The ARRL
General Class
License Manual

Edited By:

Larry D. Wolfgang, WR1B

Production Staff:

David Pingree, N1NAS, Senior Technical Illustrator:
Technical Illustrations

Jodi Morin, KA1JPA, Assistant Production Supervisor: Layout

Kathy Ford, Proofreader

Sue Fagan, Graphics Design Supervisor: Cover Design

Michelle Bloom, WB1ENT, Production Supervisor: Layout

Published By:

ARRL *The national association for* **AMATEUR RADIO**

225 Main Street, Newington, CT 06111-1494
ARRLWeb: **www.arrl.org**

This book may be used for General class license exams given beginning July 1, 2004. The General class (Element 3) question pool in this book is expected to be used for exams given until July 1, 2008. (This ending date assumes no FCC Rules changes to the licensing structure or privileges for this license class, which would force the VEC Question Pool Committee to modify the question pool.) *QST* and **ARRLWeb** (**www.arrl.org**) will have news about any rules changes.

Contents

FOREWORD

Congratulations on making the decision to study for your General class Amateur Radio license! Earning the General license will add to your enjoyment of Amateur Radio far beyond the effort required to learn the material in this book and pass your exam. You will gain access to major portions of all our amateur HF bands. You will be permitted to use a wide variety of operating modes on those bands, as well. Go ahead. Accept the challenge to earn your General license and prepare for more Amateur Radio excitement than you can imagine.

For 90 years, the ARRL has been every ham's own organization. *QST*, our monthly membership journal is chock full of articles of interest to active hams. Construction projects, news, operating hints and events, and much more fill the magazine every month. Product Reviews give detailed reports about the latest Amateur Radio equipment, and the advertising section is a showcase of the many commercial products offered. ARRL represents the interests of Amateur Radio to the FCC and the US Government as well as helping to make sure the International Telecommunication Union understands the concerns and needs of our country's Amateur Radio operators.

ARRL publishes the most widely used, most complete license preparation materials available to radio amateurs. ARRL's *Now You're Talking!* has been the standard study guide for the Technician class license for many years. ARRL's *Your Introduction to Morse Code*, available as a set of two audio CDs or cassette tapes will prepare you for the Element 1, 5 words per minute Morse code exam. For those who prefer the unlimited practice and code speeds of a computer program, we offer the MICA *Ham University* program for IBM-PC compatible computers.

In this fifth edition of *The ARRL General Class License Manual* you will find all of the material you need to pass the Element 3B written exam! There is study material to explain every question in the General question pool. Chapter 12 discusses the best way to learn and practice Morse code to prepare for the Element 1 exam.

Chapter 13 is the complete General question pool as released by the Volunteer-Examiner Coordinators' Question Pool Committee in December 2003 for use beginning July 1, 2004. It has been laid out so you can drill yourself on the actual examination questions, to check your understanding of the study material. The answer key is included on the same page as each question, for your convenience. If you do well answering the questions from the pool, you can be confident when you go to take the actual exam.

Of course you aren't studying just to pass a license exam, and we aren't satisfied just to help you pass that exam! We want you to enjoy every aspect of this avocation. That's why we have included Chapter 11, Setting Up Your HF Station. We want to help you get on the air and use those new privileges. The ARRL provides plenty of technical material and operating aids in the many books and supplies that make up our "Radio Amateur's Library." Contact the Publication Sales Office at ARRL Headquarters to request the latest publications catalog or to place an order. (You can reach us by phone — 888-277-5289; by fax — 860-594-0303; by electronic mail — **pubsales@arrl.org**; and now on the World Wide Web — **www.arrl.org**)

This fifth edition of *The ARRL General Class License Manual* is not just the product of the many ARRL staff members who have helped bring the book to you. Readers of the earlier editions sent comments and suggestions; you can, too. First, use this book to prepare for your exam. Then, write your suggestions (or any corrections you think need to be made) on the Feedback Form at the back of the book, and send the form to us. Your comments and suggestions are important to us. Thanks, and good luck!

David Sumner, K1ZZ
Executive Vice President, ARRL
Newington, Connecticut

HOW TO USE THIS BOOK

To earn a General Amateur Radio license, you will have to know some electronics theory and the rules and regulations governing the Amateur Service, as contained in Part 97 of the FCC Rules. You'll also have to be able to send and receive the international Morse code at a rate of 5 wpm. This book provides a brief description of the Amateur Service, and the General license in particular. Applicants for the General license must pass a 35-question written exam drawn from the Element 3 question pool, as released by the Volunteer-Examiner Coordinators' Question Pool Committee (QPC), and the Element 1 (5 wpm) Morse code exam.

Please note that each class of Amateur Radio license requires that you pass all of the exam elements below that class as well. So if you hold a Novice license, for example, you will also have to pass the Element 2 — Technician — exam. (If you don't already have an Amateur Radio license, you will also have to pass the Element 1 Morse code exam and the Element 2 Technician exam.) Chapter 13 of this book contains the complete Element 3 question pool and multiple-choice answers released by the VEC QPC in December 2003 for use starting July 1, 2004.

At the beginning of each chapter, you will find a list of **Key Words** that appear in that chapter, along with a simple definition for each word or phrase. As you read the text, you will find these words printed in **boldface type** the first time they appear. You may want to refer to the **Key Words** list at the beginning of the chapter when you come to a **boldface** word, to review the definition. After you have studied the material in that chapter, you may also want to review those definitions. Most of the key words are terms that will appear on exam questions, so a quick review of the definitions just before you go to take the test may also be helpful. Appendix C is a glossary of all the key words used in the book. They are arranged in alphabetical order for your convenience as you review before the exam.

The Question Pool

As you study the material, you will be instructed to turn to sections of the questions in Chapter 13. Be sure to use these questions to review your understanding of the material at the suggested times. This breaks the material into bite-sized pieces, and makes it easier for you to learn. Do not try to memorize all of the questions and answers. With over 400 questions, that will be nearly impossible! Instead, by using the questions for review, you will be familiar with them when you take the test, but you will also understand the electronics theory or Rules point behind each question.

Most people learn more easily when they are actively involved in the learning process. Study the text, rather than passively reading it. Use the questions to review your progress rather than just reading the question and looking at the correct answer letter. Fold the answer-key column under at the dashed line down the page before you begin answering questions on that page. Look at the answer key only to verify your answer. This will help you check your progress and ensure you know the answer to each question. If you missed one, go back to the supporting text and review that material. Page numbers are included with each answer letter. These indicate where to turn in the book to find the explanatory text for that question. You may have to read more than one page for the complete explanation. Paper clips make excellent place markers to help you find your spot in the text and question pool.

Other ARRL Study Materials

To help you prepare for the Morse code exam, ARRL offers a set of two audio CDs or cassette tapes to teach Morse code: *Your Introduction to Morse Code*. This package introduces each of the required characters one at a time, and drills you on each character as it is introduced. Then the character is used in words and text before proceeding to the next character. After all characters have been introduced, there is plenty of practice at 5 words per minute to help you prepare for the Element 1 (5 wpm) exam. This set includes two 74-minute audio CDs or cassette tapes. The Morse code is sent using 15-wpm characters, with extra space between characters to slow the overall code speed. This same technique is used on ARRL/VEC Morse code exams.

For those who prefer a computer program to learn and practice Morse code, ARRL offers the MICA *Ham University* program for IBM PC and compatible computers.

Even with the tapes or a computer program, you'll want to tune in the code-practice sessions transmitted by W1AW, the ARRL Headquarters station. There is a W1AW code-practice schedule on page 8 in the Introduction. You can also listen to W1AW Morse code practice anytime via the Internet by going to **www.arrl.org/w1aw/morse.html**. For more information about W1AW or how to order any ARRL publication or set of code CDs or tapes, write to ARRL Headquarters, 225 Main St, Newington, CT 06111-1494, tel 888-277-5289 (toll-free), on **ARRLWeb (www.arrl.org)** or by e-mail at **pubsales@arrl.org**.

Introduction

THE GENERAL LICENSE

If you have a Technician Amateur Radio license, upgrading to a General class license will provide a giant step in amateur privileges! After all, the General license grants high-power phone privileges on the amateur HF bands. Even if you have added the HF "Novice" privileges to your Technician license by passing the 5 wpm Morse code exam, a General class license will add significant privileges. Some people even prefer to earn the General license as their first "ticket" because of the license privileges it offers.

If you have put off earning your General license because you thought the electronics theory was too difficult, put that thought aside right now! Once you make the commitment to study and learn what it takes to pass the exam, you *will* be able to do it. Many amateurs pass the Element 3 exam on their first try, but even if it takes more than one attempt to pass, the extra privileges are well worth the effort. The key is that you must make the commitment, and be willing to study.

To help you pass the Morse code exam, ARRL offers Morse code training and practice on audio CDs as well as cassette tapes. *Your Introduction to Morse Code* will teach you all the characters you need to know, and prepare you for the 5 wpm Morse code exam, Element 1. If you want a computer program to learn Morse code, the MICA *Ham University* program is hard to beat. This program for IBM-PC and compatible computers is available from ARRL. There is more information about practicing Morse code in Chapter 12.

With this book, carefully designed to teach the required electronics theory, operating practices and FCC Rules for your Element 3 General license exam, you will soon have that higher class license.

Chapter 1 of this book, Commission's Rules covers those sections of the FCC Part 97 Rules that you will be tested on for your General license. We recommend that you also obtain a copy of *The ARRL FCC Rule Book*, which includes a com-

Figure 1 — Hank Wolcott, KA1WTS, seated at the left and Tom Dean, KB1IJI, demonstrate Amateur Radio to a group of Scouts at a Connecticut Rivers Council Boy Scout Show and campout.

plete copy of Part 97, along with detailed explanations for *all* the rules governing Amateur Radio.

Chapters 2 through 10 cover the remaining nine subelements for the General license exam. All of the questions in the General class question pool are covered in this book. Chapter 11 provides some hints and guidelines for selecting equipment and setting up and operating your HF amateur radio station. Chapter 12 has study hints and recommendations for learning and practicing Morse code. Chapter 13 contains the complete question pool, with answers, for the General license exam.

Whether you now hold a Novice, Technician or Technician Plus (Technician with credit for passing the 5 wpm Morse code exam) license, or even if you don't yet have any license, you will find the operating privileges available to a General licensee to be worth the time spent learning about your hobby. General operators are allowed to use substantial portions of every amateur high-frequency band. World-wide communication through Morse code and voice operation — as well as modes such as slow-scan TV (SSTV), radioteletype (RTTY), PSK-31 and facsimilie (fax) — is available to General class licensees.

You can see that the General license allows a wide range of transmitting privileges. It is no wonder that this is such a popular license class in the US. To earn those extra privileges, however, you will have to demonstrate that you know the international Morse code, basic electronic theory, operating practices and FCC Rules and regulations. This book is designed to teach you everything you need to know to qualify for the General Amateur Radio license.

IF YOU'RE A NEWCOMER TO AMATEUR RADIO

Earning an Amateur Radio license, at whatever level, is a special achievement. Nearly 700,000 people in the US call themselves Amateur Radio operators, or hams.

We are also part of a much larger global fraternity. There are about 3 million Amateur Radio operators around the world! Radio amateurs serve the public as a voluntary, noncommercial, communication service. This is especially true during natural disasters or other emergencies. Hams have made many important contributions to the field of electronics and communications, and this tradition continues today. Amateur Radio experimentation is yet another reason many people become part of this self-disciplined group of trained operators, technicians and electronics experts — an asset to any country. Hams pursue their hobby purely for personal enrichment in technical and operating skills, without consideration of any type of payment except the personal satisfaction they feel from a job well done!

Radio signals do not know territorial boundaries, so hams have a unique ability (and responsibility) to enhance international goodwill. Hams become ambassadors of their country every time they put their stations on the air.

Amateur Radio has been around since the early 1900s. Hams have always been at the forefront of technology. Today, hams relay signals through their own satellites in the OSCAR (Orbiting Satellite Carrying Amateur Radio) series, bounce signals off the moon, relay messages automatically through computerized radio networks and use any number of other "exotic" communications techniques. Amateurs talk from hand-held transceivers through mountaintop repeater stations that can relay their signals to other hams' cars or homes. Hams send their own pictures by television, talk with other hams around the world by voice, or, keeping alive a distinctive traditional skill, tap out messages in Morse code. The "code" represents a distinctive traditional skill, which often proves to be the only way to communicate during an emergency or when signals are weak and interference is present. When emergencies arise, radio amateurs are on the spot to relay information to and from disaster-stricken areas that have lost normal lines of communication.

The US government, through the Federal Communications Commission (FCC), grants all US Amateur Radio licenses. This licensing procedure

Figure 2 — Once you begin operating on the amateur HF bands, you will soon begin to collect many colorful QSL cards. Many of them will likely be from countries around the world.

Figure 3 — When you begin collecting DX QSL cards you are working towards the ARRL's prestigious DX Century Club award. When you have collected QSL cards from 100 "DXCC Countries" from around the world you can qualify for this award.

ensures operating skill and electronics knowledge. Without these skills, radio operators might unknowingly cause interference to other services using the radio spectrum, because of improperly adjusted equipment or neglected regulations.

Who Can Be a Ham?

The FCC doesn't care how old you are or whether you're a US citizen: If you pass the examination, the Commission will issue you an amateur license. Any person (except the agent of a foreign government) may take the exam, and, if successful, receive an amateur license. It's important to understand that if a citizen of a foreign country receives an amateur license in this manner, he or she is a US Amateur Radio operator. (This should not be confused with reciprocal operating authority, which allows visitors from certain countries who hold valid amateur licenses in their homelands to operate their own stations in the US without having to take an FCC exam.)

License Structure

By examining **Table 1**, you'll see that new amateurs can earn three license classes in the US. These are the Technician, General and Amateur Extra license. They vary in the degree of knowledge required and frequency privileges granted. Higher class licenses have more comprehensive examinations. The FCC requires proof of your ability to operate an Amateur Radio station properly. The required knowledge is in line with the privileges of the license you hold. Higher license classes require more knowledge — and offer greater operating privileges. So as you upgrade your license class, you must pass more challenging written examinations. The vast majority of beginners start with the most basic license, the Technician, although it's possible to start with any class of license.

Table 1
Amateur Operator Licenses†

Class	Code Test	Written Examination	Privileges
Technician		Basic theory and regulations. (Element 2)*	All amateur privileges above 50.0 MHz.
Technician with Morse code Credit	5 wpm (Element 1)	Basic theory and regulations. (Element 2)* all Technician privileges.	All "Novice" HF privileges in addition to
General	5 wpm (Element 1)	Basic theory and regulations; General theory and regulations. (Elements 2 and 3)	All amateur privileges except those reserved for Advanced and Amateur Extra class.
Amateur Extra	5 wpm (Element 1)	All lower exam elements, plus Extra-class theory (Elements 2, 3 and 4)	All amateur privileges.

†A licensed radio amateur will be required to pass only those elements that are not included in the examination for the amateur license currently held.
*If you have a Technician-class license issued before March 21, 1987, you also have credit for Elements 1 and 3. You must be able to prove your Technician license was issued before March 21, 1987 to claim this credit.

There are also several other amateur license classes, but the FCC is no longer issuing new licenses for these classes. The Novice license was long considered the beginner's license. Exams for this license were discontinued as of April 15, 2000. The FCC also stopped issuing new Advanced class licenses on that date. They will continue to renew previously issued licenses, however, so you will meet Novice and Advanced class operators on the air. You will also find certain frequency privileges designated for Novice and Advanced class licensees.

Technician licensees can pass an exam to demonstrate their knowledge in international Morse code at 5 words per minute. With proof of passing the Morse code exam, a Technician licensee gains some frequency privileges on four of the amateur HF bands. This license was previously called the Technician Plus license, and many amateurs will refer to it by that name. The HF privileges earned are known as the Novice privileges.

Each step up the Amateur Radio license ladder requires the applicant to pass the lower exams. So to earn a General class or even an Amateur Extra class license, you must also pass the Technician written exam. This does not mean you have to pass the Technician exam again if you already hold a Technician license! Your valid Amateur Radio license gives you credit for all the exam elements of that license when you go to upgrade. If you now hold a Technician license, you will only have to pass the Element 1 Morse code exam and the Element 3 General class written exam.

As a General, you can use a wide range of frequency bands — at least some portion of *all amateur bands*, in fact. You'll be able to talk with other hams around

Table 2
Amateur Operating Privileges

US Amateur Bands

ARRL The national association for AMATEUR RADIO

June 1, 2003   Novice, Advanced and Technician Plus Allocations

New Novice and Technician Plus licenses are no longer being issued, but *existing* Novice, Technician Plus and Advanced class licenses are unchanged. Amateurs can continue to renew these licenses. Technicians who pass the 5 wpm Morse code exam *after* that date have Technician Plus privileges, although their license says Technician. They must retain the 5 wpm Certificate of Successful Completion of Examination (CSCE) as proof. The CSCE is valid indefinitely for operating authorization, but is valid only for 365 days for upgrade credit.

160 METERS

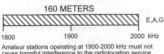

E,A,G

1800   1900   2000 kHz

Amateur stations operating at 1900-2000 kHz must not cause harmful interference to the radiolocation service and are afforded no protection from radiolocation operations.

80 METERS

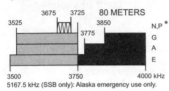

N,P *
G
A
E

3500   3750   4000 kHz
5167.5 kHz (SSB only): Alaska emergency use only.

60 METERS

General, Advanced, and Amateur Extra licensees may use the following five channels on a secondary basis with a maximum effective radiated power of 50 W PEP relative to a half wave dipole. Only upper sideband suppressed carrier voice transmissions may be used. The frequencies are 5330.5, 5346.5, 5366.5, 5371.5 and 5403.5 kHz. The occupied bandwidth is limited to 2.8 kHz centered on 5332, 5348, 5368, 5373, and 5405 kHz respectively.

40 METERS

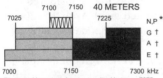

N,P *
G †
A †
E †

7000   7150   7300 kHz

† Phone and Image modes are permitted between 7075 and 7100 kHz for FCC licensed stations in ITU Regions 1 and 3 and by FCC licensed stations in ITU Region 2 West of 130 degrees West longitude or South of 20 degrees North latitude. See Sections 97.305(c) and 97.307(f)(11). Novice and Technician Plus licensees outside ITU Region 2 may use CW only between 7050 and 7075 kHz. See Section 97.301(e). These exemptions do not apply to stations in the continental US.

30 METERS

E,A,G

10,100   10,150 kHz
Maximum power on 30 meters is 200 watts PEP output. Amateurs must avoid interference to the fixed service outside the US.

20 METERS

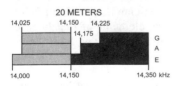

G
A
E

14,000   14,150   14,350 kHz

17 METERS

E,A,G

18,068   18,110   18,168 kHz

15 METERS

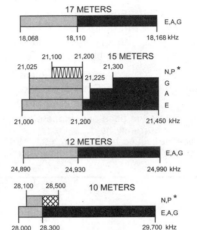

N,P *
G
A
E

21,000   21,200   21,450 kHz

12 METERS

E,A,G

24,890   24,930   24,990 kHz

10 METERS

N,P *
E,A,G

28,000   28,300   29,700 kHz
Novices and Technician Plus Licensees are limited to 200 watts PEP output on 10 meters.

6 METERS

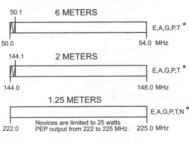

E,A,G,P,T *

50.0   54.0 MHz

2 METERS

144.1

E,A,G,P,T *

144.0   148.0 MHz

1.25 METERS

E,A,G,P,T,N *

222.0   Novices are limited to 25 watts
PEP output from 222 to 225 MHz.   225.0 MHz

70 CENTIMETERS **

E,A,G,P,T *

420.0   450.0 MHz

33 CENTIMETERS **

E,A,G,P,T *

902.0   928.0 MHz

23 CENTIMETERS **

1270   1295
N
E,A,G,P,T *

1240   1300 MHz
Novices are limited to 5 watts PEP output from 1270 to 1295 MHz.

US AMATEUR POWER LIMITS

At all times, transmitter power should be kept down to that necessary to carry out the desired communications. Power is rated in watts PEP output. Unless otherwise stated, the maximum power output is 1500 W. Power for all license classes is limited to 200 W in the 10,100-10,150 kHz band and in all Novice subbands below 28,100 kHz. Novices and Technicians are restricted to 200 W in the 28,100-28,500 kHz subbands. In addition, Novices are restricted to 25 W in the 222-225 MHz band and 5 W in the 1270-1295 MHz subband.

KEY

= CW, RTTY and data

= CW, RTTY, data, MCW, test, phone and image

= CW, phone and image

= CW and SSB phone

= CW, RTTY, data, phone, and image

= CW only

E = EXTRA CLASS
A = ADVANCED
G = GENERAL
P = TECHNICIAN PLUS
T = TECHNICIAN
N = NOVICE

*Technicians who have passed the 5 wpm Morse code exam are indicated as "P".

**Geographical and power restrictions apply to all bands with frequencies above 420 MHz. See *The ARRL FCC Rule Book* for more information about your area.

All licensees except Novices are authorized all modes on the following frequencies:
2300-2310 MHz
2390-2450 MHz
3300-3500 MHz
5650-5925 MHz
10.0-10.5 GHz
24.0-24.25 GHz
47.0-47.2 GHz
75.5-76.0, 77.0-81.0 GHz
119.98-120.02 GHz
142-149 GHz
241-250 GHz
All above 300 GHz

For band plans and sharing arrangements, see *The ARRL FCC Rule Book*.

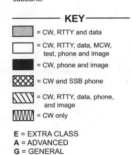

the world using voice, digital modes, slow-scan TV and Morse code on eight HF bands. (Another HF band — the 30-meter band — is reserved for Morse code and digital modes only.) See **Table 2**. You can provide public service through emergency communications and message handling.

As a General class licensee, you will experience the thrill of working (contacting) other Amateur Radio operators in just about any country in the world. There's nothing quite like making friends with other amateurs around the world.

In addition to passing the written exams, you must demonstrate an ability to send and receive international Morse code at 5 wpm for the General and Amateur Extra licenses. It's important to stress that although you may intend to use voice rather than code, this doesn't excuse you from the code test.

The FCC allows Volunteer Examiners to use a range of procedures to accommodate applicants with various disabilities. That doesn't mean you won't have to pass the test, but if you have a specific disability, you may qualify for special examination procedures for the Morse code examination or the written exams. Contact your local Volunteer Examiners or the Volunteer Examiner Coordinator responsible for the test session you will be attending if you believe you may need special testing procedures. You can contact the ARRL/VEC Office, 225 Main Street, Newington, CT 06111-1494 (860-594-0200) for assistance. Ask for more information about the special exam procedures their Volunteer Examiners will use to administer the exams to you.

Learning Morse Code

Learning the Morse code is a matter of practice. As mentioned earlier, the ARRL publishes Morse code practice audio CDs and cassette tapes to help you pass the code exam. *Your Introduction to Morse Code* teaches all the characters required by the FCC. If you prefer to use a computer program to learn Morse code, ARRL offers *Ham University* by MICA. This program for Microsoft Windows includes a versatile Morse code training section that provides unlimited practice at any Morse code speed you desire.

Chapter 12 of this book is dedicated to helping you learn Morse code. The ARRL also publishes *Morse Code: The Essential Language*, which describes much of the history of Morse code and includes plenty of hints on learning and using the code. All of these materials are available from the ARRL, 225 Main Street, Newington, CT 06111 and from many local Amateur Radio equipment dealers. To place an order, call, toll-free, **888-277-5289**. You can also send e-mail to: **pubsales@arrl.org** or check out our World Wide Web site: **www.arrl.org/** Prospective new hams can call: **800-32-NEW HAM (800-326-3942)** for additional information. You can learn more about these products and order them directly at *ARRLWeb*: **www.arrl.org**.

ARRL's Maxim Memorial Station, W1AW, transmits code practice and information bulletins of interest to all amateurs. These code-practice sessions and Morse code bulletins provide an excellent opportunity for code practice. **Table 3** is a W1AW operating schedule. When we change from standard time to daylight saving time, the same local times are used.

Station Call Signs

Many years ago, by international agreement, the nations of the world decided to allocate certain call-sign prefixes to each country. This means that if you hear a

Table 3
W1AW Schedule

W1AW Schedule

PACIFIC	MTN	CENT	EAST	MON	TUE	WED	THU	FRI
6 AM	7 AM	8 AM	9 AM		FAST CODE	SLOW CODE	FAST CODE	SLOW CODE
7 AM-1 PM	8 AM-2 PM	9 AM-3 PM	10 AM-4 PM	VISITING OPERATOR TIME (12 PM-1 PM CLOSED FOR LUNCH)				
1 PM	2 PM	3 PM	4 PM	FAST CODE	SLOW CODE	FAST CODE	SLOW CODE	FAST CODE
2 PM	3 PM	4 PM	5 PM	CODE BULLETIN				
3 PM	4 PM	5 PM	6 PM	TELEPRINTER BULLETIN				
4 PM	5 PM	6 PM	7 PM	SLOW CODE	FAST CODE	SLOW CODE	FAST CODE	SLOW CODE
5 PM	6 PM	7 PM	8 PM	CODE BULLETIN				
6 PM	7 PM	8 PM	9 PM	TELEPRINTER BULLETIN				
6⁴⁵PM	7⁴⁵PM	8⁴⁵PM	9⁴⁵PM	VOICE BULLETIN				
7 PM	8 PM	9 PM	10 PM	FAST CODE	SLOW CODE	FAST CODE	SLOW CODE	FAST CODE
8 PM	9 PM	10 PM	11 PM	CODE BULLETIN				

♦ **Morse code transmissions:**
Frequencies are 1.818, 3.5815, 7.0475, 14.0475, 18.0975, 21.0675, 28.0675 and 147.555 MHz.

Slow Code = practice sent at 5, 7$\frac{1}{2}$, 10, 13 and 15 wpm.

Fast Code = practice sent at 35, 30, 25, 20, 15, 13 and 10 wpm.

Code practice text is from the pages of *QST*. The source is given at the beginning of each practice session and alternate speeds within each session. For example, "Text is from July 2001 *QST*, pages 9 and 81," indicates that the plain text is from the article on page 9 and mixed number/letter groups are from page 81.

Code bulletins are sent at 18 wpm.

W1AW qualifying runs are sent on the same frequencies as the Morse code transmissions. West Coast qualifying runs are transmitted on approximately 3.590 MHz by K6YR.

At the beginning of each code practice session, the schedule for the next qualifying run is presented.

Underline one minute of the highest speed you copied, certify that your copy was made without aid, and send it to ARRL for grading. Please include your name, call sign (if any) and complete mailing address.

The fee structure is $10 for a certificate, and $7.50 for endorsements.

♦ **Teleprinter transmissions:**
Frequencies are 3.625, 7.095, 14.095, 18.1025, 21.095, 28.095 and 147.555 MHz.

Bulletins are sent at 45.45-baud Baudot and 100-baud AMTOR, FEC Mode B. 110-baud ASCII will be sent only as time allows.

On Tuesdays and Fridays at 6:30 PM Eastern Time, Keplerian elements for many amateur satellites are sent on the regular teleprinter frequencies.

♦ **Voice transmissions:**
Frequencies are 1.855, 3.99, 7.29, 14.29, 18.16, 21.39, 28.59 and 147.555 MHz.

♦ **Miscellany:**
On Fridays, UTC, a DX bulletin replaces the regular bulletins.

W1AW is open to visitors from 10 AM until noon and from 1 PM until 3:45 PM on Monday - Friday.

FCC licensed amateurs may operate the station during that time. Be sure to bring your current FCC amateur license or a photocopy.

In a communication emergency, monitor W1AW for special bulletins as follows: voice on the hour, teleprinter at 15 minutes past the hour, and CW on the half hour.

Headquarters and W1AW are closed on New Year's Day, President's Day, Good Friday, Memorial Day, Independence Day, Labor Day, Thanksgiving and the following Friday, and Christmas Day.

radio station call sign beginning with K, N or W, for example, you know the station is licensed by the United States. A call sign beginning with the letter G is licensed by Great Britain and a call sign beginning with VE is from Canada. *The ARRL DXCC Countries List* is an operating aid no ham who is active on the HF bands should be without. That booklet, available from ARRL, includes the common call-sign prefixes used by amateurs in virtually every location in the world. It also includes a check-off list to help keep track of the countries you contact as you work towards collecting QSL cards from 100 or more countries to earn the prestigious DX Century Club award. (DX is ham lingo for distance. On the HF bands, DX is generally taken to mean any country outside the one from which you are operating.)

The International Telecommunication Union (ITU) radio regulations outline the basic principles used in forming amateur call signs. According to these regulations, an amateur call sign must be made of one or two characters (either of which may be a numeral) as a prefix, followed by a numeral, and then a suffix of not more than three letters. The prefixes W, K, N and A are used in the United States. When the letter A is used to begin a US call sign there are two letters for the prefix, AA through AL. The continental US is divided into 10 Amateur Radio call districts (sometimes called areas), numbered 0 through 9). **Figure 4** is a map showing the US call districts.

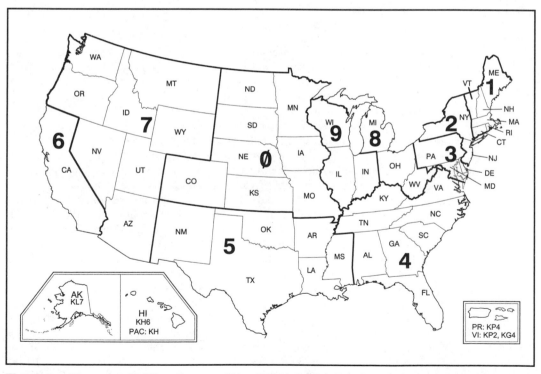

Figure 4 — There are 10 US call areas. Hawaii is part of the sixth call area, and Alaska is part of the seventh.

For information on the FCC's call-sign assignment system, and a table listing the blocks of call signs for each license class, see *The ARRL FCC Rule Book*. You may keep the same call sign when you change license class, if you wish. If you want to receive a new call sign, you must check off the "Change my station call sign systematically" box and initial the line below that box when you fill out an NCVEC Quick Form 605 to apply for the exam. (There is more about filling out the paper forms later in this chapter.) With both the check box and initials, the goal is to ensure that you really do want to change your call sign. Once done, the only way to go back to your old call sign will be through the FCC's Vanity Call Sign system. You can also change your call sign when you renew your license or simply file for a modification to your license. In that case you can use FCC Form 605, and mark the appropriate box on Schedule D of the form.

The FCC also has a vanity call sign system. Under this system the FCC will issue a call sign selected from a list of preferred available call signs. While there is no fee for an Amateur Radio license, there is a fee for the selection of a vanity call sign. The fee as of January, 2004 is $16.30 for a 10-year Amateur Radio license, paid upon application for a vanity call sign and at license renewal after that. (That fee may change as costs of administering the program change.) The latest details about the vanity call sign system are available from ARRL Regulatory Information, 225 Main Street, Newington, CT 06111-1494 and on *ARRLWeb* at **www.arrl.org/.**

EARNING A LICENSE

Forms and Procedures

To renew or modify a license, you can file a copy of FCC Form 605 or use one of the electronic filing options. Licenses are normally good for ten years. Your application for a license renewal must be submitted to the FCC no more than 90 days before the license expires. (We recommend you submit the application for renewal between 90 and 60 days before your license expires.) If the FCC receives your renewal application before the license expires, you may continue to operate until your new license arrives, even if it is past the expiration date.

If you forget to apply before your license expires, you may still be able to renew your license without taking another exam. There is a two-year grace period, during which you may apply for renewal of your expired license. Use an FCC Form 605 to apply for reinstatement (and your old call sign), or follow the electronic filing procedures. If you apply for reinstatement of your expired license under this two-year grace period, you may not operate your station until your new license is issued.

If you move or change addresses you should use an FCC Form 605 or electronic filing to notify the FCC of the change. If your license is lost or destroyed you will also have to either file an FCC Form 605 or use the electronic filing option to obtain a new copy. Use Application Purpose code DU (Duplicate License) on Form 605.

You can ask one of the Volunteer Examiner Coordinators' offices to file your renewal application electronically if you don't want to mail the form to the FCC. You must still mail the form to the VEC, however. If you are asking one of the VECs to file your application renewal or license modification for you, you should complete a copy of the NCVEC Quick Form 605 and mail it to the VEC. The Volunteer Examiner Coordinator organizations each determine their own fee structure for this service. Some VECs may not charge a fee, and others do. The ARRL/VEC Office charges anyone who is not an ARRL member, but will electronically file application forms for any ARRL member free of charge.

Electronic Filing

You can also file your license renewal or address modification using the Universal Licensing System (ULS) on the World Wide Web. To use ULS, you must have an FCC Registration Number, or FRN. Obtain your FRN by registering with the Commission Registration System, known as CORES.

Described as an agency-wide registration system for anyone filing applications with or making payments to the FCC, CORES will assign a unique 10-digit FCC Registration Number, or FRN to all registrants. All Commission systems that handle financial, authorization of service, and enforcement activities will use the FRN. The FCC says use of the FRN will allow it to more rapidly verify fee payment. Amateurs mailing payments to the FCC — for example as part of a vanity call sign application — would include their FRN on FCC Form 159.

The on-line filing system and further information about CORES is available by visiting the FCC Web site, **www.fcc.gov** and clicking on the Commission Registration System link. Follow the directions on the Web site.

When you register with CORES you must supply a Taxpayer Identification Number, or TIN. For individuals, this is usually a Social Security Number. Club stations must obtain an Assigned Taxpayer Identification Number (ATIN) before registering on CORES

Anyone can register on CORES and obtain an FRN. CORES/FRN is "entity registration." You don't need a license to be registered. It is also possible to register on CORES using a paper Form 160.

Once you have registered on CORES and obtained your FRN, you can proceed to renew or modify your license using the Universal Licensing System (ULS), also on the World Wide Web. Go to **www.fcc.gov/wtb/uls** and click on the "Online Filing" button. Follow the directions provided on the Web page to connect to the FCC's ULS database. You can also find specific instructions for the various common filing tasks with the Universal License System (ULS) at **wireless.fcc.gov/services/amateur/**.

Paper Filing

The FCC has a set of detailed instructions for the Form 605, which are included with the form. To obtain a new Form 605, call the FCC Forms Distribution Center at 800-418-3676. You can also write to: Federal Communications Commission, Forms Distribution Center, 2803 52nd Avenue, Hyattsville, MD 20781 (specify "Form 605" on the envelope). The Form 605 also is available from the FCC's fax on demand service. Call 202-418-0177 and ask for form number 000605. Form 605 also is available via the Internet. The World Wide Web location is: **www.fcc.gov/ formpage.html** or you can receive the form via ftp to: **ftp.fcc.gov/pub/Forms/ Form605**.

The ARRL/VEC has created a package that includes the portions of Form 605 that are needed for amateur applications, as well as a condensed set of instructions for completing the form. Write to: ARRL/VEC, Form 605, 225 Main Street, Newington, CT 06111-1494. (Please include a large business-sized stamped self-

Quick-Form Application for Authorization in the Ship, Aircraft, Amateur, Restricted and Commercial Operator, and General Mobile Radio Services

Approved by OMB
3060 - 0850
See instructions for
public burden estimate

1) Radio Service Code:	HA

Application Purpose (Select only one) (*MD*)

2) NE – New	RO – Renewal Only	WD – Withdrawal of Application
MD – Modification	RM – Renewal / Modification	DU – Duplicate License
AM – Amendment	CA – Cancellation of License	AU – Administrative Update

3)	If this request if for Developmental License or STA (Special Temporary Authorization) enter the appropriate code and attach the required exhibit as described in the instructions. Otherwise enter 'N' (Not Applicable).	(N) D S N/A
4)	If this request is for an Amendment or Withdrawal of Application, enter the file number of the pending application currently on file with the FCC.	File Number
5)	If this request is for a Modification, Renewal Only, Renewal / Modification, Cancellation of License, Duplicate License, or Administrative Update, enter the call sign (serial number for Commercial Operator) of the existing FCC license. If this is a request for consolidation of DO & DM Operator Licenses, enter serial number of DO. Also, if filing for a ship exemption, you must provide call sign.	Call Sign/Serial # WR1B
6)	If this request is for a New, Amendment, Renewal Only, or Renewal Modification, enter the requested expiration date of the authorization (this item is optional).	MM DD
7)	Does this filing request a Waiver of the Commission's rules? If 'Y', attach the required showing as described in the instructions.	(N) Yes No
8)	Are attachments (other than associated schedules) being filed with this application?	(N) Yes No

Applicant/Licensee Information

9) FCC Registration Number (FRN): 0003- 3573 - 99

10) Applicant /Licensee is a(n): (I) Individual   Unicorporated Association   Trust   Government Entity   Joint Venture
Corporation   Limited Liability Corporation   Partnership   Consortium

11) First Name (if individual): Larry	MI: D	Last Name: Wolfgang	Suffix:

12) Entity Name (if other than individual):

13) Attention To:

14) P.O. Box:	And/Or	15) Street Address: 225 Main Street

16) City: Newington	17) State: CT	18) Zip Code/Postal Code: 06111	19) Country: USA

20) Telephone Number: 860- 594-0200	21) FAX Number:

22) E-Mail Address: wr1b@arrl.net

Ship Applicants/Licensees Only

23) Enter new name of vessel:_____

Aircraft Applicants/Licensees Only

24) Enter the new FAA Registration Number (the N-number):_____
    NOTE: Do not enter the leading "N".

FCC 605 – Main Form
December 2003 - Page 1

Figure 5 — This sample FCC Form 605 shows the sections you should complete to notify the FCC of a change in your address.

Fee Status

25) Is the applicant/licensee exempt from FCC application Fees?	(**N**) <u>Yes</u> <u>No</u>
26) Is the applicant/licensee exempt from FCC regulatory Fees?	(**N**) <u>Yes</u> <u>No</u>

General Certification Statements

1) The applicant/licensee waives any claim to the use of any particular frequency or of the electromagnetic spectrum as against the regulatory power of the United States because of the previous use of the same, whether by license or otherwise, and requests an authorization in accordance with this application.

2) The applicant/licensee certifies that all statements made in this application and in the exhibits, attachments, or documents incorporated by reference are material, are part of this application, and are true, complete, correct, and made in good faith.

3) Neither the applicant/licensee nor any member thereof is a foreign government or a representative thereof.

4) The applicant/licensee certifies that neither the applicant/licensee nor any other party to the application is subject to a denial of Federal benefits pursuant to Section 5301 of the Anti-Drug Abuse Act of 1988, 21 U.S.C. § 862, because of a conviction for possession or distribution of a controlled substance. **This certification does not apply to applications filed in services exempted under Section 1.2002(c) of the rules, 47 CFR § 1.2002(c).** See Section 1.2002(b) of the rules, 47 CFR § 1.2002(b), for the definition of "party to the application" as used in this certification.

5) Amateur or GMRS applicant/licensee certifies that the construction of the station would NOT be an action which is likely to have a significant environmental effect (see the Commission's rules 47 CFR Sections 1.1301-1.1319 and Section 97.13(a) rules (available at web site http://wireless.fcc.gov/rules.html).

6) Amateur applicant/licensee certifies that they have READ and WILL COMPLY WITH Section 97.13(c) of the Commission's rules (available at web site http://wireless.fcc.gov/rules.html) regarding RADIOFREQUENCY (RF) RADIATION SAFETY and the amateur service section of OST/OET Bulletin Number 65 (available at web site http://www.fcc.gov/oet/info/documents/bulletins/).

Certification Statements For GMRS Applicants/Licensees

1) Applicant/Licensee certifies that he or she is claiming eligibility under Rule Section 95.5 of the Commission's rules.

2) Applicant/Licensee certifies that he or she is at least 18 years of age.

3) Applicant/Licensee certifies that he or she will comply with the requirement that use of frequencies 462.650, 467.650, 462.700 and 467.700 MHz is not permitted near the Canadian border North of Line A and East of Line C. These frequencies are used throughout Canada and harmful interference is anticipated.

4) Non-Individual applicants/licensees certify that they have NOT changed frequency or channel pairs, type of emission, antenna height, location of fixed transmitters, number of mobile units, area of mobile operation, or increase in power.

Certification Statements for Ship Applicants/Licensees (Including Ship Exemptions)

1) Applicant/Licensee certifies that they are the owner or operator of the vessel, a subsidiary communications corporation of the owner or operator of the vessel, a state or local government subdivision, or an agency of the US Government subject to Section 301 of the Communications Act.

2) This application is filed with the understanding that any action by the Commission thereon shall be limited to the voyage(s) described herein, and that apart from the provisions of the specific law from which the applicant/licensee requests an exemption, the vessel is in full compliance with all applicable statues, international agreements and regulations.

Signature

27) Typed or Printed Name of Party Authorized to Sign

First Name: *Larry*	MI: *D*	Last Name: *Wolfgang*	Suffix:
28) Title:			

Signature: *Larry D Wolfgang*	29) Date: *1/4/04*

Failure to Sign This Application May Result In Dismissal Of The Application And Forfeiture Of Any Fees Paid

WILLFUL FALSE STATEMENTS MADE ON THIS FORM OR ANY ATTACHMENTS ARE PUNISHABLE BY FINE AND/OR IMPRISONMENT (U.S. Code, Title 18, Section 1001) AND / OR REVOCATION OF ANY STATION LICENSE OR CONSTRUCTION PERMIT (U.S. Code, Title 47, Section 312(a)(1)), AND / OR FORFEITURE (U.S. Code, Title 47, Section 503).

FCC 605 – Main Form
December 2003 - Page 2

addressed envelope with your request.) **Figure 5** is a sample of those portions of an FCC Form 605 that you would complete to submit a change of address to the FCC.

Most of the form is simple to fill out. You will need to know that the Radio Service Code for box 1 is HA for Amateur Radio. (Just remember HAm radio.) You will have to include a "Taxpayer Identification Number" on the Form. This is normally your Social Security Number. If you don't want to write your Social Security Number on this form, then you can register with CORES as described above. Then you will receive your FRN from the FCC, and you can use that number instead of your Social Security Number on the Form. Of course, you will have to supply your Social Security Number to register with CORES.

The telephone number, fax number and e-mail address information is optional. The FCC will use that information to contact you in case there is a problem with your application.

Page two of the Form includes six General Certification Statements. Statement five may seem confusing. Basically, this statement means that you do not plan to install an antenna over 200 feet high, and that your permanent station location will not be in a designated wilderness area, wildlife preserve or nationally recognized scenic and recreational area.

The sixth statement indicates that you are familiar with the FCC RF Safety Rules, and that you will obey them. Chapter 10 (Subelement G0) includes exam questions and explanations about the RF Safety Rules.

Volunteer Examiner Program

Before you can take an FCC exam, you'll have to fill out a copy of the National Conference of Volunteer Examiner Coordinators' (NCVEC) Quick Form 605. This form is used as an application for a new license or an upgraded license. The NCVEC Quick Form 605 is only used at license exam sessions or for a Volunteer Examiner Coordinator organization to process a license renewal or change request with the FCC on your behalf. This form includes some information that the Volunteer Examiner Coordinator's office will need to process your application with the FCC. See **Figure 6**. You should not use an NCVEC Quick Form 605 to apply for a license renewal or modification with the FCC. *Never* mail these forms to the FCC, because that will result in a rejection of the application. Likewise, an FCC Form 605 can't be used for an exam application.

All US amateur exams are administered by Volunteer Examiners who are certified by a Volunteer-Examiner Coordinator (VEC). Program. *The ARRL FCC Rule Book* contains more details about the Volunteer-Examiner program.

To qualify for a General license you must pass Elements 1, 2 and 3. If you already hold a valid Novice license, or a Technician Plus license (which includes Morse code credit), then you have credit for passing Elements 1. If you have a Technician license you will receive credit for Element 2. If you have a Technician license issued before March 21, 1987, and you wish to upgrade to General, you also have credit for passing Element 3. In that case, all you need to do is go to an exam session with your old original license. If you no longer have the original license, you may be able to obtain suitable proof of having held such a license. Such proof can come from an old *Radio Amateurs Callbook*, the data from an old edition of the *QRZ* CD ROM (available on the Internet at **www.qrz.com**) and other sources. Contact the ARRL/VEC Office if you have questions about this. Finally, if you have a

Technician license issued after March 21, 1987, you must study the questions in the Element 3 question pool and pass that exam. See **Table 4** for details.

It is an unfortunate quirk of the FCC Rules that if you currently have a Technician license and then pass the Morse code exam, your Certificate of Successful Completion of Examination (CSCE) is valid for 365 days toward your General class upgrade. After that, you will have to take and pass the Morse code exam again to complete your upgrade. The CSCE for the code credit is valid indefinitely to use the Novice/Technician with Morse code privileges on the HF bands, but only valid for 365 days to upgrade to General. On the other hand, if you have ever held a Novice license, even if it is expired, you will receive credit for the Morse code exam. Just show proof of having held that Novice license.

The Element 3 exam consists of 35 questions taken from a pool of more than

---

Table 4
Exam Elements Needed to Qualify for a General Class License

Current License	Exam Requirements	Study Materials
None	Morse code (Element 1) Technician (Element 2)	*Your Introduction to Morse Code* *Now You're Talking!*, 5th Ed or *ARRL Technician Class Video Course*, 4th Ed or *ARRL's Tech Q & A*, 2nd Ed
	General (Element 3)	*The ARRL General Class License Manua*, 5th Ed or *ARRL's General Q & A*, 2nd Ed
Novice	Technician (Element 2)	*Now You're Talking!*, 5th Ed or *ARRL Technician Class Video Course*, 4th Ed or *ARRL's Tech Q & A*, 2nd Ed
	General (Element 3)	*The ARRL General Class License Manual*, 5th Ed or *ARRL's General Q & A*, 2nd Ed
Technician issued on or after Feb 14, 1991	Morse code (Element 1) General (Element 3)	*Your Introduction to Morse Code* *The ARRL General Class License Manual*, 5th Ed or *ARRL's General Q & A*, 2nd Ed
Technician issued before Feb 14, 1991	General (Element 3)	*The ARRL General Class License Manual*, 5th Ed *ARRL's General Q & A*, 2nd Ed
Technician Plus or Technician with	General (Element 3)	*The ARRL General Class License Manual*, 5th EdV or *ARRL's General Q & A*, 2nd Ed
Technician issued before Mar 21, 1987		None*

*Individuals who qualified for the Technician license before March 21, 1987 will be able to upgrade to General class by providing documentary proof to a Volunteer Examiner Coordinator, paying an application fee and completing NCVEC Quick Form 605.

NCVEC QUICK-FORM 605 APPLICATION FOR
AMATEUR OPERATOR/PRIMARY STATION LICENSE

SECTION 1 - TO BE COMPLETED BY APPLICANT

PRINT LAST NAME	SUFFIX	FIRST NAME	INITIAL	STATION CALL SIGN (IF ANY)
MORIN		Joanne	B	KA1JPA

MAILING ADDRESS (Number and Street or P.O. Box)
225 Man St.

SOCIAL SECURITY NUMBER (SSN) or (FRN) FCC FEDERAL REGISTRATION NUMBER
987-65-4321

CITY	STATE CODE	ZIP CODE (5 or 9 Numbers)	E-MAIL ADDRESS (OPTIONAL)
Newington	CT	06111	

DAYTIME TELEPHONE NUMBER (Include Area Code) OPTIONAL
860-594-0200

FAX NUMBER (Include Area Code) OPTIONAL

ENTITY NAME (IF CLUB, MILITARY RECREATION, RACES)

Type of Applicant: ☒ Individual ☐ Amateur Club ☐ Military Recreation ☐ RACES (Modify Only)

CLUB, MILITARY RECREATION, OR RACES CALL SIGN

SIGNATURE OF RESPONSIBLE CLUB OFFICIAL

I HEREBY APPLY FOR (Make an X in the appropriate box(es))

☐ EXAMINATION for a **new** license grant

☒ EXAMINATION for **upgrade** of my license class

☐ CHANGE my **name** on my license to my new name

Former Name: _____
(Last name) (Suffix)   (First name)   (MI)

☐ CHANGE my mailing address to **above** address

☐ CHANGE my station **call sign** systematically

Applicant's Initials: _____

☐ RENEWAL of my license grant.

Do you have another license application on file with the FCC which has not been acted upon?	PURPOSE OF OTHER APPLICATION	PENDING FILE NUMBER (FOR VEC USE ONLY)

I certify that:
* I waive any claim to the use of any particular frequency regardless of prior use by license or otherwise;
* All statements and attachments are true, complete and correct to the best of my knowledge and belief and are made in good faith;
* I am not a representative of a foreign government;
* I am not subject to a denial of Federal benefits pursuant to Section 5301of the Anti-Drug Abuse Act of 1988, 21 U.S.C. § 862;
* The construction of my station will NOT be an action which is likely to have a significant environmental effect (See 47 CFR Sections 1.301-1.319 and Section 97.13(a));
* I have read and WILL COMPLY with Section 97.13(c) of the Commission's Rules regarding RADIOFREQUENCY (RF) RADIATION SAFETY and the amateur service section of OST/OET Bulletin Number 65.

Signature of applicant (Do not print, type, or stamp. Must match applicant's name above.)

X Joanne B Morin   Date Signed: April 1, 2004

SECTION 2 - TO BE COMPLETED BY ALL ADMINISTERING VEs

Applicant is qualified for operator license class:

☐ NO NEW LICENSE OR UPGRADE WAS EARNED

☐ TECHNICIAN   Element 2

☒ GENERAL   Elements 1, 2 and 3

☐ AMATEUR EXTRA   Elements 1, 2, 3 and 4

DATE OF EXAMINATION SESSION
4 / 1 / 2004

EXAMINATION SESSION LOCATION
Newington, CT

VEC ORGANIZATION
ARRL

VEC RECEIPT DATE

I CERTIFY THAT I HAVE COMPLIED WITH THE ADMINISTERING VE REQUIRMENTS IN PART 97 OF THE COMMISSION'S RULES AND WITH THE INSTRUCTIONS PROVIDED BY THE COORDINATING VEC AND THE FCC.

1st VEs NAME (Print First, MI, Last, Suffix)	VEs STATION CALL SIGN	VEs SIGNATURE (Must match name)	DATE SIGNED
Larry D. Wolfgang	WR1B	Larry D. Wolfgang	4/1/2004
2nd VEs NAME (Print First, MI, Last, Suffix) Larry T. GREEN	WY1O	Larry T. Green	4/1/2004
3rd VEs NAME (Print First, MI, Last, Suffix) JOHN C. HENNESSEE	N1KB	John C.	4/1/2004

DO NOT SEND THIS FORM TO FCC – THIS IS NOT AN FCC FORM.
IF THIS FORM IS SENT TO FCC, FCC WILL RETURN IT TO YOU WITHOUT ACTION.

NCVEC FORM 605 - APRIL 2003
FOR VE/VEC USE ONLY - Page 1

Figure 6 — This sample NCVEC Quick Form 605 shows how your form will look after you have completed your upgrade to General.

INSTRUCTIONS FOR COMPLETING APPLICATION FORM NCVEC FORM 605

ARE WRITTEN TESTS AN FCC-LICENSE REQUIREMENT? ARE THERE EXEMPTIONS?

Beginning April 15, 2000, you may be examined on only three classes of operator licenses, each authorizing varying levels of privileges. The class for which each examinee is qualified is determined by the degree of skill and knowledge in operating a station that the examinee demonstrates to volunteer examiners (VEs) in his or her community. The demonstration of this knowledge is required in order to obtain an Amateur Operator/Primary Station License. There is no exemption from the written exam requirements for persons with difficulty in reading, writing, or because of a handicap or disability. There are exam accommodations that can be afforded examinees (see ACCOMMODATING A HANDI-CAPPED PERSON below). Most new amateur operators start at the Technician class and then advance one class at a time. The VEs give examination credit for the license class currently (and in some cases, previously) held so that examinations required for that license need not be repeated. The written examinations are constructed from question pools that have been made public (see: <http://www.arrl.org/arrlvec/pools.html>.) Helpful study guides and training courses are also widely available. To locate examination opportunities in your area, contact your local club, VE group, one of the 14 VECs or see the online listings at: <http://www.w5yi.org/vol-exam.htm> or <http://www.arrl.org/arrlvec/examsearch.phtml>.

IS KNOWLEDGE OF MORSE CODE AN FCC-LICENSE REQUIREMENT? ARE THERE EXEMPTIONS?

Some persons have difficulty in taking Morse code tests because of a handicap or disability. There is available to all otherwise qualified persons, handicapped or not, the Technician Class operator license that does not require passing a Morse code examination. Because of international regulations, however, any US FCC licensee seeking access to the HF bands (frequencies below 30 MHz) must have demonstrated proficiency in Morse code. If a US FCC licensee wishes to gain access to the HF bands, there is no exemption available from this Morse code proficiency requirement. If licensed as a Technician class, upon passing a Morse code examination operation on certain HF bands is permitted.

THE REASON FOR THE MORSE CODE EXAMINATION

Telegraphy is a method of electrical communication that the Amateur Radio Service community strongly desires to preserve. The FCC supports this objective by authorizing additional operating privileges to amateur operators who pass a Morse Code examination. Normally, to attain this skill, intense practice is required. Annually, thousands of amateur operators prove, by passing examinations, that they have acquired the skill. These examinations are prepared and administered by amateur operators in the local community who volunteer their time and effort.

THE EXAMINATION PROCEDURE

The volunteer examiners (VEs) send a short message in the Morse code. The examinee must decipher a series of audible dots and dashes into 43 different alphabetic, numeric, and punctuation characters used in the message. Usually a 10-question quiz is then administered asking questions about items contained in the message.

ACCOMMODATING A HANDICAPPED PERSON

Many handicapped persons accept and benefit from the personal challenge of passing the examination in spite of their hardships. For handicapped persons who have difficulty in proving that they can decipher messages sent in the Morse code, the VEs make exceptionally accommodative arrangements. To assist such persons, the VEs will:

- adjust the tone in frequency and volume to suit the examinee.
- administer the examination at a place convenient and comfortable to the examinee, even at bedside.
- for a deaf person, they will send the dots and dashes to a vibrating surface or flashing light.
- write the examinee's dictation.
- where warranted, they will pause in sending the message after each sentence, each phrase, each word, or in extreme cases they will pause the exam message character-by-character to allow the examinee additional time to absorb, to interpret or even to speak out what was sent.
- or they will even allow the examinee to send the message, rather than receive it.

Should you have any questions, please contact your local volunteer examiner team, or contact one of the 14 volunteer examiner coordinator (VEC) organizations. For contact information for VECs, or to contact the FCC, call 888-225-5322 (weekdays), or write to FCC, 1270 Fairfield Road, Gettysburg PA 17325-7245. Fax 717-338-2696. Also see the FCC web at: <http://www.fcc.gov/wtb/amateur/>.

RENEWING, MODIFYING OR REINSTATING YOUR AMATEUR RADIO OPERATOR/PRIMARY STATION LICENSE

RENEWING YOUR AMATEUR LICENSE

The NCVEC Form 605 may also be used to renew or modify your Amateur Radio Operator/Primary Station license. License renewal may only be completed during the final 90 days prior to license expiration, or up to two years after expiration. Changes to your mailing address, name and requests for a sequential change of your station call sign appropriate for your license class may be requested at any time. This form may not be used to apply for a specific ("Vanity") station call sign.

REINSTATING YOUR AMATEUR LICENSE

This form may also be used to reinstate your Amateur Radio Operator/Station license if it has been expired less than the two year grace period for renewal. After the two year grace period you must retake the amateur license examinations to become relicensed. You will be issued a new systematic call sign.

RENEWING OR MODIFYING YOUR LICENSE

On-line renewal: You can submit your renewal or license modifications to FCC on-line via the internet/WWW at: <http://www.fcc.gov/wtb/uls>. To do so, you must first register in ULS by following the "Register, TIN/Call Sign" link and then complete your registration information (ignore the contact person, and SGIN references). You must then choose the "File, ULS Filing" link to perform your on-line transaction with FCC. Direct any on-line filing or password questions to FCC Tech Support weekdays at tel: 202-414-1250.

Renewal by mail: If you choose to renew by mail, you can mail the "FCC Form 605" to FCC. You can obtain FCC Form 605 via the internet at <http://www.fcc.gov/formpage.html> or <ftp://ftp.fcc.gov/pub/Forms/Form605/>. It's available by fax at 202-418-0177 (request Form 000605). The FCC Forms Distribution Center will accept form orders by calling 800-418-3676. FCC Form 605 has a main form, plus a Schedule D. The main form is all that is needed for renewals. Mail FCC Form 605 to: FCC, 1270 Fairfield Rd, Gettysburg PA 17325-7245. This is a free FCC service. The NCVEC Form 605 application can be used for a license renewal, modification or reinstatement. NCVEC Form 605 can be processed by VECs, but not all VECs provide this as a routine service. ARRL Members can submit NCVEC Form 605 to the ARRL/VEC for processing. ARRL Members or others can choose to submit their NCVEC Form 605 to a local VEC (check with the VEC office before forwarding), or it can be returned with a $6.00 application fee to: The W5YI Group, Inc., P.O. Box 565101, Dallas, Texas 75356 (a portion of this fee goes to the National Conference of VECs to help defray their expenses). The NCVEC Form 605 may not be returned to the FCC since it is an internal VEC form. Once again, the service provided by FCC is free.

THE FCC APPLICATION FORM 605

The FCC version of the Form 605 may not be used for applications submitted to a VE team or a VEC since it does not request information needed by the administering VEs. The FCC Form 605 may, however, be used to routinely renew or modify your license without charge. It should be sent to the FCC, 1270 Fairfield Rd., Gettysburg PA 17325-7245.

CLUB STATION CALL SIGN ADMINISTRATORS (CSCSA)

The NCVEC Form 605 is also used for the processing of applications for Amateur Service club and military recreation station call signs and for the modification of RACES stations. No fee may be charged by an administrator for this service. The Club Station Call Sign Administrators are: ARRL/VEC (225 Main St., Newington, CT 06111), W4VEC (3504 Stonehurst Pl., High Point, NC 27265) and the W5YI-VEC (P.O. Box 565101, Dallas, TX 75356.) Please return this form to one of these three CSCSAs.

NCVEC FORM 605
FOR VE/VEC USE ONLY - Page 2

350. The question pools for all amateur exams are maintained by the Volunteer Examiner Coordinators' Question Pool Committee. The FCC allows Volunteer Examiners to select the questions for an amateur exam, but they must use the questions exactly as they are released by the VEC that coordinates the test session. If you attend a test session coordinated by the ARRL/VEC, your test will be designed by the ARRL/VEC or by a computer program designed by the VEC. The questions and answers will be exactly as they are printed in Chapter 13 of this book. Be careful, though. The ARRL/VEC and some other VECs scramble the positions of the answers within the questions.

Finding an Exam Opportunity

To determine where and when an exam will be given, contact the ARRL/VEC Office, or watch for announcements in the Hamfest Calendar and Coming Conventions columns in *QST*. Many local clubs sponsor exams, so they are another good source of information on exam opportunities. ARRL officials such as Directors, Vice Directors and Section Managers receive notices about test sessions in their area. See the latest issue of *QST* for the names and addresses of your local ARRL officials. You can also use the ARRL Exam Session Search feature of *ARRLWeb*, at: **www.arrl.org/arrlvec/examsearch.phtml**. Registration deadlines, and the time and location of the exams, are mentioned prominently in publicity releases about upcoming sessions.

Taking The Exam

By the time examination day rolls around, you should have already prepared yourself. This means getting your schedule, supplies and mental attitude ready. Plan your schedule so you'll get to the examination site with plenty of time to spare. There's no harm in being early. In fact, you might have time to discuss hamming with another applicant, which is a great way to calm pretest nerves. Try not to discuss the material that will be on the examination, as this may make you even more nervous. By this time, it's too late to study anyway!

What supplies will you need? First, be sure you bring your current *original* Amateur Radio license, if you have one. Bring along several sharpened number 2 pencils and two pens (blue or black ink). Be sure to have a good eraser. A pocket calculator may also come in handy. You may use a programmable calculator if that is the kind you have, but take it into your exam "empty" (cleared of all programs and contents in memory). Don't program equations ahead of time, because you may be asked to demonstrate that there is nothing in the calculator memory. The examining team has the right to refuse a candidate the use of any calculator that they feel may contain information for the test or could otherwise be used to cheat on the exam. If you still use a slide rule, that should also be allowed, although you will probably *not* be allowed to take math tables, such as trigonometry tables or logarithm tables into the exam with you.

The Volunteer Examiner Team is required to check two forms of identification before you enter the test room. This includes your *original* Amateur Radio license, if you have one — not a photocopy of your license. (You will need a photocopy of your license to file with your application, but only the original is valid for ID.) A photo ID of some type is best for the second type, but is not required by the FCC. Other acceptable forms of identification include a driver's license, a piece of mail addressed to you, a birth certificate, or some other such document.

The following description of the testing procedure applies to exams coordinated by the ARRL/VEC, although many other VECs use a similar procedure.

The Code Test

The 5 wpm code test is usually given before the written exams. If you don't plan to take the code exam, just sit quietly while the other candidates give it a try.

Before you take the code test, you'll be handed a piece of paper to copy the code as it is sent. The test will begin with about a minute of practice copy. Then comes the actual test: at least five minutes of Morse code. You are responsible for knowing the 26 letters of the alphabet, the numerals 0 through 9, the period, comma, question mark, and procedural signals $\overline{AR}$, $\overline{SK}$, $\overline{BT}$ and $\overline{DN}$ (/ or fraction bar, sometimes called "slant bar").

You may copy the entire text word for word, or just take notes on the content. At the end of the transmission, the examiner will hand you 10 questions about the text. Simply fill in the blanks with your answers. (You must spell each answer exactly as it was sent.) If you get at least 7 correct, you pass! Alternatively, the exam team has the option to look at your copy sheet if you fail the 10-question exam. If you have one minute of solid copy (25 characters), they can certify that you passed the test on that basis. The format of the test transmission is generally similar to one side of a normal on-the-air amateur conversation (QSO).

A sending test probably won't be required. The Commission has decided that if applicants can demonstrate receiving ability, they most likely can also send at that speed. But be prepared, just in case! Subpart 97.503(a) of the FCC Rules says, "A telegraphy examination must be sufficient to prove that the examinee has the ability to send correctly by hand and to receive correctly by ear texts in the international Morse code at not less than the prescribed speed..."

Written Tests

After the code test has been administered, you'll then take the written examination. The examiner will give each applicant a test booklet, an answer sheet and scratch paper. After that, you're on your own. The first thing to do is read the instructions. Be sure to sign your name every place it's called for. Do all of this at the beginning to get it out of the way.

Next, check the examination to see that all pages and questions are there. If not, report this to the examiner immediately. When filling in your answer sheet, make sure your answers are marked next to the numbers that correspond to each question.

Go through the entire exam, and answer the easy questions first. Next, go back to the beginning and try the harder questions. Leave the tough questions for last. Guessing can only help, as there is no additional penalty for answering incorrectly.

If you have to guess, do it intelligently: At first glance, you may find that you can eliminate one or more "distracters." Of the remaining responses, more than one may seem correct; only one is the *best* answer, however. To the applicant who is fully prepared, incorrect distracters to each question are obvious. Nothing beats preparation!

After you've finished, check the examination thoroughly. You may have read a question wrong or goofed in your arithmetic. Don't be overconfident. There's no rush, so take your time. Think, and check your answer sheet. When you feel you've done your best and can do no more, return the test booklet, answer sheet and scratch pad to the examiner.

The Volunteer Examiner team will grade the exam right away. The passing mark is 74%. (That means 26 out of 35 questions correct — or no more than 9 incorrect answers on the Element 3 exam.) You will receive a Certificate of Successful Completion of Examination (CSCE) showing all exam elements that you pass. If you are already licensed, and you pass the exam elements required to earn a higher license class, the CSCE authorizes you to operate with your new privileges immediately. When you use your new privileges, you must sign your call sign, followed by the slant mark, "/"; on voice say "stroke" or "slant" and the letters "AG." You only have to follow this special identification procedure when you use your new privileges, and only until your new license is issued by the FCC.

If you pass only some of the exam elements required for a higher class license you will still receive a CSCE. That certificate shows what exam elements you passed, and is valid for 365 days. Use it as proof that you passed those exam elements so you won't have to take them over again the next time you try for the upgrade.

AND NOW, LET'S BEGIN

The complete General question pool (Element 3) is printed in Chapter 13. Chapters 1 through 10 explain the material covered in subelements G1 through G0 of the question pool. This book provides the background in FCC Rules, operating procedures, radio-wave propagation, Amateur Radio practices, electrical principles, circuit components, practical circuits, signals and emissions, antennas and feed lines, and RF safety you will need to know to pass the Element 3 General license exam.

Table 5 shows the study guide, or syllabus, for the Element 3 exam. This study guide was released by the Volunteer Examiner Coordinators' Question Pool Committee. The syllabus lists the topics to be covered by the General exam, and so forms the basic outline for the remainder of this book.

The question numbers used in the question pool refer to this syllabus. Each question number begins with a syllabus-point number (G1C, G2A, G3B). The question numbers end with a two-digit number. For example, question G1C03 is the third question in the series about syllabus point G1C (transmitter power standards). G3B09 is the ninth question about the G3B syllabus point (maximum usable frequency; propagation "hops").

The Question Pool Committee designed the syllabus and question pool so there are the same number of points in each subelement as there are exam questions from that subelement. For example, six questions on the General exam are from the "Commission's Rules" subelement. There are six groups of questions for this subelement, G1A through G1F. Three exam questions come from the "Radio-Wave Propagation" subelement, so there are three groups for that point. These are numbered G3A, G3B and G3C. There will be five questions on your exam from the RF Safety subelement. So you will find five groups of questions (G0A through G0E) in the RF Safety section. While not a requirement of the FCC Rules, the Question Pool Committee recommends that one question be taken from each group to make the best possible General license exam.

Good luck with your studies!

Table 5
General (Element 3) Syllabus

SUBELEMENT G1 — COMMISSION'S RULES
[6 Exam Questions — 6 Groups]
G1A General control operator frequency privileges
G1B Antenna structure limitations; good engineering and good amateur practice;
beacon operation; restricted operation; retransmitting radio signals
G1C Transmitter power standards; certification of external RF- power-amplifiers;
standards for certification of external RF-power amplifiers; HF data emission standards
G1D Examination element preparation; examination administration;
temporary station identification
G1E Local control; repeater and harmful interference definitions; third party communications
G1F Certification of external RF-power-amplifiers; standards for certification of external
RF-power amplifiers; HF data emission standards

SUBELEMENT G2 — OPERATING PROCEDURES
[6 Exam Questions — 6 Groups]
G2A Phone operating procedures
G2B Operating courtesy
G2C Emergencies, including drills and emergency communications
G2D Amateur auxiliary to the FCC's Compliance and Information Bureau; antenna orientation to
minimize interference; HF operations, including logging practices
G2E Third-party communications; ITU Regions; VOX operation
G2F CW operating procedures, including procedural signals, Q signals and common abbrevia-
tions; full break-in; RTTY operating procedures, including procedural signals and common
abbreviations and operating procedures for other digital modes, such as HF packet, AMTOR,
PacTOR, G-TOR, Clover and PSK31

SUBELEMENT G3 — RADIO- WAVE PROPAGATION
[3 Exam Questions — 3 Groups]
G3A Ionospheric disturbances; sunspots and solar radiation
G3B Maximum usable frequency; propagation "hops"
G3C Height of ionospheric regions; critical angle and frequency; HF scatter

SUBELEMENT G4 — AMATEUR RADIO PRACTICES
[5 Exam Questions — 5 Groups]
G4A Two-tone test; electronic TR switch; amplifier neutralization
G4B Test equipment: oscilloscope; signal tracer; antenna noise bridge; monitoring oscilloscope;
field-strength meters
G4C Audio rectification in consumer electronics; RF ground
G4D Speech processors; PEP calculations; wire sizes and fuses
G4E Common connectors used in amateur stations: types; when to use; fastening methods;
precautions when using; HF mobile radio installations; emergency power systems;
generators; battery storage devices and charging sources including solar; wind generation

SUBELEMENT G5 — ELECTRICAL PRINCIPLES
[2 Exam Questions — 2 Groups]

G5A   Impedance, including matching; resistance, including ohm; reactance; inductance; capacitance; and metric divisions of these values

G5B   Decibel; Ohm's Law; current and voltage dividers; electrical power calculations and series and parallel components; transformers (either voltage or impedance); sine wave root-mean-square (RMS) value

SUBELEMENT G6 — CIRCUIT COMPONENTS
[1 Exam Question — 1 Group]

G6A   Resistors; capacitors; inductors; rectifiers and transistors; etc.

SUBELEMENT G7 — PRACTICAL CIRCUITS
[1 Exam Question — 1 Group]

G7A   Power supplies and filters; single-sideband transmitters and receivers

SUBELEMENT G8 — SIGNALS AND EMISSIONS
[2 Exam Questions — 2 Groups]

G8A   Signal information; AM; FM; single and double sideband and carrier; bandwidth; modulation envelope; deviation; overmodulation

G8B   Frequency mixing; multiplication; bandwidths; HF data communications

SUBELEMENT G9 — ANTENNAS AND FEED LINES
[4 Exam Questions — 4 Groups]

G9A   Yagi antennas — physical dimensions; impedance matching; radiation patterns; directivity and major lobes

G9B   Loop antennas — physical dimensions; impedance matching; radiation patterns; directivity and major lobes

G9C   Random wire antennas — physical dimensions; impedance matching; radiation patterns; directivity and major lobes; feed point impedance of ½-wavelength dipole and ¼-wavelength vertical antennas

G9D   Popular antenna feed lines — characteristic impedance and impedance matching; SWR calculations

SUBELEMENT G0 — RF SAFETY
[5 Exam Questions — 5 Groups]

G0A   RF Safety Principles

G0B   RF Safety Rules and Guidelines

G0C   Routine Station Evaluation and Measurements (FCC Part 97 refers to RF Radiation Evaluation)

G0D   Practical RF-safety applications

G0E   RF-safety solutions

CHAPTER 1
KEYWORDS
KEYWORDS
KEYWORDS

Beacon station — An amateur station transmitting communications for the purposes of observation of propagation and reception or other related experimental activities. [§97.3(a)(9)]

Broadcasting — Transmissions intended to be received by the general public, either direct or relayed. [§97.3(a)(10)] Broadcasting is prohibited on Amateur Radio.

Emission types — Any signals from a transmitter. Phone, data and CW are emission types.

Frequency privileges — The specific band segments assigned to holders of each particular license. General operators have frequency privileges on the amateur high-frequency bands.

Harmful interference — Interference that seriously degrades, obstructs or repeatedly interrupts a radiocommunication service operating in accordance with the Radio Regulations. [§97.3(a)(23)]

International Telecommunication Union (ITU) — The international organization with responsibility for dividing the range of communications frequencies between the various services for the entire world.

One-way communications — Radio signals not directed to a specific amateur station, or for which no reply is expected. The FCC Rules provide for limited types of one-way communications on the amateur bands. [§97.111(b)]

Repeater — An amateur station that automatically retransmits the signals of other amateur stations. [§97.3(a)(39)]

Third-party communication — A message from the control operator (first party) of an amateur station to another amateur station (second party) on behalf of another person (third party). [§97.3(a)(46)]

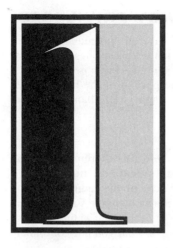

Commission's Rules

Part 97 of the Federal Communications Commission's Rules govern the Amateur Radio Service in the United States. Each Amateur Radio license exam includes questions covering sections of these rules. The General license exam includes six rules questions. In this chapter of *The ARRL General Class License Manual*, you will learn about the specific rules covered on the General (Element 3) license exam.

This book does not contain a complete listing of the Part 97 Rules. The FCC modifies Part 97 on an irregular basis, and it would be impossible to keep an up-to-date set of rules in a book designed to be revised with the release of a new General Question Pool. We do recommend, however, that every Amateur Radio operator have an up-to-date copy of the rules in their station for reference. *The FCC Rule Book*, published by the ARRL, contains the complete text of Part 97, along with detailed explanations of all the regulations. *The FCC Rule Book* is updated as necessary to keep it current with the latest rule changes. In addition, The ARRL Web site includes a listing of Part 97, with all the latest updates, at **www.arrl.org/FandES/ field/regulations/rules-regs.html**.

The six rules questions on your General license exam will come from six exam-question groups. Each group forms a subelement section on the syllabus or study guide, and the question pool. The six groups are:

G1A   General control operator frequency privileges
G1B   Antenna structure limitations; good engineering and good amateur practice; beacon operation; restricted operation; retransmitting radio signals
G1C   Transmitter power standards
G1D   Examination element preparation; examination administration; temporary station identification
G1E   Local control; repeater and harmful interference definitions; third-party communications
G1F   Certification of external RF-power-amplifiers; standards for certification of external RF-power amplifiers; HF data emission standards

GENERAL CONTROL OPERATOR FREQUENCY PRIVILEGES

As a General Amateur Radio licensee, you will have all Amateur Radio **frequency privileges** above 50 MHz, just like a Technician licensee. These frequency privileges are listed in Section 97.301(a) of the FCC Rules. (Section is often represented by the § character when referencing a specific portion of the FCC Rules.) You will also have extensive privileges on the high-frequency (HF) bands, as listed in §97.301(d). **Table 1-1** shows that section of the Rules.

What is the ITU?

When you look at Table 1-1 you will notice that there are different listings for ITU Regions 1, 2 and 3. The **International Telecommunication Union (ITU)** is the organization responsible for international regulation of the radio spectrum. The ITU regulations state that communications between amateurs in foreign countries must be limited to technical topics or "remarks of a personal nature of relative unimportance" — you can talk about your new rig or what you had for dinner, but don't get into a serious political discussion with a foreign ham.

Under the ITU, the world has been separated into three areas, or Regions. **Figure 1-1** shows a world map that indicates the general area of those three regions. Amateurs in different Regions may not be able to use all the same frequencies or operating modes. The United States, including Alaska and Hawaii, is in ITU Region 2. Region 3 comprises most of the islands in the Pacific Ocean, including American Samoa, the Northern Mariana Islands, Guam and Wake Island. (Many Pacific islands are under FCC regulation.) Also included in Region 3 are Australia, China, India and parts of the Middle East. Region 1 comprises Europe, Africa, Russia and the remainder of the Middle East.

There are two issues to consider when you look at the ITU regions. First, the

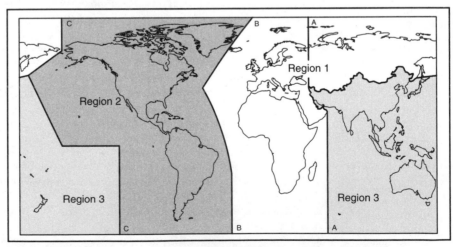

Figure 1-1 — This map shows the world divided into three ITU Regions.

Table 1-1
Authorized Frequency Bands and Sharing Requirements

§97.301 Authorized frequency bands.

The following transmitting frequency bands are available to an amateur station located within 50 km of the Earth's surface, within the specified ITU Region, and outside any area where the amateur service is regulated by any authority other than the FCC.

(d) For a station having a control operator who has been granted an operator license of General Class:[†]

Wavelength Band	ITU Region 1	ITU Region 2	ITU Region 3	Sharing requirements See §97.303, Paragraph:
MF	kHz	kHz	kHz	
160 m	1810-1850	1800-2000	1800-2000	(a), (b), (c)
HF	MHz	MHz	MHz	
80 m	3.525-3.750	3.525-3.750	3.525-3.750	(a)
75 m	—	3.85-4.00	3.85-3.90	(a)
40 m	7.025-7.100	7.025-7.150	7.025-7.100	(a)
40 m	—	7.225-7.300	—	(a)
30 m	10.10-10.15	10.10-10.15	10.10-10.15	(d)
20 m	14.025-14.150	14.025-14.150	14.025-14.150	
20 m	14.225-14.350	14.225-14.350	14.225-14.350	
17 m	18.068-18.168	18.068-18.168	18.068-18.168	
15 m	21.025-21.200	21.025-21.200	21.025-21.200	
15 m	21.30-21.45	21.30-21.45	21.30-21.45	
12 m	24.89-24.99	24.89-24.99	24.89-24.99	
10 m	28.0-29.7	28.0-29.7	28.0-29.7	

[†]These frequencies may also be used by Advanced Class and Amateur Extra Class licensees.

§97.303 Frequency sharing requirements.

The following is a summary of the frequency sharing requirements that apply to amateur station transmissions on the frequency bands specified in §97.301 of this Part. (For each ITU Region, each frequency band allocated to the amateur service is designated as either a secondary service or a primary service. A station in a secondary service must not cause harmful interference to, and must accept interference from, stations in a primary service. See §§2.105 and 2.106 of the FCC Rules, *United States Table of Frequency Allocations* for complete requirements.)

(a) Where, in adjacent ITU Regions or Subregions, a band of frequencies is allocated to different services of the same category, the basic principle is the equality of right to operate. The stations of each service in one region must operate so as not to cause harmful interference to services in the other Regions or Subregions. (See ITU *Radio Regulations*, No. 346 (Geneva, 1979).)

(b) No amateur station transmitting in the 1900-2000 kHz segment, the 70 cm band, the 33 cm band, the 13 cm band, the 9 cm band, the 5 cm band, the 3 cm band, the 24.05-24.25 GHz segment, the 76-81 GHz segment, the 144-149 GHz segment and the 241-248 GHz segment shall cause harmful interference to, nor is protected from interference due to the operation of, the Government radiolocation service.

(c) No amateur station transmitting in the 1900-2000 kHz segment, the 3 cm band, the 76-81 GHz segment, the 144-149 GHz segment and the 241-248 GHz segment shall cause harmful interference to, nor is protected from interference due to the operation of, stations in the non-Government radiolocation service.

(d) No amateur station transmitting in the 30 meter band shall cause harmful interference to stations authorized by other nations in the fixed service. The licensee of the amateur station must make all necessary adjustments, including termination of transmissions, if harmful interference is caused.

governments of the various countries of the world set the frequency limits of the frequency bands for their own amateurs, within the recommendations of the ITU Radio Regulations. Table 1-1 does not show Amateur Radio privileges for amateurs licensed in countries besides the US. The table does show the frequency limits for US licensed amateurs operating in the three regions. For example, a US licensed ham might be operating a station on board a ship registered in the US, while sailing in international waters. In that case, the US amateur must obey the frequency limits for the ITU Region in which they are operating, as listed in §97.301.

If you will be operating in an FCC-regulated area in ITU Regions 1 or 3, be sure to look in §97.301 (a) and (d) for your frequency privileges in those Regions. In addition to the land areas described, your operation is under FCC regulation if you are operating from onboard a US-registered ship sailing in international waters. In such cases, you must follow the frequency privileges authorized for use in the ITU region in which you are located.

What Modes Can I Operate?

In addition to listing the frequency privileges for the various license classes in §97.301, the FCC also specifies frequency ranges in each band for certain operating modes. On the HF bands there is one range for RTTY and data and another range for phone and image operation on each band. These mode restrictions are listed in §97.305, Authorized emission types. **Emission type** refers to any signals from a transmitter. Phone, data and CW are common emission types. It can be a bit confusing to figure out just what privileges you have as a General class licensee, because §97.305 lists the complete frequency range for the modes. You have to compare that with the frequencies for you license class, as listed in §97.301. **Figure 1-2** combines the information from both sections of the Rules to show your operating frequency and mode privileges by license class. Remember that CW is authorized on the entire range of each band. (There is one exception to that rule: the 60-meter band. We will discuss that band in a later section.)

Frequency Sharing

Many of our amateur bands are shared with other radio services. By sharing the frequencies with several other services, including the US military, hams can have the use of a greater amount of spectrum than would otherwise be possible. When a band is shared between two or more services, one service will be listed as the *primary service* for that band and the others will be listed as a *secondary service*.

If you are operating on a band on which the amateur service is the secondary service, and a station in the primary service causes interference, you should change frequency immediately. You may also be interfering with the other station, and that is prohibited by Part 97 of the FCC Rules. For example, if you are operating in the amateur 30-meter band and you receive some interference from a station in the fixed service of another country, you should change frequency because you may be interfering with the primary-service station. Table 1-1 lists sharing requirements for some of the General class frequency bands (§97.303(d)).

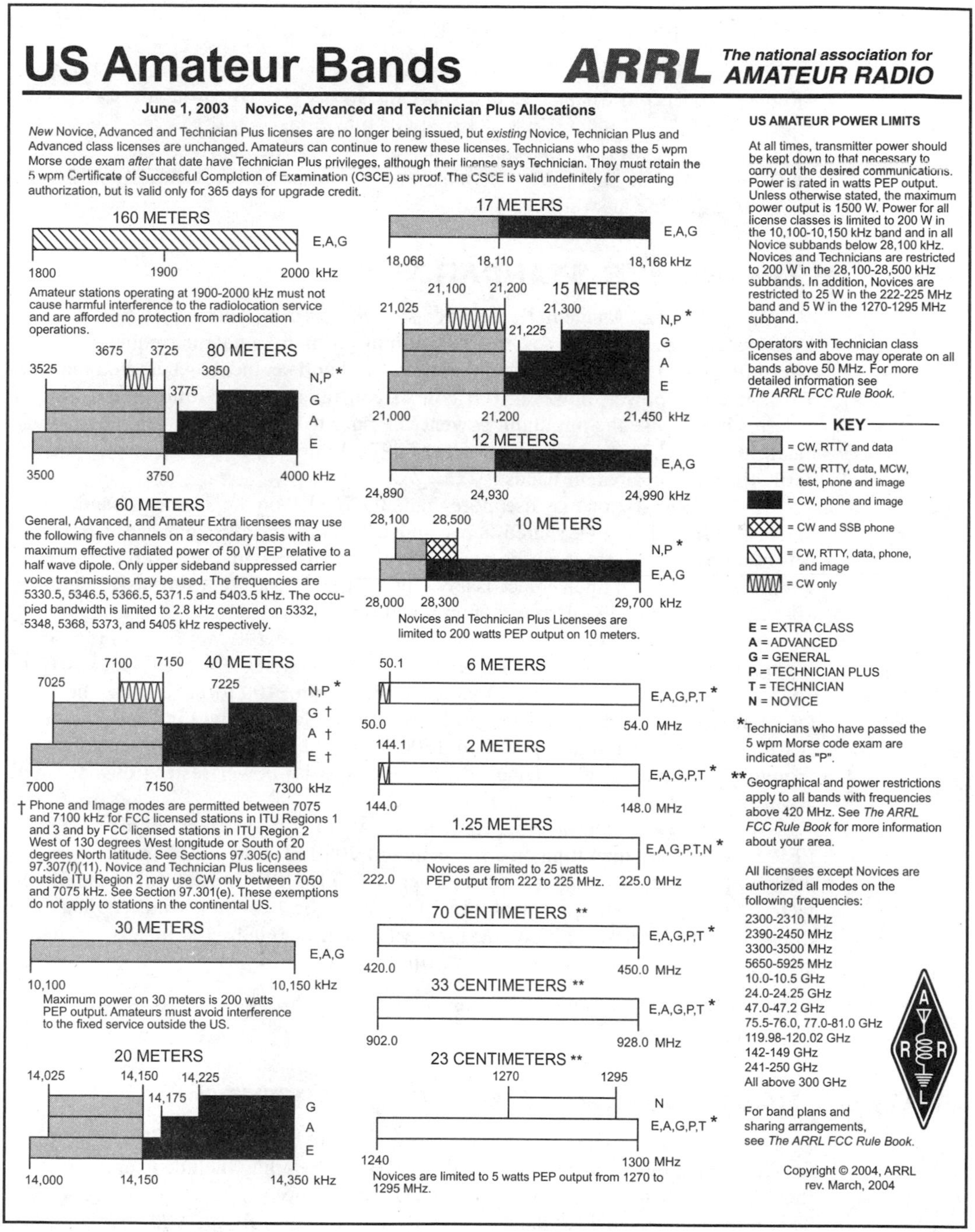

US Amateur Bands

ARRL *The national association for* **AMATEUR RADIO**

June 1, 2003   Novice, Advanced and Technician Plus Allocations

New Novice, Advanced and Technician Plus licenses are no longer being issued, but *existing* Novice, Technician Plus and Advanced class licenses are unchanged. Amateurs can continue to renew these licenses. Technicians who pass the 5 wpm Morse code exam *after* that date have Technician Plus privileges, although their license says Technician. They must retain the 5 wpm Certificate of Successful Completion of Examination (CSCE) as proof. The CSCE is valid indefinitely for operating authorization, but is valid only for 365 days for upgrade credit.

160 METERS

E,A,G

1800   1900   2000 kHz

Amateur stations operating at 1900-2000 kHz must not cause harmful interference to the radiolocation service and are afforded no protection from radiolocation operations.

80 METERS

3525   3675   3725   3850
3775

N,P *
G
A
E

3500   3750   4000 kHz

60 METERS

General, Advanced, and Amateur Extra licensees may use the following five channels on a secondary basis with a maximum effective radiated power of 50 W PEP relative to a half wave dipole. Only upper sideband suppressed carrier voice transmissions may be used. The frequencies are 5330.5, 5346.5, 5366.5, 5371.5 and 5403.5 kHz. The occupied bandwidth is limited to 2.8 kHz centered on 5332, 5348, 5368, 5373, and 5405 kHz respectively.

40 METERS

7025   7100   7150   7225

N,P *
G †
A †
E †

7000   7150   7300 kHz

† Phone and Image modes are permitted between 7075 and 7100 kHz for FCC licensed stations in ITU Regions 1 and 3 and by FCC licensed stations in ITU Region 2 West of 130 degrees West longitude or South of 20 degrees North latitude. See Sections 97.305(c) and 97.307(f)(11). Novice and Technician Plus licensees outside ITU Region 2 may use CW only between 7050 and 7075 kHz. See Section 97.301(e). These exemptions do not apply to stations in the continental US.

30 METERS

E,A,G

10,100   10,150 kHz

Maximum power on 30 meters is 200 watts PEP output. Amateurs must avoid interference to the fixed service outside the US.

20 METERS

14,025   14,150   14,225
14,175

G
A
E

14,000   14,150   14,350 kHz

17 METERS

E,A,G

18,068   18,110   18,168 kHz

15 METERS

21,100   21,200   21,300
21,025   21,225

N,P *
G
A
E

21,000   21,200   21,450 kHz

12 METERS

E,A,G

24,890   24,930   24,990 kHz

10 METERS

28,100   28,500

N,P *
E,A,G

28,000   28,300   29,700 kHz

Novices and Technician Plus Licensees are limited to 200 watts PEP output on 10 meters.

6 METERS

50.1
E,A,G,P,T *
50.0   54.0 MHz

2 METERS

144.1
E,A,G,P,T *
144.0   148.0 MHz

1.25 METERS

E,A,G,P,T,N *
222.0   225.0 MHz

Novices are limited to 25 watts PEP output from 222 to 225 MHz.

**70 CENTIMETERS ** **

E,A,G,P,T *
420.0   450.0 MHz

**33 CENTIMETERS ** **

E,A,G,P,T *
902.0   928.0 MHz

**23 CENTIMETERS ** **

1270   1295
N
E,A,G,P,T *
1240   1300 MHz

Novices are limited to 5 watts PEP output from 1270 to 1295 MHz.

US AMATEUR POWER LIMITS

At all times, transmitter power should be kept down to that necessary to carry out the desired communications. Power is rated in watts PEP output. Unless otherwise stated, the maximum power output is 1500 W. Power for all license classes is limited to 200 W in the 10,100-10,150 kHz band and in all Novice subbands below 28,100 kHz. Novices and Technicians are restricted to 200 W in the 28,100-28,500 kHz subbands. In addition, Novices are restricted to 25 W in the 222-225 MHz band and 5 W in the 1270-1295 MHz subband.

Operators with Technician class licenses and above may operate on all bands above 50 MHz. For more detailed information see *The ARRL FCC Rule Book.*

KEY

= CW, RTTY and data

= CW, RTTY, data, MCW, test, phone and image

= CW, phone and image

= CW and SSB phone

= CW, RTTY, data, phone, and image

= CW only

E = EXTRA CLASS
A = ADVANCED
G = GENERAL
P = TECHNICIAN PLUS
T = TECHNICIAN
N = NOVICE

*Technicians who have passed the 5 wpm Morse code exam are indicated as "P".

**Geographical and power restrictions apply to all bands with frequencies above 420 MHz. See *The ARRL FCC Rule Book* for more information about your area.

All licensees except Novices are authorized all modes on the following frequencies:

2300-2310 MHz
2390-2450 MHz
3300-3500 MHz
5650-5925 MHz
10.0-10.5 GHz
24.0-24.25 GHz
47.0-47.2 GHz
75.5-76.0, 77.0-81.0 GHz
119.98-120.02 GHz
142-149 GHz
241-250 GHz
All above 300 GHz

For band plans and sharing arrangements, see *The ARRL FCC Rule Book.*

Figure 1-2 — This US Amateur Bands chart shows Amateur operating privileges by band and operating mode.

The FCC Rule Book, published by ARRL explains on which bands the amateur service has primary status and which has secondary status.

[After you have learned the HF frequency privileges described in Table 1-1 and Figure 1-2, turn to Chapter 13 and study questions G1A01 through G1A11 and questions G1E07 and G1E08. Also study questions G2E05 through G2E09. Review this section of the chapter if you have difficulty with any of these questions. As you study the questions, turn the edge of the question-pool page under to hide the answer key.]

TRANSMITTER POWER STANDARDS

As a General Amateur Radio licensee, you will be authorized to use up to 1500 watts peak envelope power (PEP) output on many amateur frequencies. It is especially important to understand that there are band segments where you may not use that much power, however. (Of course you must always follow the rule that requires you to use the minimum power necessary to carry out the desired communications — §97.313(a).) The remainder of §97.313 describes maximum power standards for various amateur bands.

No amateur station may use more than 200 W PEP on the 30-meter band, 10.10 to 10.15 MHz. This is a shared band, and US amateurs must not cause **harmful interference** to stations authorized by other nations in the fixed service — §97.303(d). Harmful interference is any type of interference that seriously degrades, obstructs or repeatedly disrupts communications.

In addition, no amateur station may use more than 200 W PEP on the Novice segments of the 80, 40 and 15-meter bands. (The Novice 80-meter band is from 3675 to 3725 kHz. The Novice 40-meter band is from 7100 to 7150 kHz. The Novice 15-meter band is from 21.100 to 21.200 MHz.) General and higher class licensees may use the maximum power of 1500 W PEP on the Novice 10-meter band, though. The amateur 60-meter band has some special power restrictions. We will cover the 60-meter band later in this chapter.

When you transmit on other frequencies, you may use a maximum of 1500 W PEP. So if you are operating on a frequency of 7080 kHz on the 40-meter band, for example, you may use the maximum power of 1500 W PEP output. Likewise, if you were operating on a frequency of 24.950 MHz, in the 12-meter band, you may use the maximum 1500 W PEP output. Figure 1-2 will also help you identify frequencies that have 200 W maximum power limits.

[Turn to Chapter 13 and study questions G1C01 through G1C10. Review this section as needed.]

AMATEUR RADIO'S NEWEST BAND: 60 METERS

When the FCC decided to allow Amateur Radio operators to make use of some frequencies in the 60-meter band (in the 5 MHz range) they included some rules for the use of the band that had never been heard of in the Amateur Radio regulations. This new band opened for amateur use at midnight on July 3, 2003. General, Advanced and Amateur Extra licensees are only allowed to operate upper sideband voice mode on five specific channels — 5332 kHz, 5348 kHz, 5368 kHz, 5373 kHz

Table 1-2

Operating on the Amateur 60-Meter Band

§97.303 Frequency Sharing Requirements

The following is a summary of the frequency sharing requirements that apply to amateur station transmissions on the frequency bands specified in §97.301 of this Part. (For each ITU Region, each frequency band allocated to the amateur service is designated as either a secondary service or a primary service. A station in a secondary service must not cause harmful interference to, and must accept interference from, stations in a primary service. See §§2.105 and 2.106 of the FCC Rules, United States Table of Frequency Allocations for complete requirements.) (a) Where, in adjacent ITU Regions or Subregions, a band of frequencies is allocated to different services of the same category, the basic principle is the equality of right to operate. The stations of each service in one region must operate so as not to cause harmful interference to services in the other Regions or Subregions. (See ITU Radio Regulations, No. 346 (Geneva, 1979).)

(s) An amateur station having an operator holding a General, Advanced or Amateur Extra Class license may only transmit single sideband, suppressed carrier, (emission type 2K8J3E) upper sideband on the channels 5332 kHz, 5348 kHz, 5368 kHz, 5373 kHz, and 5405 kHz. Amateur stations shall ensure that their transmission occupies only the 2.8 kHz centered around each of these frequencies. Transmissions shall not exceed an effective radiated power (e.r.p.) of 50 W PEP. For the purpose of computing e.r.p. the transmitter PEP will be multiplied with the antenna gain relative to a dipole or equivalent calculation in decibels. A half wave dipole antenna will be presumed to have a gain of 0 dBd. Licenses using other antennas must maintain in their records either the manufacturer data on the antenna gain or calculations of the antenna gain. No amateur station may cause harmful interference to stations authorized in the mobile and fixed services; nor is any amateur station protected from interference due to the operation of any such station.

and 5405 kHz. Transmitted signals must occupy a bandwidth of no more than 2.8 kHz centered on one of these channel frequencies. In addition, transmitted signals are limited to an effective radiated power (ERP) of no more than 50 W PEP. ERP is specified as the transmitted power relative to a dipole antenna.

If you are using a dipole antenna, you can use the full 50 W of transmitter power. But if you are using an antenna that has some gain compared to a dipole you will have to reduce your transmitter power accordingly. For example, if your antenna has a gain of 3 dBd (3 dB of gain compared to a dipole) then you will have to reduce your transmitter power by 3 dB below the 50 W level. (3 dB of gain represents a doubling of the effective radiated power, so you would have to reduce your transmitter power by one half, to 25 W.)

If you are using an antenna that has some gain relative to a dipole, then you must keep station records with either the manufacturer's data on the antenna gain or calculations of the antenna gain. Amateurs are limited to no more than 50 W PEP ERP.

Amateurs are secondary users on this band. As such, we cannot cause any harmful interference to stations in the mobile and fixed services, who are primary users on this band. We are not protected from interference caused by stations in those services, either. **Table 1-2** lists the complete FCC Rules for amateur operation on this band.

[Turn to Chapter 13 and study questions G1C11, G1E10 and G1E11. Also study question G2D12. Review this section as needed.]

SOME DEFINITIONS

A **repeater** is an amateur station that automatically retransmits the signals of other amateur stations (§97.3(a)(39)). Section 97.205 of the FCC Rules outlines some restrictions and standards for repeaters.

One common question about repeater operation is whether an operator may use a repeater if the repeater output is on a frequency the operator is not allowed to use. For example, may a Technician operator use a 2-meter repeater that has an output on the 10-meter band? There are two tests to determine if such operation is legal. First, the operator in question must be transmitting to the repeater on a frequency on which he or she is authorized to operate. In our example, the Technician operator is transmitting to the repeater on 2 meters, so this part is legal. The second test is that the repeater control operator must have privileges on the repeater output frequency §97.205(a). So in this example, the repeater control operator must have a General or higher license. As long as both conditions are met, the operation is completely legal.

Another common question about repeaters involves frequency coordination. Volunteer frequency coordinators help ensure that repeaters in their area use frequencies that will not tend to interfere with each other. The FCC encourages frequency coordination, but the process is organized and run by hams and groups of hams who use repeaters.

The FCC has ruled in favor of coordinated repeaters, if there is harmful interference between two repeaters. In such a case, if a frequency coordinator has coordinated one but not the other, the licensee of the *uncoordinated* repeater has the primary responsibility for solving the interference problem. If both repeaters are coordinated, or if neither is, then both licensees are equally responsible for resolving the interference.

Section 97.3(a)(23) defines **harmful interference** as "interference which endangers the functioning of a radionavigation service or of other safety services or seriously degrades, obstructs or repeatedly interrupts a radiocommunication service operating in accordance with the Radio Regulations." In other words, any type of signal that causes interference to any type of radio service produces harmful interference. Your amateur station may receive harmful interference from other amateurs or from other radio signals. This could happen, for example, if a nearby business-radio transmitter produces an unwanted signal on an amateur band. Your amateur station might also cause harmful interference to another radio signal. Any time you receive a report that your signal may be interfering with another radio service you should investigate the possible causes of this interference and take every means possible to ensure that your station does not cause continued interference.

Third-party communication is defined in §97.3(a)(46). This definition says, "A message from the control operator (first party) of an amateur station to another amateur station control operator (second party) on behalf of another person (third party)." There are further details about such communications in §97.115 and §97.117 of the FCC Rules. See **Table 1-3**.

While these rules can seem confusing at times, basically they mean you can allow an unlicensed person to talk over the air on your Amateur Radio station, provided you remain at the radio controls. They also mean you can send a message on behalf of another person, or to an unlicensed person. This is the basic idea of the

Table 1-3
Third Party and International Communications

§97.115 Third party communications.

(a) An amateur station may transmit messages for a third party to:
  (1) Any station within the jurisdiction of the United States.
  (2) Any station within the jurisdiction of any foreign government whose administration has made arrangements with the United States to allow amateur stations to be used for transmitting international communications on behalf of third parties. No station shall transmit messages for a third party to any station within the jurisdiction of any foreign government whose administration has not made such an arrangement. This prohibition does not apply to a message for any third party who is eligible to be a control operator of the station.

(b) The third party may participate in stating the message where:
  (1) The control operator is present at the control point and is continuously monitoring and supervising the third party's participation; and
  (2) The third party is not a prior amateur service licensee whose license was revoked; suspended for less than the balance of the license term and the suspension is still in effect; suspended for the balance of the license term and relicensing has not taken place; or surrendered for cancellation following notice of revocation, suspension or monetary forfeiture proceedings. The third party may not be the subject of a cease and desist order which relates to amateur service operation and which is still in effect.

(c) At the end of an exchange of international third party communications, the station must also transmit in the station identification procedure the call sign of the station with which a third party message was exchanged.

§97.117 International communications.

Transmissions to a different country, where permitted, shall be made in plain language and shall be limited to messages of a technical nature relating to tests, and, to remarks of a personal character for which, by reason of their unimportance, recourse to the public telecommunications service is not justified.

ARRL's National Traffic System, designed to relay messages around the US and to many parts of the world.

When you are transmitting third-party traffic your station must be under local or remote control. An amateur station cannot transmit third-party traffic while it is under automatic control. The exception to this rule is that a station can be operated under automatic control while it is relaying third-party traffic using RTTY or data modes. Such a station might be part of a network of stations used to relay such messages. In that case, the third-party messages must originate at a station that is under local or remote control.

There are two important rule points to keep in mind when you are conducting any type of third-party communication with another country. First, the US and the other government must have made prior arrangements to allow such third-party communications. This agreement is called a "third-party agreement." **Table 1-4** is a list of countries with which the US currently has such an agreement. This list changes

Table 1-4
Third-Party Traffic Agreements List

Countries that share third-party traffic agreements with the US:

Occasionally, DX stations may ask you to pass a third-party message to a friend or relative in the States. This is all right as long as the US has signed an official third-party traffic agreement with that particular country, or the third party is a licensed amateur. The traffic must be noncommercial and of a personal, unimportant nature. During an emergency, the US State Department will often work out a special temporary agreement with the country involved. But in normal times, never handle traffic without first making sure it is legally permitted.

US Amateurs May Handle Third-Party Traffic With:

C5	The Gambia	J7	Dominica	VE	Canada
CE	Chile	J8	St Vincent and the	VK	Australia
CO	Cuba		Grenadines	VP6/VR6**	Pitcairn Island
CP	Bolivia	JY	Jordan	XE	Mexico
CX	Uruguay	LU	Argentina	YN	Nicaragua
D6	Federal Islamic	OA	Peru	YS	El Salvador
	Rep. of the Comoros	PY	Brazil	YV	Venezuela
DU	Philippines	TA	Turkey	ZP	Paraguay
EL	Liberia	TG	Guatemala	ZS	South Africa
GB*	United Kingdom	TI	Costa Rica	3DAØ	Swaziland
HC	Ecuador	T9	Bosnia-Herzegovina	4U1ITU	ITU Geneva
HH	Haiti	V2	Antigua and Barbuda	4U1VIC	VIC, Vienna
HI	Dominican Republic	V3	Belize	4X	Israel
HK	Colombia	V4	St Christopher and	6Y	Jamaica
HP	Panama		Nevis	8R	Guyana
HR	Honduras	V6	Federated States of	9G	Ghana
J3	Grenada		Micronesia	9L	Sierra Leone
J6	St Lucia	V7	Marshall Islands	9Y	Trinidad and Tobago

Notes:

*Third-party traffic permitted between US amateurs and special-events stations in the United Kingdom having the prefix GB only, with the exception that GB3 stations are not included in this agreement.

**Since 1970, there has been an informal agreement between the United Kingdom and the US, permitting Pitcairn and US amateurs to exchange messages concerning medical emergencies, urgent need for equipment or supplies, and private or personal matters of island residents.

Please note that the Region 2 Division of the International Amateur Radio Union (IARU) has recommended that international traffic on the 20 and 15-meter bands be conducted on the following frequencies:

14.100-14.150 MHz

14.250-14.350 MHz

The IARU is the alliance of Amateur Radio societies from around the world; Region 2 comprises member-societies in North, South and Central America, and the Caribbean.

Note: At the end of an exchange of third-party traffic with a station located in a foreign country, an FCC-licensed amateur must also transmit the call sign of the foreign station as well as his own call sign.

Current as of January 2004

from time to time because of additions or deletions. Occasionally there are temporary agreements, such as during some special event such as the Olympic Games, or during certain times of the year, such as the Christmas/New Year Holiday season.

Many governments operate the public telephone and other telecommunications services in their country. They do not look kindly on losing the fees that would be paid for such services because Amateur Radio operators bypassed the system. For this reason, all international communications are limited to comments of a technical nature related to Amateur Radio experiments and personal remarks. Such communications are considered relatively unimportant because you wouldn't normally use the public telecommunications system for those messages.

[Now turn to Chapter 13 and study questions G1E02 through G1E06 and question G1E12. You should also study question G2E01. If you have any difficulty with these questions, review the "Some Definitions" section of the text.]

GOOD ENGINEERING AND GOOD AMATEUR PRACTICE

The Amateur Service is a radiocommunication service intended to encourage Amateur Radio operators to train themselves in the technical and operating phases of electronics and communications technology by experimenting and communicating with other amateurs. Amateur Radio operators should strive always to maintain and operate their stations in a manner that will meet the highest standards.

Part 97 of the FCC Rules spells out many technical and operating standards that all US Amateur Radio stations must meet. The Rules describe various operating modes and set minimum standards for transmitted signal quality. The Rules must also be sufficiently vague, however, so as to encourage experimentation and the development of new communications techniques.

The FCC can't anticipate every question amateurs may have about how to set up or operate their stations, nor should we expect them to. After all, that's what experimentation is all about. So what should we do when faced with those unanswered questions about our stations? One place to look is in §97.101(a), under General standards, where it says, "In all respects not specifically covered by FCC Rules each amateur station must be operated in accordance with good engineering and good amateur practice." This means you should comply with the conventions that have evolved within the amateur community to promote efficient use of our limited radio spectrum. It means cooperating with other amateurs who have interests in diverse operating modes and activities. Band plans and "gentlemen's agreements" *may not* be specifically addressed in the Rules, but following them is part of operating your station "in accordance with good engineering and good amateur practice."

[Turn to the question pool in Chapter 13 and study question G1B02 before going on. Review this section if you do not understand this question.]

PROHIBITED TRANSMISSIONS

The FCC Rules are generally intended to allow flexibility and room to try new types of communication. Still, there are some things the Rules clearly prohibit. Some of these prohibitions provide for your safety, or the safety of others. Some

Table 1-5
What Isn't Allowed?

§97.113 Prohibited transmissions.

(a) No amateur station shall transmit:

    (1) Communications specifically prohibited elsewhere in this Part;

    (2) Communications for hire or for material compensation, direct or indirect, paid or promised, except as otherwise provided in these rules;

    (3) Communications in which the station licensee or control operator has a pecuniary interest, including communications on behalf of an employer. Amateur operators may, however, notify other amateur operators of the availability for sale or trade of apparatus normally used in an amateur station, provided that such activity is not conducted on a regular basis;

    (4) Music using a phone emission except as specifically provided elsewhere in this Section; communications intended to facilitate a criminal act; messages in codes or ciphers intended to obscure the meaning thereof, except as otherwise provided herein; obscene or indecent words or language; or false or deceptive messages, signals or identification;

    (5) Communications, on a regular basis, which could reasonably be furnished alternatively through other radio services.

(b) An amateur station shall not engage in any form of broadcasting, nor may an amateur station transmit one-way communications except as specifically provided in these rules; nor shall an amateur station engage in any activity related to program production or news gathering for broadcasting purposes, except that communications directly related to the immediate safety of human life or the protection of property may be provided by amateur stations to broadcasters for dissemination to the public where no other means of communication is reasonably available before or at the time of the event.

(c) A control operator may accept compensation as an incident of a teaching position during periods of time when an amateur station is used by that teacher as a part of classroom instruction at an educational institution.

(d) The control operator of a club station may accept compensation for the periods of time when the station is transmitting telegraphy practice or information bulletins, provided that the station transmits such telegraphy practice and bulletins for at least 40 hours per week; schedules operations on at least six amateur service MF and HF bands using reasonable measures to maximize coverage; where the schedule of normal operating times and frequencies is published at least 30 days in advance of the actual transmissions; and where the control operator does not accept any direct or indirect compensation for any other service as a control operator.

(e) No station shall retransmit programs or signals emanating from any type of radio station other than an amateur station, except propagation and weather forecast information intended for use by the general public and originated from United States Government stations and communications, including incidental music, originating on United States Government frequencies between a space shuttle and its associated Earth stations. Prior approval for shuttle retransmissions must be obtained from the National Aeronautics and Space Administration. Such retransmissions must be for the exclusive use of amateur operators. Propagation, weather forecasts, and shuttle retransmissions may not be conducted on a regular basis, but only occasionally, as an incident of normal amateur radio communications.

(f) No amateur station, except an auxiliary, repeater or space station, may automatically retransmit the radio signals of other amateur stations.

restrictions are required to preserve the very nature of Amateur Radio. Several questions in the General question pool are about these restrictions. Section 97.113 lists specific types of prohibited transmissions. **See Table 1-5**.

No amateur station may transmit music. The FCC doesn't want the general public tuning in to your Amateur Radio transmissions to listen to music. This would be a form of **broadcasting**, also forbidden on the amateur bands. The rules specifically state that transmitting music using a phone emission is prohibited. If you wanted to exchange digital files containing music scores with another ham, that would be okay. Likewise, if you wanted to experiment with sending MIDI control codes as part of a digital transmission, you could do that. Just don't play the actual composition over the air.

The prohibition against transmitting music also includes background music. If you have a radio or recorded music playing in the background while you are talking into the microphone of your transmitter, you could be in violation of this rule. Turn down the background audio before transmitting, to keep your station from transmitting music.

There is one exception to the "no music" rule. Amateur stations are permitted to retransmit signals that originate on US Government frequencies between a space shuttle and its associated Earth stations, with prior approval from NASA. Under these conditions, if there is incidental music transmitted between the shuttle and the Earth station, the amateur station is allowed to continue the retransmission.

In general, an amateur station may not retransmit signals from any other radio service. The exception to this rule is the retransmission of propagation and weather forecast information transmitted from US Government stations, and the space shuttle communications just discussed.

No amateur station may transmit communications intended to facilitate a criminal act. Amateur stations may not transmit obscene or indecent words, nor may they transmit false or deceptive messages. Amateur stations may not transmit any false signals or false identification.

No amateur station in two-way communication may transmit messages using codes or ciphers that are intended to obscure the meaning of the communications. It doesn't matter if the communication is domestic (between US amateurs) or international (between a US amateur and an amateur in another country). Procedural signals and common abbreviations are not considered to be codes or ciphers under this rule. So it is okay to use Q signals, procedural signals (prosigns) and other abbreviations that you commonly hear over the air. You and your friend may not make up your own "secret code," however, in order to keep your communications "private."

A station designated as a telecommand station of an amateur space station (such as a satellite) may transmit special telecommand messages to the space station, however. Those telecommand messages may use special codes intended to obscure the meaning of the message. This rule allows the use of special codes so only properly authorized stations can transmit control codes to the space station. Telecommand signals transmitted to control a model craft are also not considered to be codes or ciphers intended to obscure the meaning of the communication.

One important restriction designed to protect others is an antenna-height limitation. You may not build an antenna structure (including the radiating elements, tower, supports or any other attachments) that is over 200 feet high, unless you obtain prior FCC approval. In addition, if you live within 4 miles of an airport the height of your antenna may be further limited by the Rules. Basically, the closer you are to the airport, the lower your antenna must be. Section 97.15 and Part 17 of the FCC Rules contain the details of these antenna-height limitations. Check *The FCC Rule Book*, published by ARRL, if you think you may be affected by the airport limitation.

[Now turn to Chapter 13 and study questions G1B01 and G1B05 through G1B09. Review this section if any of these questions give you difficulty.]

BROADCASTING

No amateur station may engage in any type of **broadcasting**. Broadcasting generally refers to any type of communication intended to be received by the general public. Broadcasts would often be transmitted as a form of **one-way communication**, because they are not transmitted specifically to another amateur.

According to the Rules, "…an amateur station may transmit the following types of one-way communications: (1) Brief transmissions necessary to make adjustments to the station; (2) Brief transmissions necessary to establishing two-way communications with other stations; (3) Telecommand; (4) Transmissions necessary to providing emergency communications; (5) Transmissions necessary to assisting persons learning, or improving proficiency in, the international Morse code; (6) Transmissions necessary to disseminate information bulletins; (7) Transmissions of telemetry." [§97.111(b)] The information bulletins mentioned in (6) refer to information bulletins of interest to amateurs, not information intended for the general public.

In addition to this section, §97.203(g) says, "A beacon may transmit one-way communications." A **beacon station** is "an amateur station transmitting communications for the purposes of observation of propagation and reception or other related experimental activities." [§97.3(a)(9)] Beacon stations are limited to no more than 100 W output. Beacon stations are only allowed to transmit one signal per band from any given location. Any Amateur Radio operator with a license other than a Novice license may be the control operator of a beacon station. There are no operating mode restrictions on beacon stations, however.

When a natural disaster such as an earthquake, flood or severe storm hits an area, normal means of communication are often wiped out. Amateur Radio plays a significant role in providing communications with the disaster area because many hams maintain complete emergency stations that can be put on the air without commercial power or other services. Amateur Radio operators practice their communications skills under these difficult conditions during the annual ARRL Field Day and other emergency drills.

During such disasters, the broadcast news media may be cut off from their normal news-gathering methods. They often turn to Amateur Radio operators for news and information. Amateurs must be careful not to become part of the news media in such cases. In their eagerness for information, the media may take advantage of a well-meaning amateur's eagerness to help. They may request the amateur to ask specific questions of stations in the disaster area, and even attempt to conduct "interviews" over Amateur Radio. The FCC Rules state some limited conditions under which an amateur station may assist in such "news-gathering" activities.

These are spelled out in §97.113(b). (See Table 1-5 for the complete text.) The communications must be directly related to the immediate safety of human life or the protection of property and there must be no other means of communication reasonably available before or at the time of the event. Of course the news information must be directly related to the event. The news media is interested in obtaining the information by the fastest method possible, but that isn't reason enough to use Amateur Radio.

The real test here is that you, as the amateur operator, must feel confident that what you are doing falls within the Rules. Don't be swayed into asking questions for (or turning the mike over to) an eager reporter. If the FCC decides the news-gathering was inappropriate, you are the one left to face the consequences!

[Now turn to Chapter 13 and study questions G1B03, G1B04 and G1B10. If you are uncertain about the answers, review this section.]

RF POWER AMPLIFIERS

RF power amplifiers capable of operating on frequencies below 144 MHz may require FCC certification. Sections 97.315 and 97.317 describe the conditions under which certification is required, and set out the standards to be met for certification. See **Table 1-6**.

Many of these rules apply to manufacturers of amplifiers or kits, but several points are important for individual amateurs. Amateurs may build their own amplifiers or modify amplifiers for use in their own station without concern for the FCC certification rules. If you build or modify an amplifier for use by another amateur operator, then §97.315(a) applies. This rule says you cannot build or modify more than one amplifier of any particular model during any calendar year without obtaining a grant of certification from the FCC.

To receive a grant of certification, an amplifier must satisfy the spurious emission standards specified in §97.307(d) or (e) when operated at full power output. The amplifier must not be capable of reaching its designed output power when driven with less than 50 watts. The amplifier must also not be capable of operation on any frequency between 24 and 35 MHz. (This is to prevent the amplifier from being used illegally on the Citizen's Band frequencies.) The amplifier may be capable of operation on all amateur bands with frequencies below 24 MHz, however. An amplifier is considered incapable of operating between 24 and 35 MHz if it has no more than 6 dB of gain between 24 and 26 MHz and between 28 and 35 MHz. In that case it must also have no gain (0 dB) between 26 and 28 MHz.

If there is any wiring that can be simply altered to allow the amplifier to operate outside of the amateur bands, or if there are instructions describing how to modify or operate the amplifier in a way that violates FCC Rules, the amplifier would not be granted FCC certification. The main reason for these rules is to prevent amateur equipment from being used illegally on the Citizen's Band.

[Now turn to Chapter 13 and study questions G1F01 through G1F04. Also study questions G1F10 and G1F11. If any of these questions give you difficulty, review this section.]

Table 1-6
RF Power Amplifiers For The HF Bands

§97.315 Certification of external RF power amplifiers.

(a) No more than 1 unit of 1 model of an external RF power amplifier capable of operation below 144 MHz may be constructed or modified during any calendar year by an amateur operator for use at a station without a grant of certification. No amplifier capable of operation below 144 MHz may be constructed or modified by a non-amateur operator without a grant of certification from the FCC.

(b) Any external RF power amplifier or external RF power amplifier kit (see §2.815 of the FCC Rules), manufactured, imported or modified for use in a station or attached at any station must be certificated for use in the amateur service in accordance with Subpart J of Part 2 of the FCC Rules. This requirement does not apply if one or more of the following conditions are met:

  (1) The amplifier is not capable of operation on frequencies below 144 MHz. For the purpose of this part, an amplifier will be deemed to be incapable of operation below 144 MHz if it is not capable of being easily modified to increase its amplification characteristics below 120 MHz and either:

    (i) The mean output power of the amplifier decreases, as frequency decreases from 144 MHz, to a point where 0 dB or less gain is exhibited at 120 MHz; or

    (ii) The amplifier is not capable of amplifying signals below 120 MHz even for brief periods without sustaining permanent damage to its amplification circuitry.

  (2) The amplifier was manufactured before April 28, 1978, and has been issued a marketing waiver by the FCC, or the amplifier was purchased before April 28, 1978, by an amateur operator for use at that amateur operator's station.

  (3) The amplifier was:

    (i) Constructed by the licensee, not from an external RF power amplifier kit, for use at the licensee's station; or]

    (ii) Modified by the licensee for use at the licensee's station.

  (4) The amplifier is sold by an amateur operator to another amateur operator or to a dealer.

  (5) The amplifier is purchased in used condition by an equipment dealer from an amateur operator and the amplifier is further sold to another amateur operator for use at that operator's station.

(c) Any external RF power amplifier appearing in the Commission's database as certificated for use in the amateur service may be marketed for use in the amateur service.

§97.317 Standards for certification of external RF power amplifiers.

(a) To receive a grant of certification, the amplifier must satisfy the spurious emission standards of §97.307(d) or (e) of this Part, as applicable, when the amplifier is:

  (1) Operated at its full output power;

  (2) Placed in the "standby" or "off" positions, but still connected to the transmitter; and

  (3) Driven with at least 50 W mean RF input power (unless higher drive level is specified).

(b) To receive a grant of certification, the amplifier must not be capable of operation on any frequency

---

EXAM ELEMENT PREPARATION AND ADMINISTRATION

As a General Amateur Radio operator, you will earn some important new privileges that have nothing to do with on-the-air operation. A General licensee is eligible to be certified as a Volunteer Examiner (VE) by a Volunteer-Examiner Coordinator (VEC). This privilege is very important, and one you should take very seriously. You will be responsible for maintaining the high standards set for the Amateur Radio examination process.

Part 97, Subpart F of the FCC Rules describes the Qualifying Examination Sys-

or frequencies between 24 MHz and 35 MHz. The amplifier will be deemed incapable of such operation if it:

(1) Exhibits no more than 6 dB gain between 24 MHz and 26 MHz and between 28 MHz and 35 MHz. (This gain will be determined by the ratio of the input RF driving signal (mean power measurement) to the mean RF output power of the amplifier); and

(2) Exhibits no amplification (0 dB gain) between 26 MHz and 28 MHz.

(c) Certification may be denied when denial would prevent the use of these amplifiers in services other than the amateur service. The following features will result in dismissal or denial of an application for certification:

(1) Any accessible wiring which, when altered, would permit operation of the amplifier in a manner contrary to the FCC Rules;

(2) Circuit boards or similar circuitry to facilitate the addition of components to change the amplifier's operating characteristics in a manner contrary to the FCC Rules;

(3) Instructions for operation or modification of the amplifier in a manner contrary to the FCC Rules;

(4) Any internal or external controls or adjustments to facilitate operation of the amplifier in a manner contrary to the FCC Rules;

(5) Any internal RF sensing circuitry or any external switch, the purpose of which is to place the amplifier in the transmit mode;

(6) The incorporation of more gain in the amplifier than is necessary to operate in the amateur service; for purposes of this paragraph, the amplifier must:

  (i) Not be capable of achieving designed output power when driven with less than 50 W mean RF input power;

  (ii) Not be capable of amplifying the input RF driving signal by more than 15 dB, unless the amplifier has a designed transmitter power of less than 1.5 kW (in such a case, gain must be reduced by the same number of dB as the transmitter power relationship to 1.5 kW; This gain limitation is determined by the ratio of the input RF driving signal to the RF output power of the amplifier where both signals are expressed in peak envelope power or mean power);

  (iii) Not exhibit more gain than permitted by paragraph (c)(6)(ii) of this Section when driven by an RF input signal of less than 50 W mean power; and

  (iv) Be capable of sustained operation at its designed power level.

(7) Any attenuation in the input of the amplifier which, when removed or modified, would permit the amplifier to function at its designed transmitter power when driven by an RF frequency input signal of less than 50 W mean power; or

(8) Any other features designed to facilitate operation in a telecommunication service other than the Amateur Radio Services, such as the Citizens Band (CB) Radio Service.

tems of the amateur service. This section of the Rules describes the VEC system and outlines the procedures used to make Amateur Radio exams and administer them. The FCC's basic principle for preparing or administering exams is that a Volunteer Examiner can only prepare or administer those exam elements one license grade below the VE's own license. An Amateur Extra licensee may prepare or administer all exam elements, but a General class licensee may only work with the exam elements for Technician and Technician with credit for passing the 5 wpm Morse code exam.

A General licensee may prepare or administer the Element 1 telegraphy exami-

Table 1-7
License Examination Elements

§97.501 Qualifying for an amateur operator license.

Each applicant must pass an examination for a new amateur operator license grant and for each change in operator class. Each applicant for the class of operator license grant specified below must pass, or otherwise receive examination credit for, the following examination elements:

(a) Amateur Extra Class operator: Elements 1, 2, 3, and 4;

(b) General Class operator: Elements 1, 2, and 3;

(c) Technician Class operator: Element 2.

§97.503 Element standards.

(a) A telegraphy examination must be sufficient to prove that the examinee has the ability to send correctly by hand and to receive correctly by ear texts in the international Morse code at not less than the prescribed speed, using all the letters of the alphabet, numerals 0-9, period, comma, question mark, slant mark and prosigns $\overline{AR}$, $\overline{BT}$ and $\overline{SK}$.

Element 1: 5 words per minute.

(b) A written examination must be such as to prove that the examinee possesses the operational and technical qualifications required to perform properly the duties of an amateur service licensee. Each written examination must be comprised of a question set as follows:

(1) Element 2: 35 questions concerning the privileges of a Technician Class operator license. The minimum passing score is 26 questions answered correctly.

(2) Element 3: 35 questions concerning the privileges of a General Class operator license. The minimum passing score is 26 questions answered correctly.

(3) Element 4: 50 questions concerning the privileges of an Amateur Extra Class operator license. The minimum passing score is 37 questions answered correctly.

---

nation, which is the 5 words-per-minute Morse code exam. A General licensee may prepare or administer the Element 2 written exam. Element 2 is the 35-question Technician written exam. If you are preparing one of these exam elements you must select questions from the appropriate current question pool, as maintained by the Question Pool Committee of the Volunteer-Examiner Coordinators. These question pools are released into the public domain and are published by the ARRL and other exam-preparation publishers in the applicable license study guides.

Section 97.501 of the Rules specifies the minimum examination elements an applicant must pass to qualify for any particular Amateur Radio license. Section 97.503 specifies the standards for the various Amateur Radio exam elements. **Table 1-7** lists Sections 97.501 and 97.503. For a Technician license, the applicant must pass Exam Element 2. By passing Exam Element 1 a Technician licensee earns the Novice license privileges on the HF bands.

The rules for administering all license exams are the same. Sections 97.509 and 97.511 describe how to conduct an exam session. Section 97.509 lists some requirements for being a Volunteer Examiner. A VE must be at least 18 years of age, and be accredited by the coordinating VEC. No VE may administer an examination to any member of their family, including in-laws. If your amateur station license or operator license has ever been revoked or suspended by the FCC, you are not eligible to be a VE.

If you meet all these qualifications, you can apply for accreditation with a Volunteer-Examiner Coordinator. When you are accredited by a VEC, you will receive an ID badge or other form of accreditation. If you would like more information about becoming a Volunteer Examiner, contact the ARRL/VEC Office, 225 Main Street, Newington, CT 06111-1494. As soon as the FCC grants your General class or higher license and you receive your accreditation from the VEC, you can begin helping administer Technician license exams.

Every exam session must be coordinated by a VEC and conducted by three VEs accredited by that VEC. All three VEs must hold the proper license for the exam elements being administered at that session. So to conduct a Technician exam you will need three VEs who hold at least General licenses.

[Now it is time to review a few more questions from the question pool printed in Chapter 13. Study questions G1D01 through G1D06 and question G1D11. Review this section if any of these questions give you difficulty.]

STATION IDENTIFICATION

At the end of your General license exam you will receive a Certificate of Successful Completion of Examination (CSCE) showing those exam elements you passed. This certificate is valid as proof that you passed those elements, and is valid for 365 days. So if you don't pass all the exam elements to earn your new license, you can retake only the ones you didn't pass to earn the license. You can also use this certificate to show that you passed certain elements if you want to go back to another exam session before your license actually arrives from the FCC. If you want to upgrade as fast as possible, you might go to another exam session the following month (or sooner) and take the Amateur Extra exam.

If you pass all the exam elements for the General license, and have a Technician license, you may begin using your new General license privileges on the air immediately. Your CSCE serves as a temporary "endorsement" to your license, until the FCC grants your new license. You must follow a special identification procedure when you use these new privileges with the CSCE, however. In §97.119(f) we read that "When the control operator who is exercising the rights and privileges authorized by §97.9(b) of this Part, an indicator must be included after the call sign as follows: ... (2) For a control operator who has requested a license modification from Novice or Technician Class to General Class: AG." Section 97.9(b) is the Rule point that allows you to use your new privileges after an upgrade, as long as you have a CSCE to prove your new privileges, until the upgraded license is granted by the FCC.

There is a bit more detail about how to use this special identifier in §97.119(c): "One or more indicators may be included with the call sign. Each indicator must be separated from the call sign by the slant mark (/) or by any suitable word that denotes the slant mark..." So if you are operating on the 20-meter band, transmitting phone emissions on 14.325 MHz, for example, you would give your station identification with your call sign, some word to denote the slant mark, and the letters AG. This would sound like "This is WB3IOS slant AG." You may also hear stations use words like "stroke," "interim," or "temporary" instead of "slant."

Remember that you only have to give this temporary identifier when you operate using your new privileges. If you previously had a Technician license and are operating on the 2-meter band, you don't have to say "temporary AG."

Every amateur station must have a control operator. When the station is being *locally controlled*, the control operator must be at the control point (§97.109(b)). Sometimes the control operator may have license privileges that exceed those of the station licensee. For example, an operator with a General Amateur Radio license may operate a station licensed to a Technician or Technician operator with Morse code credit. If the General licensee is operating the station outside of the other licensee's privileges, such as on 7250 kHz, special identification procedures apply. According to §97.119(e), the General operator would use the station licensee's call sign followed by a slant mark (/) or by a word that denotes the slant mark, and the control operator's call sign.

You can use any language you want to communicate with other amateurs. Amateur Radio gives you an excellent opportunity to practice your "foreign" language skills with other hams who speak the language you are learning. They will be very helpful with questions you may have, and you will find they are quite pleased that you would try to speak with them in their native language. When you give your station identification, though, you must use English. It's also a good idea to use a phonetic alphabet when you give your station identification. Using words that are internationally recognized substitutes for letters, such as the ITU phonetic alphabet, makes it easier for everyone to understand the letters. Of course, you can always use Morse code to transmit your station identification, no matter what frequency or mode you are using.

[Congratulations! You have completed your study of the Commission's Rules section of the question pool. Before you go on to the next chapter, though, you should turn to Chapter 13 and study questions G1D07 through G1D10 and G1D12. Also study questions G1E01 and G1E09. Review this last section of Chapter 1 if you have any difficulty with these questions.]

CHAPTER 2
KEYWORDS
KEYWORDS
KEYWORDS

Amateur Auxiliary — A voluntary organization, administered by ARRL. The primary objectives are to foster amateur self-regulation and compliance with the rules.

Amateur Teleprinting Over Radio (AMTOR) — AMTOR provides error-correcting capabilities. See Automatic Repeat Request and Forward Error Correction.

American National Standard Code for Information Interchange (ASCII) — A seven-bit digital code used in computer and radioteleprinter applications.

Automatic Repeat Request (ARQ) — One of two AMTOR communications modes. In ARQ, also called Mode A, the two stations are constantly confirming each other's transmissions. If information is lost, it is repeated until the receiving station confirms reception.

Azimuthal-equidistant projection map — A map made with its center at one geographic location and the rest of the continents projected from that point. Also called a great-circle map, this map is the most useful type for determining where to point a directional antenna to communicate with a specific location.

Band plan – An agreement for operating within a certain portion of the radio spectrum. Band plans set aside certain frequencies for each different mode of amateur operation, such as CW, SSB, RTTY, SSTV, FM, repeaters and simplex.

Baudot — A five-bit digital code used in teleprinter application.

Directional antenna — An antenna that concentrates more of the transmitted energy in a particular direction. A directional antenna must be turned to provide the best signal coverage in various directions. Such antennas also receive signals better from the direction they are pointed. Also called a beam antenna.

Effective radiated power (ERP) — The relative amount of power radiated in the direction of maximum signal by an antenna, as compared to a dipole. ERP takes system gains and losses into account.

Emergency communications — Communications involving the immediate safety of human life and immediate protection of property when normal communication systems are not available.

Forward Error Correction (FEC) — One of two AMTOR communications modes. In the FEC mode, each character is sent twice. The receiving station checks for errors in the mark/space ratio. If an error is detected, a space is printed to show that an incorrect character was received. Also called Mode B.

Fox hunt — A friendly Amateur Radio competition to locate a hidden transmitter. Amateurs practice their direction-finding skills, which can be useful in tracking down interference sources.

Full break-in (QSK) — With QSK, an amateur can hear signals between code characters. This allows another amateur to break into the communication without waiting for the transmitting station to finish.

Great-circle path — Either one of two direct paths between two points on the surface of the Earth. One of the great-circle paths is the shortest distance between those two points. Great-circle paths can be visualized if you think of a globe with a rubber band stretched around it, connecting the two points.

Long-path communication — Communication made by pointing beam antennas in the directions indicated by the longer **great-circle path** between the stations. To work each other by long-path, an amateur in Hawaii would point his antenna west and an amateur in Florida would aim east.

Net — A group of amateurs who meet at regular times on a specific frequency to share common interests.

Radio Amateur Civil Emergency Service (RACES) — Part of the amateur service that provides radio communications only for civil defense purposes.

Radioteletype (RTTY) — Radio signals sent from one teleprinter machine to another machine. Anything that one operator types on a teleprinter will be printed on the other machine when the two operators are communicating with each other. Also known as narrow-band direct-printing telegraphy.

Short-path communication — Communication made by pointing beam antennas in the direction indicated by the shorter **great-circle path**.

Single sideband (SSB) — The type of voice (phone) operation used most often on the amateur high-frequency (HF) bands. One sideband and the carrier are removed from a double-sideband amplitude modulated (AM) signal, leaving only a single sideband to convey the voice information.

Voice-Operated Transmit (VOX) — Circuitry that activates the transmitter when the operator speaks into the microphone.

Operating Procedures

On-the-air operation is the thrill of Amateur Radio. There are many different operating modes available, and true enjoyment of the hobby comes from taking advantage of this diversity. This chapter covers rules and procedures normally followed on the air when using some of the modes that you, as a General licensee, are authorized. Your General exam will have six questions from the material covered in this chapter. Those questions come from the following syllabus groups:

G2A   Phone operating procedures
G2B   Operating courtesy
G2C   Emergencies, including drills and emergency communications
G2D   Amateur auxiliary to the FCC's Compliance and Information Bureau; antenna orientation to minimize interference; HF operations, including logging practices
G2E   Third-party communications; ITU Regions; VOX operation
G2F   CW operating procedures, including procedural signals, Q signals and common abbreviations; full break-in; RTTY operating procedures, including procedural signals and common abbreviations and operating procedures for other digital modes, such as HF packet, AMTOR, PacTOR, G-TOR, Clover and PSK31

TELEPHONY

Radiotelephony is probably the most natural communication mode of all. Although "phone" operation appears to be as simple as grabbing a microphone and talking, there is a little more to it. There are a few rules and guidelines to keep in mind for courteous and legal on-the-air operation.

Most phone operation on the high frequency (HF) bands uses **single sideband**

(SSB) for a variety of reasons. Single-sideband transmissions require less spectrum space, they're more power efficient and no carrier is transmitted. A double-sideband, full-carrier, amplitude-modulated (DSB AM) signal takes twice the spectrum bandwidth necessary to convey the information, since both sidebands contain the same information, as mirror images of each other. (Chapter 8 gives more information about this signal type and how it relates to SSB signals.)

An SSB signal transmits only one of the sidebands; the other one is suppressed. But which one is transmitted? The lower sideband is normally used on the 160-meter band (1800 to 2000 kHz), the 75-meter band (3850 to 4000 kHz) and the 40-meter band (7225 to 7300 kHz). The lower sideband is given this name because it is *lower* in frequency than the carrier. The upper sideband is *higher* in frequency than the carrier. The upper sideband is normally used on the 20-meter band (14,225 to 14,350 kHz), the 17-meter band (18,110 to 18,168 kHz), the 15-meter band (21,300 to 21,450 kHz), the 12-meter band (24,930 to 24,990 kHz) and the 10-meter band (28,300 to 29,700 kHz). While neither of these sideband choices is specified in the FCC Rules, they are common practice on the amateur bands. You might decide to use the opposite sideband for some reason, but you would not be able to understand the other stations on the band, and no one would be able to understand what you were saying unless they also switched to the opposite sideband.

In general, when you operate voice modes, you should use plain language and communicate clearly. While some operators like to use Q signals and other phrases and abbreviations that were originally intended for Morse code operation or other modes, it is hard to beat plain language for clear communication.

Clear communication can be very important when you are passing formal written messages. The ARRL National Traffic System has established specific procedures for passing formal written messages by Amateur Radio. Even if you don't participate in traffic nets, it is a good idea to be familiar with the procedures for handling such messages. It can be especially helpful in an emergency for a number of reasons. A written message is more likely to include all the necessary information, and by following the standard procedures it is more likely that an emergency message will be transmitted (and received) correctly. After sending the message preamble, the address, message body and signature, the phrase "End of message" clearly shows that all the information has been sent. When the receiving station has accurately recorded the entire message, they will acknowledge receipt of the message.

[Now turn to the question pool in Chapter 13 and study questions G2A01 through G2A11. Also study question G2E13. Review this section if you have difficulty with any of these questions.]

OPERATING COURTESY

On today's crowded ham bands, operating courtesy becomes extremely important. Operating courtesy is really nothing more than common sense; a good operator sets an example for others to follow. A courteous amateur operator always listens before transmitting so as not to interfere with stations already using the frequency. If the frequency is occupied, move to another frequency far enough away to prevent interference (QRM). Know how much bandwidth each mode occupies and act accordingly. A good rule of thumb is to leave 150 to 500 Hz between CW signals, approximately 3 kHz between the suppressed carriers of single-sideband signals and 250 to 500 Hz center-to-center for RTTY signals.

A good amateur operator transmits only what is necessary to accomplish his or her purpose and keeps the contents of the transmission within the bounds of propriety and good taste. Also, a good operator always reduces power to the minimum necessary to carry out the desired communications. This is not only operating courtesy; it's an FCC Rule! §97.313(a) states "An amateur station must use the minimum transmitter power necessary to carry out the desired communications." (A complete copy of Part 97 appears in *The ARRL FCC Rule Book*.)

It shows inexcusably bad manners to interfere with a station thousands of miles away because you are talking to a buddy across town on a DX band. Judicious selection of a band is important to reduce unintentional QRM. Use VHF and UHF bands for local communications. Sometimes propagation conditions may change while you are talking with another station. If you become aware of increasing interference from other activity on the frequency you are using, you are probably also causing interference to the other conversation. The most courteous thing to do in such a case is to move your contact to another (clear) frequency.

Nets are groups of amateurs who meet at regular times on a specific frequency to share common interests. Nets have many purposes. Some nets are part of the ARRL National Traffic System, designed to relay messages across the United States. Other nets provide Morse code practice, a chance to discuss shared interests with other hams or are just a gathering place for Amateur Radio friends to chat.

Nets help reduce congestion on the ham bands because many operators gather on a single frequency. A little courtesy goes a long way to reduce frequency conflicts that sometimes arise.

Suppose you are in contact with another station and become aware that a net is about to start, and they normally use the frequency on which you are operating. What should you do? The most courteous action you can take is to move your contact to another (unused) frequency. It is true that you have just as much right to a frequency as the net participants, but think about the difficulty of moving a large group of operators to another frequency. It is much easier for two or three operators to change frequency, and you will do the net participants a great favor by allowing them to use their predetermined frequency. Of course, you might also want to stand by and simply join the net. You will be most welcome to participate with the group, and you may even become one of the "regulars."

From the other side, consider what you should do as the net control station when the normal net frequency is occupied at net time. The net doesn't own the frequency either, and you have no right to demand the other stations to give up the frequency. Most net participants understand that the announced frequency is just a starting point, so you may be able to find a clear frequency 3 to 5 kHz away. Sometimes you may even have to move a little farther than that. You might also join the ongoing conversation and make a polite announcement that it is time for a net that normally meets on the frequency. This will often lead the other stations to move their conversation to another frequency, or perhaps join your net.

Testing and "loading up" a transmitter should be done into a dummy antenna (or dummy load) without putting a signal on the air. On-the-air tests should be made only when necessary and kept as brief as possible while conducting the test. If your amateur station is within a mile of an FCC monitoring station or you are planning portable operation near a monitoring station, you should consult with the staff at the monitoring station to ensure that your amateur operations do not cause them any harmful interference. Always be more courteous than the other person — you contribute to the image of the Amateur Radio Service as being a self-disciplined, self-policing service with high standards.

Besides SSB voice operation, you can find AM voice, CW, slow-scan TV (SSTV) and satellite operation, as well as RTTY, packet and other digital communications modes on the HF bands. On the amateur bands between 3.5 and 148 MHz, FCC-mandated subband restrictions apply, but different operating modes must still share spectrum space (RTTY and CW operators share the same FCC subbands, and slow-scan television and phone operators also share the same subbands, for example). Voluntary **Band Plans** help minimize conflicts. More information on band plans can be found in *The ARRL Operating Manual*. VHF and UHF band plans are covered in *The ARRL Repeater Directory*. A **Band Plan** refers to an agreement between concerned radio amateurs for operating within a certain portion of the radio spectrum. The goal is to minimize interference between the various modes sharing each band, by setting aside certain sections of a band for each different operating mode. Band plans are more specific about operating modes and frequencies than the FCC Rules.

You should follow a few simple steps to select an operating frequency, no matter what mode you plan to operate. First, if necessary, review the FCC Rules for your license class in regard to the frequency band and mode you wish to operate. Then check the band plan to see what frequencies are suggested for that mode. Finally, listen before you transmit, to be sure you don't interfere with an ongoing communication.

When considerate operators are going to use a phone emission mode, they will follow the steps listed above, and then say "Is the frequency in use? This is" and give their call sign. CW operators will listen on frequency and then send "QRL? de" and then send their call sign. QRL? asks "Are you busy?" or in this case, "Is the frequency in use?" This procedure will help insure that you are not interrupting an ongoing conversation before you start calling CQ or calling another station.

[Before going to the next section, turn to Chapter 13 and study questions G2B01 through G2B13. Review this section if you have difficulty with any of these questions.]

EMERGENCY COMMUNICATIONS

While no one *wants* to be involved with an emergency situation, it is something with which every amateur should be prepared to deal. If you are in need of some emergency help, your Amateur Radio station may be the best (or only) means to obtain that help. You may also find yourself on the receiving end of such a communication. In §97.101 of the Rules, we find "(c) At all times and on all frequencies, each control operator must give priority to stations providing emergency communications..." **Emergency communications** is a serious obligation, and every amateur should be aware of the conditions that define an emergency as well as what to do under those conditions.

Suppose you hear an emergency call for help while you are tuning around one of the ham bands, or a station breaks into your conversation with another amateur to make an emergency call. What should you do? Immediately acknowledge that you heard the call for help and stand by to receive the location of the emergency station and what assistance may be needed. Then you should relay the information to the proper authorities, and stay on frequency with the emergency station for further information, or until help arrives. Your responsibility in this case is to react to the call for help, and to do your best to obtain help for the station in distress.

Before you can think about how *you* should react in an emergency, let's define what a real emergency is. You may *think* you have an emergency situation, when all you really have is a minor inconvenience.

In discussing emergency communications in §97.403, the FCC Rules define an emergency as "the immediate safety of human life and immediate protection of property when normal communication systems are not available." So it seems that the best test of a real emergency is a life-threatening situation. A car accident with injuries could be life threatening, but a flat tire or other breakdown with the car safely off the side of the road probably is not. An earthquake or flood would be an emergency situation, but an inch of water in your basement again after a heavy rain probably is not.

You can use Amateur Radio to call for help even if the situation is not life threatening. You should not go outside your normal amateur privileges to call for help, however.

If you do have a *real* emergency, you can take any means at your disposal to attract attention, make known your condition and location, and obtain assistance — §97.405. This means you can use any frequency, any mode with any amount of power necessary to conduct the communications. **Table 2-1** is a copy of Subpart E — Providing Emergency Communications from the FCC Rules. This includes §97.401 through §97.407. You should select a frequency that has the best chance of reaching another operator, so you can communicate the emergency message. When another station answers your call for help, be sure to give your location and the nature of the emergency, so they will be able to send the help you need. If you are answering another operator who has an emergency, remain calm and do your best to get an accurate description of the other operator's location as well as a clear picture of their problem.

Section 97.111(2) even says you can make "Transmissions necessary to exchange messages with a station in another FCC-regulated service while providing emergency communications." You had better be certain you really do have a life-threatening emergency and no other communications means at your disposal before making such a call, though!

When normal communications systems are overloaded, damaged or disrupted because of a disaster in the US, an amateur station may make transmissions necessary to meet essential communications needs and assist relief operations — §97.401(a).

When a disaster disrupts normal communication systems in a particular area, the FCC may declare a temporary state of communication emergency. The declaration will set forth any special conditions and special rules to be observed by stations during the communication emergency — §97.401(c). The FCC Engineer in Charge of the particular area will issue such a declaration if he or she feels it is appropriate.

Table 2-1

Part 97 Subpart E — Providing Emergency Communications

§97.401 Operation during a disaster.

(a) When normal communication systems are overloaded, damaged or disrupted because a disaster has occurred, or is likely to occur, in an area where the amateur service is regulated by the FCC, an amateur station may make transmissions necessary to meet essential communication needs and facilitate relief actions.

(b) When normal communication systems are overloaded, damaged or disrupted because a natural disaster has occurred, or is likely to occur, in an area where the amateur service is not regulated by the FCC, a station assisting in meeting essential communication needs and facilitating relief actions may do so only in accord with ITU Resolution No. 640 (Geneva, 1979). The 80 m, 75 m, 40 m, 30 m, 20 m, 17 m, 15 m, 12 m, and 2 m bands may be used for these purposes.

(c) When a disaster disrupts normal communication systems in a particular area, the FCC may declare a temporary state of communication emergency. The declaration will set forth any special conditions and special rules to be observed by stations during the communication emergency. A request for a declaration of a temporary state of emergency should be directed to the EIC in the area concerned.

(d) A station in, or within 92.6 km of, Alaska may transmit emissions J3E and R3E on the channel at 5.1675 MHz for emergency communications. The channel must be shared with stations licensed in the Alaska-private fixed service. The transmitter power must not exceed 150 W.

§97.403 Safety of life and protection of property.

No provision of these rules prevents the use by an amateur station of any means of radiocommunication at its disposal to provide essential communication needs in connection with the immediate safety of human life and immediate protection of property when normal communication systems are not available.

§97.405 Station in distress.

(a) No provision of these rules prevents the use by an amateur station in distress of any means at its disposal to attract attention, make known its condition and location, and obtain assistance.

(b) No provision of these rules prevents the use by a station, in the exceptional circumstances described in paragraph (a), of any means of radiocommunications at its disposal to assist a station in distress.

§97.407 Radio amateur civil emergency service.

(a) No station may transmit in RACES unless it is an FCC-licensed primary, club, or military recreation station and it is certified by a civil defense organization as registered with that organization, or it is an FCC-licensed RACES station. No person may be the control operator of a RACES station, or may be the control operator of an amateur station transmitting in RACES unless that person holds a FCC-issued amateur operator license and is certified by a civil defense organization as enrolled in that organization.

(b) The frequency bands and segments and emissions authorized to the control operator are available to stations transmitting communications in RACES on a shared basis with the amateur service. In the event of an emergency which necessitates the invoking of the President's War Emergency Powers under the provisions of §706 of the Communications Act of 1934, as amended, 47 U.S.C. §606, RACES stations and amateur stations participating in RACES may only transmit on the following frequencies:

    (1) The 1800-1825 kHz, 1975-2000 kHz, 3.50-3.55 MHz, 3.93-3.98 MHz, 3.984-4.000 MHz, 7.079-7.125 MHz, 7.245-7.255 MHz, 10.10-10.15 MHz, 14.047-14.053 MHz, 14.22-14.23 MHz, 14.331-14.350 MHz, 21.047-21.053 MHz, 21.228-21.267 MHz, 28.55-28.75 MHz, 29.237-29.273 MHz, 29.45-29.65 MHz, 50.35-50.75 MHz, 52-54 MHz, 144.50-145.71 MHz, 146-148 MHz, 2390-2450 MHz segments;

    (2) The 1.25 m, 70 cm and 23 cm bands; and

    (3) The channels at 3.997 MHz and 53.30 MHz may be used in emergency areas when required to make initial contact with a military unit and for communications with military stations on matters requiring coordination.

(c) A RACES station may only communicate with:

    (1) Another RACES station;

    (2) An amateur station registered with a civil defense organization;

    (3) A United States Government station authorized by the responsible agency to communicate with RACES stations;

    (4) A station in a service regulated by the FCC whenever such communication is authorized by the FCC.

(d) An amateur station registered with a civil defense organization may only communicate with:

    (1) A RACES station licensed to the civil defense organization with which the amateur station is registered;

    (2) The following stations upon authorization of the responsible civil defense official for the organization with which the amateur station is registered:

        (i) A RACES station licensed to another civil defense organization;

        (ii) An amateur station registered with the same or another civil defense organization;

        (iii) A United States Government station authorized by the responsible agency to communicate with RACES stations; and

        (iv) A station in a service regulated by the FCC whenever such communication is authorized by the FCC.

(e) All communications transmitted in RACES must be specifically authorized by the civil defense organization for the area served. Only civil defense communications of the following types may be transmitted:

    (1) Messages concerning impending or actual conditions jeopardizing the public safety, or affecting the national defense or security during periods of local, regional, or national civil emergencies;

    (2) Messages directly concerning the immediate safety of life of individuals, the immediate protection of property, maintenance of law and order, alleviation of human suffering and need, and the combating of armed attack or sabotage;

    (3) Messages directly concerning the accumulation and dissemination of public information or instructions to the civilian population essential to the activities of the civil defense organization or other authorized governmental or relief agencies; and

    (4) Communications for RACES training drills and tests necessary to ensure the establishment and maintenance of orderly and efficient operation of the RACES as ordered by the responsible civil defense organizations served. Such drills and tests may not exceed a total time of 1 hour per week. With the approval of the chief officer for emergency planning the applicable State, Commonwealth, District or territory, however, such tests and drills may be conducted for a period not to exceed 72 hours no more than twice in any calendar year.

RACES

The **Radio Amateur Civil Emergency Service (RACES)** is part of the amateur service that provides radio communications only for civil defense purposes. It is active *only* during periods of local, regional or national civil emergencies.

You must register with the responsible civil defense organization in your area to operate as a RACES station. RACES stations may not communicate with amateurs not operating in a RACES capacity. Restrictions do not apply when stations are operating in a non-RACES amateur capacity, such as ARES, the Amateur Radio Emergency Service. (ARES is part of the ARRL Field Organization.)

Proper operating procedures are important for efficient handling of any emergency. This is especially true of a civil emergency that would result in RACES operation. Unless there is some way to practice, RACES operators may not know the proper procedures, however. The FCC Rules provide for RACES drills to assist RACES operators in practicing orderly and efficient operations for the civil defense organization they serve.

Only civil-preparedness communications can be transmitted during RACES operation. These are defined in §97.407 of the FCC regulations. (See Table 2-1.) The Rules permit tests and drills for a maximum of one hour per week. All test and drill messages must be clearly identified as such.

[Now turn to Chapter 13 and study questions G2C01 through G2C13. If you are uncertain about the answers to any of these questions, review this section.]

THE AMATEUR AUXILIARY

An important aspect of the Communications Amendments Act of 1982, commonly known as Public Law 97-259, is one that authorized the FCC to formally enlist the use of amateur volunteers for monitoring the airwaves for rules violations. The ARRL has taken a leadership role in the volunteer monitoring function, through the **Amateur Auxiliary** to the FCC's Compliance and Information Bureau. The primary objectives of the Amateur Auxiliary are to foster amateur self regulation and compliance with the Rules.

The Amateur Auxiliary is concerned with both maintenance monitoring and amateur-to-amateur interference. Maintenance monitoring is a term that refers to a program of formally enlisted amateurs monitoring the transmissions from other amateur stations for compliance with the FCC Rules and technical standards. Maintenance monitoring is conducted through the ARRL Official Observer program, while amateur-to-amateur interference is handled by specifically authorized interference committees.

Surprisingly, a popular ham radio activity is becoming increasingly important to the Amateur Auxiliary. In addition to providing fun and friendship for participants of all ages, Amateur Radio **fox hunts**—friendly hidden transmitter competitions—help participants to practice direction-finding skills that may one day be useful in tracking down sources of harmful interference. Competing on a variety of amateur bands, the best "fox hunters" augment their skills and equipment by keeping detailed notes on signal-strength and compass bearing while each fox (hidden transmitter) is active. When tracking down a real interference source, such notes will be very helpful in documenting the problem and reporting it to the proper authorities. A good direction-finding team can pinpoint the location of an interference source. Such fox hunts also help make everyone aware that there is a plan in place to find and eliminate an interference source.

[Turn to Chapter 13 and study questions G2D01 through G2D03. Review any material you are not sure of before proceeding.]

AIMING DIRECTIONAL ANTENNAS

Many Amateur Radio operators use a **directional antenna** or beam antenna on the HF bands, especially for the 20-meter band through the 10-meter band. You can take greatest advantage of the changing radio-wave-propagation conditions if you have a directional antenna, and can rotate it to point at various parts of the world.

A beam antenna concentrates most of your transmitted signal in the desired direction, while reducing the amount of signal transmitted in other directions. Although there is always some signal going in other directions, a beam antenna is sometimes called a *unidirectional antenna*. You can use such an antenna to your advantage to reduce or minimize interference with other stations. Beam antennas also *receive* signals better from one direction than other directions, so you can point the antenna away from, or to the side of an interfering station, reducing received interference as well.

The best propagation path between any two points on the Earth's surface is usually by the shortest direct route — the **great-circle path** that can be found by stretching a string tightly between the two points on a globe. If we replace the string with a rubber band going around the globe, we can see another, longer great-circle path. With the proper conditions, **long-path communication** is possible. To attempt communication by long-path propagation, you must rotate your antenna 180° from the "normal" (**short-path**) direction. For instance, using long-path propagation, an amateur in eastern Pennsylvania could work a station in Hawaii by pointing her beam a few degrees north of east — the Hawaiian amateur would point his beam a few degrees south of west. It is interesting to note that because Antarctica is at the South Pole, the short-path heading from anywhere to Antarctica is south, or 180° and the long-path heading is north, or 0°.

The long path seems to work best for communication between two places when the path is mostly over water. Since sunlight is an important ingredient for good ionospheric propagation on frequencies above 7.5 MHz, long-path communication also seems to work best when the signals will travel mostly through sunlight. From the East Coast of the United States, for example, you might try the long path to reach Hawaii just before your sunrise. If you are in Hawaii, you might try pointing your antenna to the west around sunset and shortly after that to contact stations on the East Coast. It is important that stations on both ends of a long-path circuit point their antennas in the long-path direction.

You can understand several aspects of long-path propagation if you become accustomed to thinking of the Earth as a ball. This is easy if you use a globe frequently. If you must use a flat map to find great-circle paths, an **azimuthal-equidistant projection map** (also called a great-circle map) is a useful substitute. See **Figure 2-1**. The azimuthal map centered on or near your location will show you beam headings for both long and short paths to other locations on the surface of the Earth.

Figure 2-1 — N5KR's computer-generated azimuthal-equidistant projection map centered on Newington, Connecticut. Information showing long paths to Perth and Tokyo have been added. Notice that the paths in both cases lie almost entirely over water, rather than over land masses.

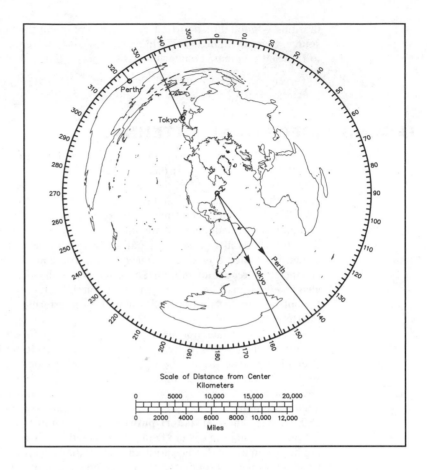

Scale of Distance from Center
Kilometers

0   5000   10,000   15,000   20,000

0   2000   4000   6000   8000   10,000   12,000
Miles

[Turn to Chapter 13 and study questions G2D04 through G2D06. Also study question G2D11. Review this section as needed.]

STATION RECORDS

If you allow another licensed ham to operate your station, you are still responsible for its proper operation. You are always responsible for the proper operation of your station. Your primary responsibility as the station licensee is to ensure the proper operation of your station. You can give your permission to another licensed amateur, allowing them to be control operator of your station. In that case, both of you will be responsible for the proper operation of the station. Unless your station records show that another amateur was the control operator of your station at a certain time, the FCC will assume that you were the only control operator.

What kind of station records should you keep? With only a few exceptions, the FCC does not *require* you to keep any particular information about the operation of your station. Many amateurs find it helpful to keep a station logbook with certain information about the operation of their station. Logbooks are useful for recording dates, times, callsigns, names and locations of those stations you contact. Many hams also include the operating band or frequency and signal reports exchanged. When you confirm contacts by sending QSL cards, a logbook is a convenient way to keep track of these exchanges. Your log will provide a useful and interesting history, if you choose to keep one. It will also prove to be a valuable record if you have to respond to a question from the FCC about your station operating times and frequencies.

Your records of operating times, frequencies and transmitter power can also be very helpful when you are trying to resolve an interference complaint. Suppose your neighbor says you were interfering with a TV or telephone at a certain time. If your log shows you were not operating at that time, then you will know you were not the cause of the interference. Your log also might help you determine that the interference only occurs when you operate on a certain band. That can be very helpful as you try to solve the problem.

One exception to the statement that the FCC does not require you to keep station records has to do with operation on one of the channels in Amateur Radio's newest band, 60 meters. In Chapter 1 you learned that you must keep records about antenna gain if you are using an antenna with more gain than a dipole antenna.

There are commercially prepared logbooks, such as the *ARRL Log Book*, or you can use a notebook or other form that you prepare for your own use. Many amateurs find it helpful to use one side of a page to log amateur contacts and the back of the page to record information about station equipment — changes and other information.

It isn't necessary to use a station logbook. If you don't, though, it will not be possible to show when someone else has been the control operator of your station. Whenever you give someone else permission to be the control operator at your station, enter the other person's name and call sign, and the time, date and frequencies used, in your logbook.

Many amateurs use computer logging programs to maintain records of their station operation. This can be convenient, especially for contest operating. Many radios have a data port that allows them to communicate with a computer, and in that case the computer logging program may be able to automatically record the date, time, operating frequency and mode of the contact. Of course there are also some disadvantages to computer logging. You will have to wait for the computer and program to start before you can read or record any data in your computer log. Computer data files can become damaged or corrupted. For these reasons, it may be a good idea to keep a paper log as a backup to your computer log.

[Now it is time for another break from the text. Turn to Chapter 13 and study questions G2D07 through G2D10. Review this section if you have difficulty with any of these questions.]

Voice-Operated Transmit (VOX)

The purpose of a **voice-operated transmit (VOX)** circuit is to provide automatic TR (transmit/receive) switching within an amateur station. When using VOX, the transmitter switches on, the antenna is connected to the transmitter, and the receiver is muted as soon as you begin to speak; when you stop talking, the transmitter is turned off and the antenna is connected to the receiver. VOX circuits can be used with separate transmitter/receiver combinations, and most modern trans-ceivers have a VOX circuit built in. VOX allows hands-free operation. That means your hands are free to write notes, log data or type entries on a computer keyboard. This can be a great advantage for contest operating.

Three controls must be adjusted before using VOX: VOX sensitivity, anti-VOX and VOX delay. These adjustments will vary with different pieces of equipment, and even with different operators using the same gear. Since VOX activates the transmitter, any VOX-circuit control adjustments should be made with the trans-mitter connected to a dummy load.

Normally, VOX sensitivity is adjusted first. The setting of this control deter-mines the level of audio input necessary to activate the transmitter. With the VOX delay and anti-VOX controls set to the middle of their adjustment ranges, turn the receiver volume to minimum. Speak into the microphone and slowly increase the setting of the VOX sensitivity control until the transmitter activates. Stop talking and allow the transmit relay to switch back to receive. Speak again, and "fine-tune" the sensitivity adjustment: your voice should activate the transmitter instantly, but background noise in the shack shouldn't.

Next, adjust the VOX delay. This part of the VOX circuit controls the amount of time the transmitter remains activated after you stop talking. Set it so the trans-mitter deactivates ("drops") when you stop to take a breath between thoughts. That way, a radiotelephone contact can approach a face-to-face conversation.

The anti-VOX control is adjusted last. Anti-VOX prevents receiver audio from tripping the transmitter. It accomplishes this by applying a portion of the receiver audio to the microphone amplifier circuit 180° out of phase with the actual receiver audio. The anti-VOX control varies the feedback level. If the anti-VOX control is set too high, you'll have trouble activating the VOX circuit when you speak into the microphone. Insufficient anti-VOX will allow receiver audio to trip the transmitter. The VOX sensitivity and anti-VOX control settings may interact slightly, so you may need to readjust the VOX sensitivity after setting the anti-VOX level.

Many Amateur Radio operators find that VOX operation provides a safe, hands free way to control a mobile station. Simply speaking into the microphone activates the transmitter, allowing the operator to keep both hands on the steering wheel. A headset with an attached microphone is one popular way to set up such a station. Of course it may not be the best idea to wear a headset that covers both ears and pre-vents you from hearing outside sounds! In many states it is illegal to drive a vehicle with earphones covering both ears, and in some locations it may even be illegal to drive while wearing a single earphone. Several manufacturers offer single-ear head-sets that are ideally suited to this type of operation.

This is a good way to provide safer, hands-free operation of the station while you are driving. Of course if you operate a radio while you are driving you must follow other safety procedures. For example, be sure to maintain your concentra-tion on road conditions and your driving.

For CW (Morse code) operation, many transceivers use the VOX circuitry to key the transmitter when you begin sending, and to shut it off when you stop. With proper adjustment, you can hear other signals between words in your transmission. This is called semi-break-in operation. With this type of circuit, the VOX delay is set to hold the transmitter on during the transmission of entire words. Nevertheless, the receiver can come back on between words, so at speeds above about 15 wpm the effect is almost the same as **full-break-in (QSK)** operation. Full-break-in operation on CW allows you to listen to your transmit frequency between character elements (dots and dashes) and between words.

QSK operation is especially useful when you are sending message traffic — the receiving station can send a few dots and "break" (stop) the transmitting station for repeats (fills) of missed words. The transmitting station can also hear if another station comes on frequency and interferes with the QSO. If you are sending formal written messages using CW, the procedural signal (prosign) $\overline{AR}$ is sent to indicate the end of each message.

[Now turn to Chapter 13 and study questions G2E02, G2E03, G2E04, G2E10 through G2E12. Also study questions G2F01 and G2F08. Review this section as needed.]

DIGITAL COMMUNICATIONS

Amateur Radio digital communications normally involve the use of the **ASCII**, **Baudot** or **AMTOR** codes, as well as the original digital code, Morse. The Baudot code, also known as the International Telegraph Alphabet Number 2 (or ITA2) was the only digital code (other than Morse) allowed on the amateur bands until 1980. The Baudot code uses five information bits, with additional bits (called "start" and "stop" bits) to indicate the beginning and end of the character.

ASCII stands for American National Standard Code for Information Interchange; ASCII is the code commonly used for computer systems. The ASCII code uses seven information bits, so more characters can be defined than with the five-bit Baudot code.

AMTOR, which stands for *Am*ateur *T*eleprinting *O*ver *R*adio, combines a modified Baudot code with error correcting capabilities. As with radiotelephone operation, there are a few standard operating practices for the digital modes. We'll cover a few of the main points here. For more details, consult *The ARRL Operating Manual*.

Radioteletype

Radioteletype (RTTY) means radio signals that are sent from one teletype machine to another. Also known as narrow-band direct-printing telegraphy, RTTY is a communications method that is intended to be copied and printed by *machine*. For most operators today that machine is a computer system and radio modem of some type. Originally, the machine was a noisy, oily, teletype machine. When most Amateur Radio operators refer to RTTY, they mean operation using the Baudot code, although any digital communications method intended for direct printing from one machine to another is a form of radioteletype. The Baudot code is a five-bit code, with additional start and stop bits to synchronize the transmitting and receiving machines.

Radioteleprinter operation uses *frequency-shift keying (FSK)* to convey information. This shift is between two frequencies, called "mark" and "space." On HF, the mark and space are normally 170 Hz apart. On VHF, there is a variety of mark and space frequency shifts. (Some common shifts on VHF are 170 Hz, 850 Hz and 1 kHz. Presently, the most common shift is 170 Hz.) Common speeds for HF RTTY operation are 60, 75 and 100 WPM (45.45, 56 and 75 bauds). Always answer a RTTY CQ call at the same speed as the calling station is using. Most computerized

Table 2-2

Suggested US RTTY Operating Frequencies (kHz)

3590 RTTY DX
3580-3620
7040 RTTY DX
7080-7100
10,130-10,140
14,070-14,095
18,100-18,105
21,070-21,095
24,920-24,925
28,070-28,120

RTTY stations will automatically set the transmit and receive speeds the same.

On the HF bands, most of the RTTY operation is found at the top of the CW portion of each band. Traditionally, the 20-meter RTTY subband has received the heaviest use during daylight hours; the 80-meter RTTY subband is active at night. Recently, the increased popularity of RTTY has brought about greater use of the 40-meter RTTY subband.

The recommended frequencies for RTTY operation on the 80-meter band are between 3580 kHz and 3620 kHz. On the 20-meter band, the Band Plan calls for RTTY operation to be between 14.070 MHz and 14.095 MHz. Most operators try to keep their RTTY operations inside these limits. Suggested operating frequencies for RTTY on the HF bands are shown in **Table 2-2**.

To call CQ on RTTY, you might use this standard calling sequence:

CQ CQ CQ DE W1AW W1AW W1AW
CQ CQ CQ DE W1AW W1AW W1AW
CQ CQ CQ DE W1AW W1AW W1AW
THIS IS JOE IN NEWINGTON CT
K

If you copied this CQ and you want to answer the station, here's the accepted method:

W1AW W1AW W1AW DE WB8IMY WB8IMY WB8IMY K

RTTY operators will occasionally send a string of the letters "R" and "Y": RYRYRYRY. The letter "R" in the Baudot code has a space for bits 1, 3 and 5 and a mark for the other two bits; "Y" is just the opposite, with a space in bits 2 and 4 and mark in bits 1, 3 and 5. A string of "RYs" alternates between the mark and space bits. This was quite common in the days of mechanical teleprinter machines because it was useful for testing equipment. It can help other amateurs tune in a station. In these days of computerized RTTY stations and modern transceivers it isn't needed for equipment testing. You should keep your use of RYRY to one or two short lines before a CQ call. Better yet, skip the RYs altogether.

There are only 32 possible character combinations with the Baudot code, so all letters are sent as upper case. One character is called the FIGS shift and one is the LTRS shift, allowing each character to serve two purposes. When you send the FIGS character, everything you type after that will be received as numbers, punctuation or other control codes until you send a LTRS character. The FIGS and LTRS characters select between two different character sets on the printer.

The reason for this limitation is that each character consists of five information bits, and each bit can have one of two possible conditions: mark or space (often represented by 1 and 0). The total number of possible characters is defined by the calculation $2^5 = 32$.

When sending formal written messages using RTTY or other digital modes, the letters NNNN are sent to indicate the end of each message. RTTY operators often use Q signals and other abbreviations commonly used on CW. For example, the letter K at the end of a transmission is an invitation for another station to transmit. SK means "signing off" or "end of contact" and CL means "clear" or "closing" as in, "I am shutting down my station."

The maximum data symbol rate for RTTY and other digital modes such as packet radio on HF (frequencies below 28 MHz) is 300 bauds. Higher data rates are used on higher-frequency bands. On 28 MHz (10 meters) and higher frequencies, for instance, the FCC permits use of symbol rates of up to 1200 bauds. As the speed of an RTTY transmission is increased, the bandwidth required for that transmission also increases. More spectrum space is available on the VHF and UHF bands, so the FCC permits the use of speeds up to 19.6 kilobauds on frequencies above 50 MHz (on the 6 and 2-meter bands) and up to 56 kilobauds above 220 MHz.

AMTOR

The benefit of using AMTOR is its error detection and correction properties. There are two AMTOR modes: Mode A, called **Automatic Repeat Request (ARQ)**, and Mode B, called **Forward Error Correction (FEC)**. Since the AMTOR code is a variation of the Baudot code, there are 32 possible character combinations, with FIGS and LTRS shift characters to change between two characters sets on the printer.

When two stations are in AMTOR Mode A communication, they are constantly confirming each other's transmissions: If a piece of information is missed or lost (because of interference or any other reason), it is repeated until the receiving station confirms reception. The transmitting station sends three characters, then pauses for the receiving station to confirm or deny correct reception. The same block of three characters is repeated or the next three are sent, depending on the reply. This interactive "handshaking" takes 450 milliseconds for every cycle between the two stations.

Mode B AMTOR is a "broadcast" mode, in which the transmitting station sends each character twice. The receiving station examines each character as it is received, looking for any errors in the mark/space ratio (each character contains four mark elements and three space elements). The first character of the pair is stored in memory as it is received. When the second character is received, the character pair is checked for errors. If one or both characters contain the proper mark/space ratio, the character is printed on the screen. If neither character has the correct ratio, the receiving station records an error signal, signified by a space.

To establish an AMTOR contact, get on a calling frequency and make a general CQ call in Mode B. (The Band Plans put all of these digital modes in the same band segments as the Baudot RTTY operation.) Once contact is established, switch to Mode A.

ASCII

ASCII is a coded character set used for computer systems and related equipment. ASCII differs from the Baudot code in that it has a larger character set. The seven information bits in each character make it possible for the ASCII character set to provide upper- and lower-case letters, numbers, punctuation and special characters. (With seven information bits, each with two possible values — 1 and 0 — the ASCII character set has $2^7 = 128$ character combinations.) In addition to the seven information bits, ASCII characters include start, stop and parity bits. To establish a contact using ASCII, use the same procedure described for Baudot RTTY.

HF ASCII transmissions are normally sent at 110 or 300 bauds, using 170-Hz shift. Higher data rates are used on higher-frequency bands. On 28 MHz (10 meters) and higher frequencies, for instance, the FCC permits use of symbol rates of up to 1200 bauds.

When you are transmitting a digital code other than one of those specified in Section 97.309 of the Rules, there is a bandwidth limitation imposed. Between 50 and 222 MHz (on the 6 and 2-meter bands), the authorized bandwidth of a RTTY, data or multiplexed emission using an unspecified digital code is 20 kHz. Between 222 and 450 MHz (on the $1^1/_4$-meter and 70-cm bands), the authorized bandwidth of a RTTY, data or multiplexed emission using an unspecified digital code is 100 kHz.

ASCII is also used for packet-radio communication. Packet radio is a communication mode that allows transmission of short bursts (packets) of information at high speed. The format (protocol) of information packets allows several stations to hold conversations on the same frequency at the same time without interfering with each other. The information to be transmitted is broken into data packets. Each data packet contains a *header* that includes the routing and handling information. Each data packet can move independently through a system of relay stations, with the data being reassembled into the original message at the receiving station. For more information on packet radio, see *The ARRL Handbook*, *The ARRL Operating Manual* or any of several other books about packet radio and digital communications published by the ARRL.

PACTOR, PSK31 and other Hybrid Modes

Amateur Radio operators are always experimenting and trying to develop more efficient means of communication. Continued amateur innovation has produced advanced HF digital modes such as G-TOR, CLOVER and PACTOR II. The operating procedures used with these modes are generally the same as for the digital modes already discussed. PSK31 is another digital mode that has become very popular. The name comes from the *phase shift keying* used to modulate the transmitter, and the 31.25-baud character transmission rate. PSK31 is a software-based system. Most of the software uses the digital signal processing capabilities of a computer sound card to modulate and demodulate the audio signals to and from the radio. One advantage of PSK31 is that the signal bandwidth is 100 Hz or less. This means you can use a very narrow bandwidth receive filter.

One of the interesting properties of the PSK31 radioteletype mode is that it uses a character code called *Varicode*. Developed by Peter Martinez, G3PLX, *Varicode* uses shorter character lengths for the more common characters, and longer codes for less common characters. The number of data bits per character varies, depending on which character is being sent. This is quite different from other RTTY modes. For example, every Baudot character has five data bits and every AMTOR character has seven data bits.

To learn more about the use of these modes, consult one of the ARRL publications mentioned earlier.

[Congratulations. You have studied all of the operating procedures information for your General class exam. Before you go on to Chapter 3, turn to Chapter 13 and study questions G1F05 through G1F09. Also study questions G2F02 through G2F07 and G2F09 through G2F11. Review this section if you have difficulty with any of these questions.]

**CHAPTER 3
KEYWORDS
KEYWORDS
KEYWORDS**

A-Index — A daily measurement for the state of activity of the Earth's magnetic field. It is based on the eight **K-index** readings from the previous day, so the A-index tells you mainly how yesterday was. The A-index is given on a scale of values from 0 to 400, to indicate the range of geomagnetic field disturbance.

Beacon — An amateur station transmitting communications for the purposes of observing propagation and reception or other related experimental activities.

Critical angle — If radio waves leave an antenna at an angle greater than the critical angle for that frequency they will pass through the ionosphere instead of returning to Earth.

Critical frequency — The highest frequency at which a vertically incident radio wave will return from the ionosphere. Above the critical frequency radio signals pass through the ionosphere instead of returning to the Earth.

D region — The lowest region of the ionosphere. The D region contributes very little to short-wave radio propagation, acting mainly to absorb energy from radio waves below about 7.5 MHz during daylight.

E region — The second lowest ionospheric region, the E region exists only during the day, and under certain conditions may refract radio waves enough to return them to Earth.

F region – The highest ionospheric region, the F region refracts radio waves and returns them to Earth. The height of the F region varies greatly depending on the time of day, season of the year and amount of sunspot activity. During the day this region often splits into two regions, called the F1 and F2 regions.

Geomagnetic disturbance — A dramatic change in the Earth's magnetic field that occurs over a short time.

Ionosphere — A region in the atmosphere about 30 to 260 miles above the Earth. The ionosphere is made up of charged particles, or ions.

K-Index — Readings of the Earth's geomagnetic field, updated every 3 hours at Boulder, Colorado. K-index values, given on a scale of 0 to 9, indicate the stability of the Earth's geomagnetic field.

Lowest Usable Frequency (LUF) — The lower limit to the range of frequencies that will provide useful communications between two locations, using **sky-wave** propagation.

Maximum usable frequency (MUF) — The highest frequency that allows a radio wave to reach a desired destination.

Propagation — The means by which radio waves travel from one place to another.

Refract — To bend. Electromagnetic energy is refracted when it passes through a boundary between different types of material. Light is refracted as it travels from air into water or from water into air.

Scatter — Several factors may cause some energy from a radio signal to follow a path other than the idealized "straight line" shown on diagrams like Figure 3-1. Scattering can take place from Earth's ionospheric and atmospheric layers as well as objects in the wave path.

Sky waves — Radio waves that travel from an antenna upward to the ionosphere, where they either pass through the ionosphere into space or are refracted back to Earth.

Solar flare — A large eruption of energy and solar material from the surface of the sun.

Solar flux — Radio energy coming from the sun.

Solar flux index — A measure of solar activity. The solar flux is a measure of the radio noise on 2800 MHz.

Sudden Ionospheric Disturbance (SID) — A blackout of HF sky-wave communications that occurs after a solar flare.

Sunspots — Dark blotches that appear on the surface of the sun.

Radio-Wave Propagation

Radio amateurs have been interested in the study of radio **propagation** since the early days of radio. Amateurs have sought to understand how different factors affect radio waves as they travel from one place to another. The height of your antenna above ground, the type of antenna used, the frequency of operation, the terrain, the weather and the height and density of the **ionosphere** (a region of charged particles high above the surface of the Earth) all affect how radio waves travel. It is important to understand the nature of radio waves and how their behavior is affected by the medium in which they travel.

There will be three questions from this Radio-Wave Propagation chapter on your General license exam. These questions will come from the three groups in Subelement G3 of the General Class Question Pool:

G3A   Ionospheric disturbances; sunspots and solar radiation.
G3B   Maximum usable frequency; propagation "hops."
G3C   Height of ionospheric regions; critical angle and frequency; HF scatter.

IONOSPHERIC PROPAGATION

Nearly all amateur communication on frequencies below 30 MHz is by means of **sky waves**. After leaving the transmitting antenna, this type of wave travels from the Earth's surface at an angle that would send it out into space if its path were not bent enough to bring it back to Earth. As the radio wave travels outward from the Earth, it encounters a region of ionized particles in the atmosphere. This region, called the **ionosphere**, begins about 30 miles from the surface of the Earth, and extends to about 300 miles. See **Figure 3-1**. The ionosphere **refracts** (or bends)

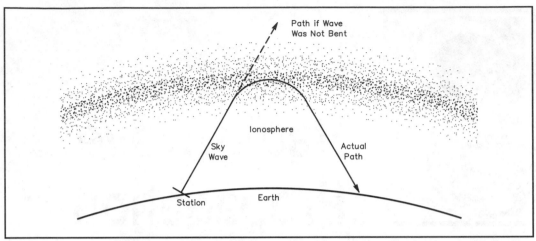

Figure 3-1 — Radio waves are bent in the ionosphere, so they return to Earth far from their origin. Without refraction (bending) in the ionosphere, radio waves would pass into space.

radio waves, and at some frequencies the radio waves are refracted enough so they return to Earth at a point far from the originating station.

The Earth's upper atmosphere is composed mainly of oxygen and nitrogen with traces of hydrogen, helium and several other gases. The atoms that make up these gases are electrically neutral — they have no charge and exhibit no electrical force outside their own structure. When the gas atoms absorb ultraviolet radiation from the Sun, however, electrons are knocked free and the atoms become positively charged. These positively charged atoms are called ions, and the process by which they are formed is called ionization. Several ionized regions are created at different heights in the atmosphere. Each region has a central region where the ionization is greatest. The intensity of the ionization decreases above and below this central region in each region. These regions have been given letter designations, as shown in **Figure 3-2**.

The Ionosphere: A Closer Look

D Region: The lowest region of the ionosphere is called the D region. This region is in a relatively dense part of the atmosphere about 30 to 60 miles above the Earth. The ions formed when sunlight is absorbed by the atmosphere are very short-lived in this region. The positive and negative ions quickly revert to their neutral atomic form. The amount of ionization in this region varies widely depending on how much sunlight hits the region. Around noon local standard time, D-region ionization overhead is maximum or very close to it. By sunset, this ionization disappears.

The D region is actually ineffective in bending high-frequency signals back to Earth. The major effect the D region has on long-distance communication is to absorb energy from radio waves. As radio waves pass through the ionosphere, they give up energy and set some of the ionized particles into motion. Lower frequencies are affected more by absorption than higher frequencies. Absorption also increases

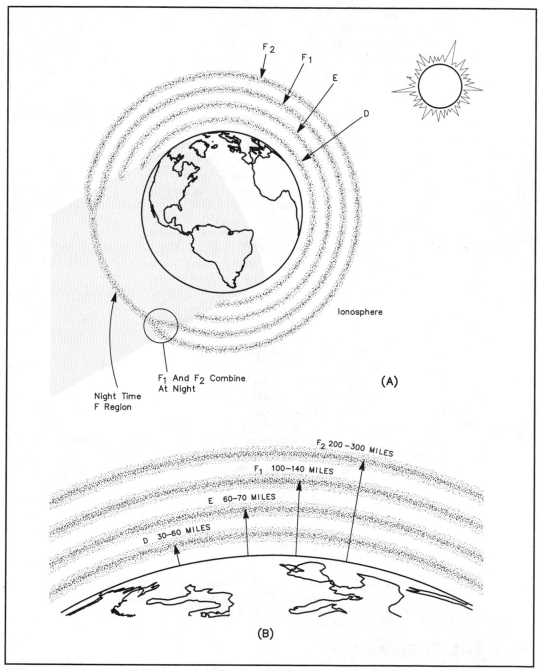

Figure 3-2 — The ionosphere consists of several regions of ionized particles at different heights above the Earth. At night, the D and E regions disappear and the F1 and F2 regions combine to form a single F region.

proportionally with the amount of ionization: the more ionization, the more energy the radio waves will lose as they pass through the ionosphere. D-region absorption is most pronounced at midday, and it is responsible for the short daytime communication ranges on the lower amateur frequencies. Long-distance propagation of signals on the 160 and 80-meter bands will nearly always disappear shortly after sunrise and not return until near sunset. Long-distance signals on the 40-meter band will fade away as daytime ionization of the D region increases. This signal fading is the result of signal absorption in the D region. Long-distance signals will appear on the 40-meter band in late afternoon, earlier than signals on the 80 and 160-meter bands.

E region: The E region appears at an altitude of about 60 to 70 miles above the Earth. At this height, the atmosphere is still dense enough so that ionization produced by sunlight does not last very long. This makes the E region useful for bending radio waves only when it is in sunlight. Like the D region, the E region reaches maximum ionization around midday, and by early evening the ionization level is very low. The ionization level reaches a minimum just before sunrise, local time. Using the E region, a radio signal can travel a maximum distance of about 1200 miles in one hop.

F region: The region of the ionosphere most responsible for long-distance amateur communication is called the F region. This region is actually a very large region ranging from about 100 to as much as 300 miles above the Earth. The height depends on the season of the year, latitude, time of day and solar activity. Ionization reaches a maximum shortly after noon local standard time, but tapers off very gradually toward sunset. At this altitude, the positive ions and electrons recombine very slowly, so the F region remains ionized throughout the night, reaching a minimum just before sunrise. After sunrise, ionization increases rapidly for the first few hours. Then it increases slowly to its noontime maximum.

During the day, the F region splits into two parts, called the F1 and F2 regions. The central area of the F1 region forms at an altitude of about 140 miles. For the F2 region, the central area forms at about 200 miles above the Earth. These altitudes vary with the season of the year and other factors. At noon in the summer, the F2 region can reach an altitude of 300 miles. At night, the two regions recombine to form a single F region slightly below the higher altitude.

The F1 region does not have much to do with long-distance communications. Its effects are similar to those caused by the E region. The F2 region is responsible for almost all long-distance communication on the amateur HF bands because it is the highest region. Signals bent back to Earth from this layer return at the greatest distance from the transmitter. A one-hop radio transmission can travel a maximum distance of about 2500 miles using the F2 region.

[Turn to Chapter 13 now and study questions G3B09 and G3B10. Also study questions G3C01 through G3C03, G3C05 and G3C12. Review this section if you have difficulty with any of these questions.]

CRITICAL FREQUENCY

Radio signals traveling through the ionosphere are gradually bent. If they are bent enough they will return to Earth. An ionospheric region is a region of consid-

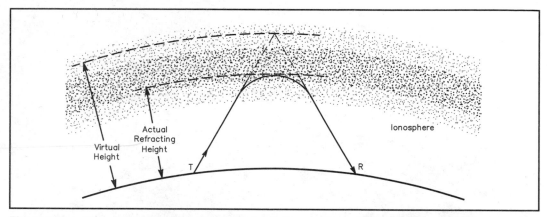

Figure 3-3 — The virtual height of a region in the ionosphere is the height at which a simple reflection would return the wave to the same point as the gradual bending that actually takes place.

erable depth, but for practical purposes it is sometimes convenient to think of each region as having a definite height. If we think of the ionosphere as a reflecting surface at a certain height, then we can picture the radio waves bouncing off that surface and returning to Earth. The height of this imagined reflecting surface is called the *virtual height* of the region. See **Figure 3-3**.

The virtual height of an ionospheric region for various frequencies is determined with a variable-frequency sounding device (called an ionosonde) that directs energy vertically and measures the time required for the round-trip path. As the frequency is increased, a point is reached where no energy will be returned from the ionosphere. The highest frequency at which a vertically incident radio wave will return from the ionosphere is called the **critical frequency**. Signals at higher frequencies pass through the ionosphere into space.

Radiation Angle and Skip Distance

A radio wave at a low angle above the horizon requires less refraction (bending) to bring the wave back to Earth. This is why we want antennas with low radiation angles to work DX.

Figure 3-4 illustrates some of the effects of radiation angle. The high-angle waves are bent only slightly in the ionosphere, and so pass through it. The wave at the somewhat lower angle is just able to be returned from the ionosphere. In daylight, it might be returned from the E region. The point of return for high-angle waves is relatively close to the transmitting station. (See point A on Figure 3-4.) The lowest-angle wave returns farther away, at point B of Figure 3-4.

The highest radiation angle that will return a radio wave of a certain frequency to Earth under specific ionospheric conditions is called the **critical angle**. Waves meeting the ionosphere at greater than the critical angle will pass through the ionosphere into space. Note that the critical angle will be different for different frequencies.

Normally, you will expect radio signals to arrive at your station by following

Radio-Wave Propagation   3-5

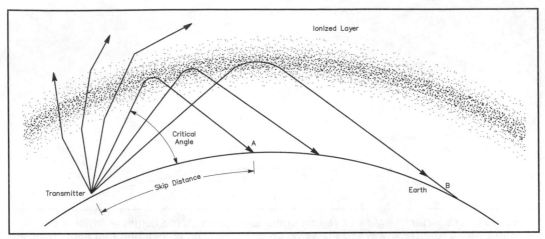

Figure 3-4 – This drawing illustrates ionospheric propagation. Waves that leave the transmitter above the critical angle are refracted in the ionosphere, but not enough to return to Earth. The wave at the critical angle will return to Earth. The lowest-angle wave will return to Earth farther away than the wave at the critical angle. This explains the emphasis on low radiation angles for DX work.

the shortest possible path between you and the transmitting station. This is called short-path propagation. Signals that might have arrived from the opposite direction, 180° different from the short-path signals are normally so weak that you would probably not hear them. Signals that arrive 180° from the short path are called long-path signals. When propagation conditions are suitable, the long-path signals may be strong enough to support communication. In fact, there are times when the long-path propagation may be even better than the short-path propagation. Stations with directional antennas can point their antennas directly away from each other to communicate. (This is not simply communicating "off the back" of the antennas.) See **Figure 3-5**.

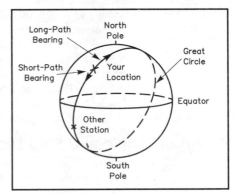

Figure 3-5 — This sketch of the Earth shows a great circle drawn between two stations. The short-path and long-path bearings are shown from the Northern Hemisphere station.

If you are listening to signals on your receiver and you hear a well-defined echo, even if it is a weak echo, the chances are you are hearing signals arrive at your station over the long path. The slightly longer time it takes the signals to travel the longer distance around the Earth results in a slight delay when compared to the direct, short-path signals. This is a good indication that you may be able to point your antenna away from the received station to communicate.

MAXIMUM USABLE FREQUENCY

There is little doubt that the critical frequency is important to amateur communication. Radio amateurs, however, are more interested in the frequency range over which communication can be carried via the ionosphere. Most amateurs want to know the **maximum usable frequency** (abbreviated **MUF**) for a particular propagation path at the time of day when communication is desired. The MUF is the highest frequency that allows a radio wave to reach the desired destination, using E or F-region propagation. The MUF is subject to seasonal variations as well as changes throughout the day, based on the amount of ultraviolet solar radiation. The MUF also varies with the distance between stations as well as their locations and the direction of the path between them.

Radio waves with frequencies below the maximum usable frequency are usually bent back to Earth, and return some distance away. The best chance of contacting another station using sky-wave propagation will be to pick a frequency as close as possible to the MUF, while staying on the low-frequency side of the MUF. Ionospheric absorption will be a minimum at frequencies near the maximum useable frequency, and that is another reason to select a frequency close to the MUF for the propagation path to the area you want to contact.

The vertical incidence critical frequency measurement can help determine the MUF for long-distance communication at the time of the measurement. Most amateurs would use computer software or read published charts to help them determine the MUF for a certain propagation path, however.

Why is the MUF so important? If we know the MUF, we can make an accurate prediction as to which amateur band will give us the best chance for communication via a particular path. For example, suppose an amateur in Minnesota wants to contact a station in France. Using a propagation-prediction computer program or propagation charts, the ham estimates the MUF to be 24 MHz at the time of the desired contact. The closest amateur band below that frequency (the 15-meter, or 21-MHz band in this case) offers the best chance for contact. If the MUF is predicted to be 17 MHz on a path between Ohio and Germany, the 20-meter (14-MHz) band probably offers the best chance of a successful contact.

There is no single MUF for a given location at any one time — the MUF will vary depending on the direction and distance to the station we wish to contact. Understanding and using the MUF is just one of the many ways we can change the study of propagation from guesswork into a science.

As the MUF for a given path increases, the single-hop distance at lower frequencies will decrease. Suppose you are operating on the 10-meter band, and are contacting stations that are 800 to 1000 miles away. After making a few more contacts you start to notice that you are contacting stations only about 500 miles away, and then you notice that you are contacting stations even closer, perhaps only out to a few hundred miles. This can be an excellent indication that the MUF for the longer-path stations has moved up to a higher frequency, perhaps even above 50 MHz. It is a good time to check for a band opening on 6 meters!

LOWEST USABLE FREQUENCY

The **lowest usable frequency** (**LUF**) is the lower limit to the range of frequencies that will provide useful communications between two locations. (This applies to communications by way of the ionosphere.) Any signals at frequencies below the LUF will be absorbed in the ionosphere rather than returning to Earth. Occasionally the LUF may be higher than the maximum usable frequency (MUF). This means that for the highest possible frequency that will propagate through the ionosphere for that path, the signal absorption is so large that even signals at the MUF are absorbed. Under these conditions it is impossible to establish sky-wave communication between those two points no matter what frequency is used! (It would normally be possible, however, to communicate between either location and other locations on some frequency using sky-wave signals.)

[Before going on, turn to Chapter 13 and study questions G3B01, G3B02, G3B05 and G3B11 through G3B14. Also study questions G3C04 and G3C11. Review this section if you have difficulty with any of these questions.]

SOLAR ACTIVITY AND RADIO-WAVE PROPAGATION

From the discussion in the previous sections, it should be clear that the Sun plays a major role in the ionization of the ionosphere. In fact, the Sun is the dominant factor in sky-wave communication. If we want to communicate over long distances, we must understand how solar conditions will eventually affect our radio signals.

You know about the day-to-day Sun-related cycle on Earth, related to the time of day. You also know about the annual cycle, related to the seasons of the year. Conditions affecting radio communication vary with these same cycles. Long and short-term solar cycles also influence propagation in ways that are not so obvious. The condition of the Sun at any given moment has a very large effect on long-distance radio communication, and the Sun is what makes propagation prediction an inexact science.

Man's interest in the Sun is older than recorded history. **Sunspots** are dark regions that appear on the surface of the Sun. They were observed and described thousands of years ago. Observers noted that the number of sunspots increased and decreased in cycles. The solar observatory in Zurich, Switzerland, has been recording solar data on a regular basis since 1749. Accordingly, the solar cycle that began in 1755 was designated Cycle 1. The low sunspot numbers marking the end of Cycle 22 and the beginning of Cycle 23 occurred in October 1996. Sunspot cycles average roughly 11 years in length, but some cycles have been as short as 9 years, and some as long as 13 years.

Several sunspot measurements are recorded and used to predict the level of solar activity and the effects on radio-wave propagation. The sunspot number is a daily index of sunspot activity. Daily values are averaged for an entire month to arrive at the montly mean sunspot value. (*Mean* is a mathematical term for average.) These monthly mean sunspot values are then averaged over a 13-month period centered around the month of interest to find the smoothed sunspot number (SSN) that is often reported in propagation studies. For example, to find the smoothed sunspot number for November 2003, you would use the monthly mean values from May 2003 through May 2004. (Only half the mean values for the first and thirteenth months are used.) The monthly values are added and the total is divided by 12 to find the smoothed sunspot number for the month of interest. As you can see, the smoothed sunspot number for a particular month cannot be determined until six months after that month.

Two separate peaks were observed in Cycle 23. The first peaked during April of 2000 with a smoothed sunspot number of 120.8. A second lower peak occurred in November of 2001 when the smoothed sunspot number was 115.6. Cycle 23 is expected to bottom out sometime between the end of 2006 and the beginning of 2007, and cycle 24 will begin to increase from that point.

The highs and lows of individual sunspot cycles also vary a great deal. For example, Cycle 19 peaked with a record-high smoothed mean (average) sunspot number of over 200, yet Cycle 14 peaked at only 60 and Cycle 20 peaked at just over 100. Cycles 21 and 22 both peaked at a smoothed mean sunspot number of just over 150.

Solar activity seems to influence all radio communication beyond ground-wave or line-of-sight ranges. Maximum ionization occurs during a sunspot cycle peak. High sunspot numbers usually indicate good worldwide radio communications on the higher-frequency amateur HF bands, such as the 15, 12 and 10-meter bands. Long-distance communication can even be enhanced on the lower VHF bands, especially the 6-meter band and sometimes the 2-meter band. During a peak in the sunspot cycle, the 20-meter amateur band is open to distant parts of the world almost continuously. Even during periods of low solar activity, however, the 20-meter band will usually provide worldwide propagation during daylight hours. When sunspot numbers are high, frequencies up to 60 MHz or higher are often usable for long-distance communication. During times of low solar activity, frequencies above 20 MHz (such as the 15-meter band and shorter wavelengths) are not as likely to provide reliable worldwide propagation.

The Cycle 19 peak in 1957 and 1958 was responsible for the best propagation conditions in the history of radio. Sunspot cycles do not follow consistent patterns — there can be highs that seem to come from nowhere during periods of relatively low sunspot activity.

An important clue for anticipating variations in solar radiation levels, and radio propagation changes resulting from them, is the time it takes the Sun to rotate on its axis, approximately 28 days. Active areas on the Sun capable of influencing propagation may recur at four-week intervals for four or five solar rotations. If the MUF is high and propagation conditions are good for several days, you can expect similar conditions to develop approximately 28 days later.

Another useful indication of solar activity is **solar flux**, or radio energy coming from the Sun. Increased solar activity produces higher levels of solar energy, which produces greater ionization in the ionosphere. Sophisticated receiving equipment and large antennas that can be pointed at the Sun are required to measure solar flux. When this measurement is made, a number called the **solar-flux index** is given to represent the amount of solar flux. The solar-flux index is gradually replacing the sunspot number as a means of predicting radio-wave propagation.

The solar-flux measurement is taken daily on 2800 MHz (10.7 centimeters). The information is transmitted by the US National Institute of Standards and Technology station WWV in Fort Collins, Colorado, and WWVH in Hawaii. Both the sunspot number and the solar-flux measurements tell us similar things about solar activity, but to get the sunspot number the Sun must be visible. The solar-flux measurement may be taken under any weather conditions. In addition, sunspot numbers are usually given as a smoothed average value, based on the previous six months of data.

Using the Solar Flux

We can use the solar flux numbers to make general predictions about band conditions from day to day. The solar flux varies directly with the activity on the Sun. Values range from around 60 to 250 or so. Flux values in the 60s and 70s generally indicate fair to poor propagation conditions on the 14-MHz (20-meter) band and higher frequencies. Values from 90 to 110 or so indicate good conditions up to about 24 MHz (12 meters), and values over 120 indicate very good conditions on 28 MHz (10 meters), and even up to 50 MHz (6 meters) or higher at times. The higher the frequency, the higher the flux value that will be required for good propagation. On the 6-meter band, for example, the flux values must be in the 200s for reliable long-range communication.

Propagation Beacons

Section 97.3(a)(9) of the FCC Rules defines a **beacon** as "an amateur station transmitting communications for the purposes of observation of propagation and reception or other related experimental activities." In the US, automatic beacon stations are permitted on 28.2 to 28.3 MHz as well as the 6, 2 and $1^1/_4$-meter bands, and the 70-cm band. (There are several US beacons on 14.10 MHz on the 20-meter band, but these are not operated under automatic control.) Several beacons are also operated at 14.1 MHz on the 20-meter band from other countries around the world.

There are numerous beacons operating on the 10-meter band between 28.2 and 28.3 MHz. You can determine 10-meter propagation conditions to various parts of the world by tuning this segment of the band listening for beacon stations. You can find the latest information about active beacons on the Northern California DX Foundation World Wide Web page (**www.ncdxf.org/beacon.htm/**) The NCDXF page includes a listing of the known beacons with call signs, locations and transmitter frequency. Other interesting information is also included. By listening on the beacon frequencies, you can estimate propagation conditions on the band of interest.

[Turn to Chapter 13 and study questions G3A04, G3A05, G3A09, G3A10 and G3A11. Also study questions G3B03, G3B04, G3B06, G3B07 and G3B08. Review this section of the text if any of these questions give you difficulty.]

SUDDEN IONOSPHERIC DISTURBANCES

One solar phenomenon that can severely disrupt HF sky-wave radio communication is a solar flare. A **solar flare** is a large eruption of energy and solar material from the surface of the Sun. A large amount of ultraviolet and X-ray radiation travels from the Sun at the speed of light, reaching the Earth approximately 8 minutes after the flare begins. When the radiation reaches the Earth, the level of ionization in the ionosphere increases rapidly and D-region absorption is greatly enhanced. This is called a **sudden ionospheric disturbance (SID)**. See **Figure 3-6**.

If the Earth and the solar flare producing the radiation are directly in line, the SID will be most severe. The D region may be so heavily ionized that HF sky-wave communications are completely blacked out. If the area of the solar flare is off to

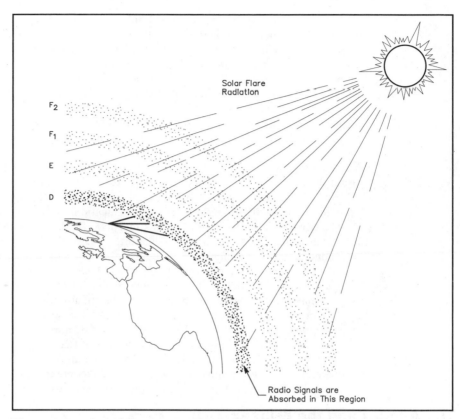

Figure 3-6 — Approximately 8 minutes after a solar flare occurs on the Sun, the ultraviolet and X-ray radiation released by the flare reaches the Earth. This radiation causes increased ionization and radio-wave absorption in the D region.

the side of the Sun, the SID will be less severe. Since D-region ionization is responsible for the disruption of HF communications during an SID, the lower frequency bands will be affected first. Communications may still be possible on a higher frequency band. Because an SID only affects the side of the Earth facing the Sun, dark-path communications will be relatively unaffected. An SID may last from a few minutes to a few hours, with conditions gradually returning to normal.

GEOMAGNETIC DISTURBANCES

Another result of a solar flare is a large increase in emission of charged particles from the Sun. These charged particles, along with the electromagnetic radiation from the Sun, influence radio-wave propagation conditions.

The corona is the Sun's outer layer. Temperatures in the corona are typically about two million degrees celsius, but can be more than four million degrees celsius over an active sunspot region. A coronal hole is an area of somewhat lower temperature. Matter ejected through such a "hole" takes the form of a plasma, a highly ionized gas made up of electrons, protons and neutral particles. This is sometimes called a *coronal mass ejection* (CME). The plasma travels at speeds up to two million miles per hour, and if the "blast" is directed toward the Earth it can result in an ionospheric storm on Earth. This will disrupt HF communications.

Because the particles travel much more slowly than the speed of light, it can take as much as 20 to 40 hours for the the plasma to reach the Earth. When the charged particles come close to the Earth, they can be trapped by the Earth's magnetic field in the north and south polar regions. The particles cause increased ionization of the gases in the E region and higher parts of the atmosphere. This produces auroral displays. The aurora conditions will also reflect radio waves above about 20 MHz. VHF signals, particularly on 6 and 2 meters, can be bounced off an aurora to provide some exciting long-distance communications.

The particles also disrupt the Earth's magnetic field. A **geomagnetic disturbance** is a sudden, dramatic change in the Earth's magnetic field over a short time. Radio communications along higher-latitude paths (North or South latitudes greater than about 45°) will be more affected than those closer to the equator.

These charged particles affect the upper regions of the ionosphere first — the F region may seem to disappear or even split into many regions. This means the higher-frequency bands are most affected. The highest frequency that will be returned from the ionosphere may be only half what it would be normally. The charged particles streaming toward the poles also increase auroral activity. In general, a geomagnetic disturbance will degrade long-distance radio communications. Under extreme conditions, the geomagnetic disturbance may completely black out long-distance radio communication, especially on paths that pass near the Earth's poles.

The A-Index and the K-Index

The **K-index** represents readings of the Earth's geomagnetic field, updated every 3 hours at Boulder, Colorado. K-index values, given on a scale of 0 to 9, indicate the stability of the Earth's geomagnetic field. Steady values indicate a stable geomagnetic field, while rising values indicate an active geomagnetic field. The K-index trends are important indicators of changing propagation conditions. Rising K-index values are generally bad news for HF propagation and falling values are good news, especially for propagation paths involving latitudes above 30° north. Values of 4 and rising warn of conditions associated with auroras and degraded HF propagation.

The **A-index** is a daily figure for the state of activity of the Earth's magnetic field. It is based on the eight K-index readings from the previous day, so the A-index tells you mainly how yesterday was. It is very revealing when charted regularly, because geomagnetic disturbances nearly always recur at four-week intervals. (It takes the sun 28 days to rotate once on its axis.) The A index is given on a scale of values from 0 to 400, to indicate the range of geomagnetic field disturbance.

[Now it is time to study some questions in Chapter 13. Study questions G3A01, G3A02 and G3A03. Also study questions G3A06, G3A07, G3A08 and G3A12 through G3A16. Review this section as needed.]

THE SCATTER MODES

When we consider radio propagation, it is convenient to look first at what happens under ideal conditions. All electromagnetic-wave propagation is subject to scattering influences that alter idealized patterns to a great degree, however. The Earth's atmosphere, ionospheric regions and any objects in the path of the radio signal act to **scatter** the energy. If we understand how this scattering takes place, we can use the phenomenon to our advantage.

Forward Scatter

There is an area between the outer limit of ground-wave propagation and the point where the first signals are returned from the ionosphere, as shown in **Figure 3-7**. We call this area the *skip zone*. This zone is often described as if communications between stations in each other's skip zone were impossible. Actually, because some of the transmitted signal is scattered in the atmosphere, the transmitted signal can be heard over much of the skip zone, if sufficiently sensitive receiving devices and techniques are used.

Ionospheric scatter, mostly from the height of the E region, is most marked at frequencies up to about 60 or 70 MHz, and this type of forward scatter may be discernible in the skip zone out to about 1200 miles. Ionospheric scatter propagation is most noticeable on frequencies above the MUF.

Another means of ionospheric scatter is provided by meteors entering Earth's atmosphere. As the meteor passes through the ionosphere, a column of ionized particles is formed. These particles can act to scatter radio energy. See **Figure 3-8**. This ionization is a short-lived phenomenon that can show up as short bursts of little communication value or as sustained periods of usable signal level, lasting up to a minute or more. Most common between midnight and dawn, and peaking

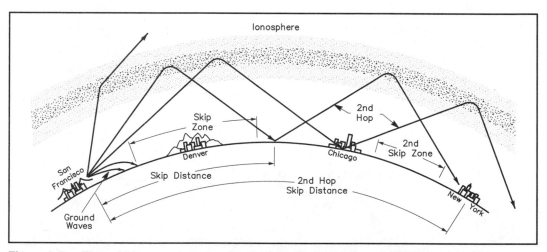

Figure 3-7 — There is an area between the farthest reaches of ground-wave propagation and the closest return of sky waves from the ionosphere. This region is known as the skip zone. Signals scattered in the ionosphere can help "fill in" the skip zone with weak signals.

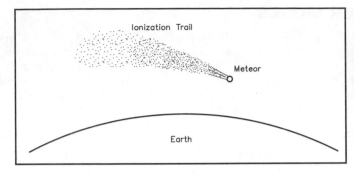

Figure 3-8 — Meteors passing through the atmosphere create trails of ionized gas. These ionized trails can be used for short-duration communications.

between 5 and 7 AM local time, meteor scatter can be an interesting adjunct to amateur communication at 21 MHz or higher, especially during periods of low solar activity.

Scatter signals are generally rather weak because only a small portion of the radio-signal energy is scattered into the area where it is being received. Scatter signals often sound distorted because the signal may arrive at the receiver from many different directions. When a signal arrives from several directions, the signal fading is called multipath interference. This may result in a signal with a fluttering, or wavering sound. With optimum equipment, scatter is usable at distances from just beyond the reliable local range out to several hundred miles.

A complex form of scatter is readily observed when you are working very near the maximum usable frequency. The transmitted wave is refracted back to Earth at some distant point, which may be an ocean area or land mass. A small portion of the transmitted signal may be reflected back into the ionosphere toward the transmitter when it reaches the Earth. This is sometimes called *backscatter*. The reflected

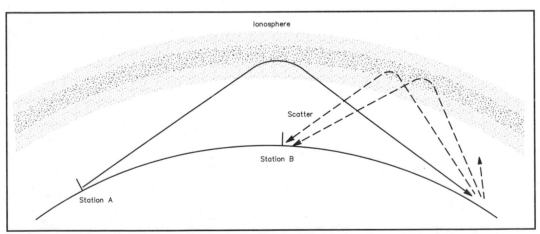

Figure 3-9 — When radio waves strike the ground after passing through the ionosphere, some of the signal may reflect back into the ionosphere toward the transmitting station. Some of this energy may be scattered back into the skip zone.

wave helps fill in the skip zone, as shown in **Figure 3-9**. The received signal will be at a distance that is too far for ground-wave reception, but may seem to be at a distance too close for normal sky-wave propagation. Under ideal conditions, backscatter is possible over 3000 miles or more, though the term "sidescatter" is probably more descriptive of what most likely happens on such long paths.

[Before going on to the next chapter, turn to Chapter 13 and study questions G3C06 through G3C10. Review this section as needed.]

CHAPTER 4
KEYWORDS
KEYWORDS
KEYWORDS

Audio rectification — Interference to electronic devices caused by a strong RF field that is rectified and amplified in the device.

Cathode-ray tube (CRT) — A vacuum tube with a phosphor coating on the inside of the face. CRTs are used in oscilloscopes and as the "picture tube" in television receivers.

Digital oscilloscope — An oscilloscope that uses digital circuits to store, change and display waveforms, and compare a waveform with one stored earlier.

Field-strength meter — A simple test instrument used to show the presence of RF energy and the relative strength of the RF field.

Monitor oscilloscope — A test instrument connected to an amateur transmitter and used to observe the shape of the transmitted-signal waveform.

Negative feedback — The process in which a portion of the amplifier output is returned to the input, 180° out of phase with the input signal.

Neutralization — Feeding part of the output signal from an amplifier back to the input so it arrives out of phase with the input. This negative feedback neutralizes the effect of positive feedback caused by coupling between the input and output circuits in the amplifier.

Noise bridge — A test instrument used to determine the impedance of an antenna system.

Oscilloscope — An electronic test instrument used to observe waveforms and voltages on a cathode-ray tube.

Peak envelope power (PEP) — The average power of the RF cycle having the greatest amplitude. (This occurs during a modulation peak.)

Photovoltaic cell – A wafer of semiconductor material that produces electricity when light shines on it. Sometimes called a *solar cell*, each cell produces about $\frac{1}{2}$ volt when fully illuminated. Cells are connected in series to increase the voltage. The size or surface area of the cell determines the maximum current that the cell can supply. An array of cells forms a *solar panel* that can be used to charge a lead-acid storage battery.

Photovoltaic conversion – The process by which a semiconductor PN junction changes sunlight directly into electricity.

Power — The rate at which energy is consumed. In an electric circuit, power is found by multiplying the voltage applied to the circuit by the current through the circuit.

Radio-Frequency Interference (RFI) — Interference to an electronic device (radio, TV, stereo) caused by RF energy from an amateur transmitter or other source.

S meter — A meter in a receiver that shows the relative strength of a received signal.

Signal tracer — A test instrument that shows the presence of RF or AF energy in a circuit. The signal tracer is used to trace the flow of a signal through a multistage circuit.

Speech processor — A device used to increase the average power contained in a speech waveform. Proper use of a speech processor can greatly improve the readability of a voice signal.

Transmit-receive (TR) switch — A mechanical switch relay or electronic circuit used to switch an antenna between a receiver and transmitter in an amateur station.

Two-tone test — Problems in a sideband transmitter can be detected by feeding two audio tones into the microphone input of the transmitter and observing the output on an oscilloscope.

Amateur Radio Practices

It's easy to form good operating habits in Amateur Radio. By understanding the fundamentals of radio, you should be able to maintain your station so that its operation is in accordance with FCC Rules and Regulations. In addition, you'll want to put out the best possible signal. If you know the principles behind the operation of your station, correctly adjust and measure its performance, and take advantage of available test equipment, you should have no problem keeping your station "up to par."

There are five questions on the General license exam from the Amateur Radio Practice subelement. These questions will come from the five syllabus groups for this subelement:

G4A Two-tone test; electronic TR switch; amplifier neutralization.

G4B Test equipment: oscilloscope; signal tracer; antenna noise bridge; monitoring oscilloscope; field-strength meters.

G4C Audio rectification in consumer electronics, RF ground.

G4D Speech processors; PEP calculations; wire sizes and fuses.

G4E Common connectors used in amateur stations: types; when to use; fastening methods; precautions when using; HF mobile radio installations; emergency power systems; generators; battery storage devices and charging sources including solar; wind generation.

TEST EQUIPMENT

You will find various types of test equipment helpful in maintaining your Amateur Radio station and ensuring your equipment operates properly. In this section you will learn about the use of several types of test equipment for your General exam.

The Oscilloscope

The **oscilloscope** (sometimes called a *scope*), is probably the most versatile of all test instruments. Oscilloscopes measure instantaneous voltage by the vertical deflection of an electron beam as it sweeps horizontally across the face of a **cathode-ray tube (CRT)**. One signal is applied to the oscilloscope vertical channel, and a second input is applied to the horizontal channel.

An oscilloscope has amplifiers in line with its horizontal and vertical-channel inputs and responds to a wide range of frequencies. Expensive oscilloscopes can handle frequencies over 30 MHz while the scopes most amateurs use only respond well up to 5 MHz or so. This lets you display audio- or low radio-frequency signals.

Oscilloscopes contain *sweep circuits*, which are oscillators used to draw the trace horizontally across the screen. The internal sweep circuits supply a signal to the horizontal channel. The sweep-oscillator frequency must be adjusted to the proper range to display various signal frequencies. For Amateur Radio use, a properly adjusted scope can display the output waveform of your transmitted signal as shown in **Figure 4-1**. This can be especially useful if you wish to check the modulation level of your signal. It can also show if there is carrier present on your single-sideband (SSB) signal. Another use for the oscilloscope is to display circuit frequency response.

A useful comparison between signals can be made using a dual-trace oscilloscope. This type of scope has two separate vertical-input channels. The two channels can be swept in the horizontal direction at the same rate and if two different signals are applied to the separate vertical inputs, both signals can be viewed on the screen at the same time. This function is especially useful for comparing the characteristics of a signal before and after passing through a stage in a circuit. **Figure 4-2** shows how a dual-trace oscilloscope can display the Morse code keying characteristics of a transmitter. The top trace shows the key closure or keyed line from an electronic keyer. The lower trace shows the actual transmitter CW output that results from the keying signal. This display shows a slight delay between the actual key closure and the start of the dot and another slight delay between opening the key and the end of the dot. It also gives a graphic demonstration of the rise and

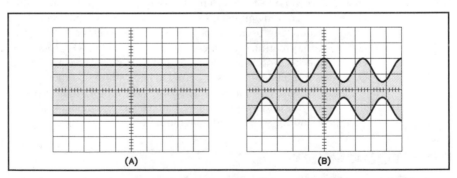

Figure 4-1 — Oscilloscope displays of RF signals. At A is an unmodulated carrier. The signal at B is from a full-carrier AM transmitter modulated with a single-frequency sine wave.

Figure 4-2 – A dual-trace oscilloscope is useful to display two waveforms for comparison. The top trace here is the key closure on the line used to key a CW transmitter. The lower trace shows the resulting CW waveform. Note the rounded corners and the smooth rise and decay times on the keyed output. Five milliseconds is generally agreed to be the ideal rise and fall time of a CW envelope. The horizontal divisions for this display each represent 5 milliseconds. If the CW signal rises too sharply, or drops off too quickly, key clicks or other keying problems can result.

fall times of the transmitted waveform from a properly adjusted CW transmitter.

There is a variety of controls on any oscilloscope and you should follow the instructions in the manual that comes with the scope when adjusting them. The focus control should be carefully adjusted for a sharp fine line on the CRT. The intensity control sets the brightness of the trace. Setting the brightness too high can damage the phosphor on the face of the oscilloscope CRT.

A **monitor oscilloscope** is used to show the signal quality of a transmitter. It lets you know if your AM signal is overmodulated, if your SSB signal is flattopping, what your CW dot-dash ratio is and the shape of your CW signal waveform.

To check the modulation of an AM or single sideband (SSB) transmitter, connect the RF output to the vertical amplifier input. The RF output also goes to the external trigger input on the oscilloscope so that the horizontal sweep signal will synchronize with the transmitter output. Set the internal sweep oscillator to a frequency that is about twice the highest modulating frequency, to provide a clear display of the modulated signal.

Oscilloscopes have been built using analog technology for many years. (Analog means the circuits measure and display continuously variable physical quantities.) Modern electronics is making more use of digital circuits, however. (Digital circuits represent the measured physical quantities with a series of numbers, or digits. Values are rounded off to the nearest allowable whole digit, so values between those digits are not allowed.) Using inexpensive digital integrated circuits, a **digital oscilloscope** can store, change and compare waveforms.

Some inexpensive digital scopes do not include a CRT display. They are designed for use with a personal computer. Measurement parameters are adjusted through the computer rather than by front-panel knobs and switches. The computer displays the measured waveform.

Analog-to-digital converter circuits measure the analog signal and change the measured amplitude values into digital data for processing, storage and display. The sampling rate of the converter limits the highest frequency signals that a digital scope can accurately display. Unfortunately, the digital circuits may process and display a waveform that does not represent the actual waveform if you try to measure a signal with a frequency that is too high for the converter. The signal must be sampled more than twice during each cycle of the highest frequency signal. So a digital-to-analog circuit that samples the input signal at a rate of 10 MHz will only be able to accurately display signal frequencies up to a little less than 5 MHz.

[Now turn to Chapter 13 and study questions G4B01, G4B02, G4B05, G4B06 and G4B14. Review this section if you have difficulty with any of these questions.]

The Signal Tracer

With today's multistage rigs, most problems that come about are a result of a defective component in one stage. Isolating the problem stage is the first step in the repair process. Armed with a block diagram of your radio and a **signal tracer**, you can proceed to locate the trouble spot in a straightforward manner.

A signal tracer, as the name implies, is a device that allows you to trace the path of a signal through the various stages in your rig until it disappears or begins to show unusual characteristics. This helps you identify the stage that isn't operating properly, so you can concentrate your repair efforts in that section. Any device that reacts to RF or AF energy can be used as a signal tracer. A typical signal tracer consists of a diode detector and a high-gain audio amplifier, which detects amplitude variations in RF signals. An oscilloscope or a VTVM can also be used to isolate trouble in your ham equipment.

The Antenna Noise Bridge

A properly operating antenna system is essential for a top-quality amateur station. If you build your own antennas, you must be able to tune them for maximum operating efficiency. Even if you buy commercial antennas, there are tuning adjustments that must be made because of differences in mounting location. Height above ground and proximity to buildings and trees will have some effect on antenna operating characteristics.

You must be able to measure your antenna's performance. After installation, you should be able to periodically monitor your antenna system for signs of problems, and troubleshoot failures as necessary.

An antenna **noise bridge** is a device that allows you to measure the impedance of antennas and other electrical circuits. The bridge produces a wide-band noise signal that is applied to the circuit under test. The noise bridge is connected between a receiver and an antenna of unknown impedance. (You can also connect other circuits with an unknown impedance to a noise bridge and receiver, but antenna measurements are the most common.) The receiver is tuned to the desired operating frequency and then you adjust the bridge controls until the noise heard in the receiver is nulled (decreased to a minimum). By noting the settings of the controls on the noise bridge, and then calculating the reactance of the indicated capacitance or inductance, you can determine the impedance of the unknown circuit.

One way to make use of a noise bridge is to set the controls for a 50-Ω resistance and zero reactance, and then adjust an antenna tuner for minimum received noise. This will help you adjust the antenna tuner for approximately the right settings before applying any transmitter power to the antenna tuner and antenna.

You can also use a noise bridge to measure the characteristic impedance of a piece of unknown transmission line. The noise bridge won't provide any information about the velocity factor of the transmission line, nor the signal loss of the line, though. A noise bridge does not give any direct information about the reflection coefficient of the line, either.

The Field-Strength Meter

A **field-strength meter** is a simple instrument that measures the relative strength of an RF field. This makes it useful for monitoring the radiated signal strength during antenna and transmitter adjustments. A field-strength meter usually contains a diode detector and a milliameter (to indicate signal strength). An antenna is attached to the detector input. By placing a field-strength meter a fixed distance from a rotatable antenna, you can get a relative indication of the antenna's horizontal radiation pattern by observing variations in signal strength as the antenna is rotated.

It is important to note that the simple field-strength meter described here is used to make *relative* measurements. You can observe changes in strength, but this instrument is not designed to measure actual electric or magnetic field strengths. This simple instrument won't measure the RF radiated field strengths from your station for the purposes of meeting the FCC RF radiation exposure limits. **Figure 4-3** shows a circuit diagram of a very simple field-strength meter. The values for L1, C2 and C3 are suitable for use on the 80-meter band. Choose smaller component values for the higher-frequency bands. The 20th edition of *The ARRL Antenna Book* includes a more elaborate band-switched portable linear field-strength meter project.

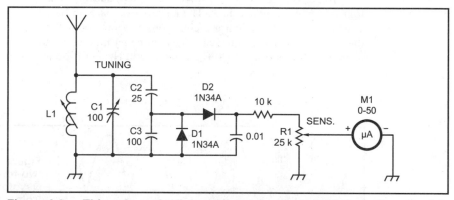

Figure 4-3 — This schematic diagram shows a simple field-strength meter. L1, C2 and C3 are selected to form a resonant circuit for the desired band of operation. The values shown will be suitable for the 80-meter band. Smaller values of L and C will be needed for the higher-frequency bands.

If you are doing some radio direction finding (RDF), you may find that a field strength meter is a handy piece of equipment to use as you get close to the transmitter source. A strong signal may drive your receiver's S-meter off scale, and this would require a variable attenuator to be used between the directional antenna and receiver as you get closer to the transmitter. A field-strength meter could replace the normal RDF equipment under those conditions, and you can use your body to shield the field-strength meter to find the direction to the transmitter.

[Before going on to the next section turn to Chapter 13 and study questions G4B03, G4B04, G4B07, G4B08 and G4B10 through G4B13. Review this section as needed.]

TESTING TRANSMITTER PERFORMANCE

Some of the most important tests and measurements you make as an amateur are those that tell you about the quality of your transmitted signal. Other hams form their first opinions of you on the air from the quality of your signal. Improperly adjusted amateur transmitters wreak havoc all over the bands and make operating unpleasant for everyone. An improperly adjusted transmitter can radiate energy all over the spectrum — not just in the amateur bands! Unless you know how to monitor your transmitted signal, you are asking for problems with radio-frequency interference and television interference (TVI).

Two-Tone Tests

The **two-tone test** is used to check the amplitude linearity of an SSB transmitter. Instead of the continuously varying pattern produced by a voice signal, a two-tone test produces a stationary pattern that can be examined on an oscilloscope. **Figure 4-4A** shows a pattern that is representative of a properly operating SSB transmitter. Figure 4-4B represents a signal that exhibits flattopping and crossover distortion. In **Figure 4-5**, a single cycle of a sine-wave signal is compared with a cycle of a signal that exhibits flattopping and crossover distortion.

To perform a two-tone test, you inject two audio signals of equal level into the transmitter microphone jack. The signals used should be about 1 kHz apart for a good display on the monitor oscilloscope and must be within the audio passband of the transmitter. The frequencies must not be harmonically related, so don't use frequencies like 1000 and 2000 Hz. Tones at 700 Hz and 1900 Hz are used in the ARRL lab for equipment testing.

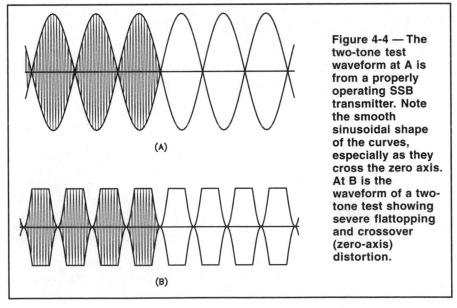

(A)

(B)

Figure 4-4 — The two-tone test waveform at A is from a properly operating SSB transmitter. Note the smooth sinusoidal shape of the curves, especially as they cross the zero axis. At B is the waveform of a two-tone test showing severe flattopping and crossover (zero-axis) distortion.

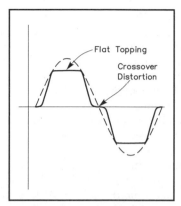

Figure 4-5 — A comparison of a sine-wave signal with a signal that exhibits flattopping and crossover distortion.

In a "clean" transmitter, this two-tone input will produce an RF signal at the output of the transmitter that contains only the two input signals. No amplifier is perfectly linear, so some mixing of the two input signals will take place. The sum and difference products produced by this mixing should be very weak in comparison with the main output. They should be weak enough that they cannot be detected on an oscilloscope pattern. What you will see on an oscilloscope is the pattern of two sine-wave signals as they add and subtract, forming peaks and valleys, as shown in Figure 4-4A.

A spectrum analyzer will tell you much more about your transmitter output, but such instruments are expensive, and few amateurs have access to one for transmitter testing. Oscilloscopes are relatively inexpensive and readily available. A two-tone test pattern observed on a scope will show only major defects in the output of an SSB transmitter, but it is still a very useful indication of how the transmitter is operating. If you keep a monitor oscilloscope connected to your transmitter output at all times you will be able to quickly notice any changes in operation. The speech pattern from a correctly adjusted SSB transmitter is shown in **Figure 4-6A**. The top and bottom halves of the oscilloscope display produce mirror images of the input waveform. Notice that the peaks are smoothly rounded. Overdriving the transmitter will cause the waveform to look more like the one shown in Figure 4-6B, with the peaks flattened, or clipped.

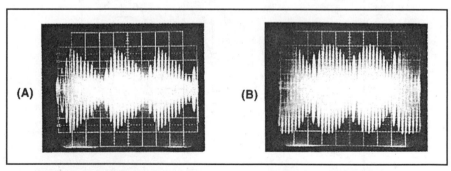

Figure 4-6 — The speech pattern from a correctly adjusted sideband transmitter is shown at A. Part B shows the same transmitter with excessive drive, causing distortion in the final amplifier.

[Turn to Chapter 13 and study questions G4A01 through G4A03 now. Also study questions G4A11 and G4A12. Review this section if you have any difficulty with these questions.]

Neutralization of Power Amplifiers

Most RF amplifiers operate with their input and output circuits tuned to the same frequency. Unless the circuit is carefully designed, some of the amplified output signal may be fed back to the input. If the signal is fed back "in phase" (so it adds to the input signal), the amplifier will oscillate. This is sometimes called *self oscillation* because there is no feedback path designed into the circuit to produce these oscillations. Care should be used in arranging components and wiring of the input and output circuits so that coupling is kept to a minimum. Keep all RF leads as short as possible, and pay particular attention to the RF return paths from input and output tank circuits to the emitter, drain or cathode.

Even with proper shielding and careful design, interelectrode capacitance in vacuum-tube amplifiers can cause oscillation. The internal elements are so close in modern vacuum tubes that signals can couple from the plate back to the grid, causing the amplifier to oscillate. This interelectrode capacitance is represented by C_{GP} in **Figure 4-7**. With tetrode and pentode tubes, interelectrode capacitance is greatly reduced by the addition of a screen grid between the control grid and the plate. Nevertheless, the power sensitivity of these tubes is so great that only a small amount of feedback is necessary to start oscillation.

In many cases, a technique called **neutralization** is used to provide stable amplifier operation. Neutralization is the process of deliberately feeding a portion of the amplifier output back to the input, 180° out of phase with the input. (This is called **negative feedback** because the feedback signal subtracts from the input.)

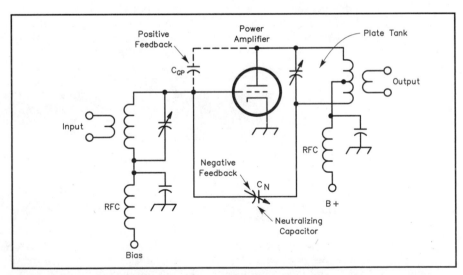

Figure 4-7 — In most vacuum tube amplifiers, some form of neutralization must be used to cancel positive feedback from the plate to the grid. In this circuit, a neutralizing capacitor (C_N) is added to counteract the effects of the grid-to-plate capacitance (represented by C_{GP}). The signal at the bottom of the tank circuit is 180° out of phase with the signal at the top of the tank, and careful adjustment of C_N will effectively cancel the positive feedback.

The feedback is usually accomplished by connecting a neutralizing capacitor to the tank circuit (C_N in Figure 4-7). If done correctly, neutralization will cancel any positive feedback in the plate-to-grid capacitance. This will render the amplifier much more stable and make it less likely to have self oscillations.

Neutralizing a Screen-Grid Amplifier Stage

One way to accomplish neutralization is to reduce to a minimum the RF-driver voltage fed from the input of the amplifier to its output circuit through the grid-to-plate capacitance of the tube.

If the tube is operated with grid current, the grid-current meter can be used to indicate neutralization. With this technique the plate voltage may remain on, but the screen voltage must be zero during the neutralization process. The dc circuit must be completed between the screen and cathode. There will be a change in grid current as the unloaded plate tank circuit is tuned through resonance. The neutralizing capacitor (or inductor) should be adjusted until this deflection is brought to a minimum. A minimum change in the grid-current reading as the output-circuit tuning is changed will indicate that you have almost achieved proper neutralization.

As a final adjustment, screen voltage should be reapplied and the neutralizing adjustment continued to the point where minimum plate current, maximum grid current and maximum screen current occur simultaneously. An increase in grid current when the plate tank circuit is tuned slightly to the high-frequency side of resonance indicates that the neutralizing capacitance is too small. If the increase is on the low frequency side, the neutralizing capacitance is too large. When neutralization is complete, there should be a slight decrease in grid current on either side of resonance.

If the screen-grid tube is operated without grid current, a sensitive output indicator can be used to indicate neutralization. With this technique, both screen and plate voltages must be removed from the tube(s). The dc circuits from the plate and screen to the cathode are left intact. The neutralizing capacitor or link coils are then carefully adjusted until the output indicator reads minimum.

[Now turn to Chapter 13 and study questions G4A06 through G4A10. Review this section if you have difficulty with any of these questions.]

SSB POWER MEASUREMENT

In a single-sideband, suppressed carrier transmission, there is no signal when there is no modulation applied to the transmitter. When the transmitter is modulated, a constantly varying RF envelope is produced, like the one shown in **Figure 4-8**. Two amplitude values associated with the wave are of particular interest. One is the peak envelope voltage, the greatest amplitude reached by the envelope at any time. The other is the RMS voltage of the envelope, as read on an RMS-reading voltmeter with an RF probe. See **Figure 4-9**. Chapter 5 includes additional information about RMS and peak voltage readings. There is also more information about power calculations in Chapter 5.

The **power** measurement used to rate transmitter output is **peak envelope power** (**PEP**). This is the *average* power of the wave at a *modulation peak*. One way to calculate PEP is to measure the peak envelope voltage, using an oscilloscope. Then

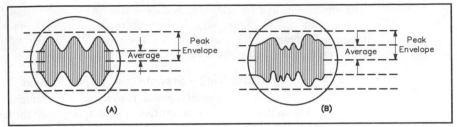

Figure 4-8 — Two RF-envelope patterns that show the difference between average amplitude and peak amplitude. In B, the average amplitude has been increased, but the peak amplitude remains constant.

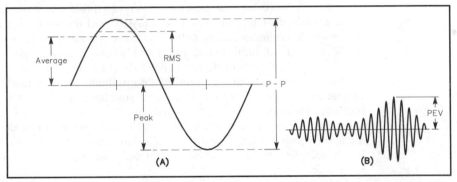

Figure 4-9 — Alternating voltage and current values used in making power measurements. The sine-wave parameters are illustrated at A, while B shows the peak envelope voltage (PEV) for a composite waveform.

multiply this value by 0.707 to calculate the RMS value of the peak envelope voltage. Remember that power equals current times voltage:

$$P = I \times E \qquad \text{(Equation 4-1)}$$

We don't have to calculate current if we know the load impedance, however. For most amateur antenna systems we want a load impedance of about 50 ohms. From Ohm's Law, current equals voltage divided by resistance:

$$I = \frac{E}{R} \qquad \text{(Equation 4-2)}$$

So we can substitute for I in the power equation:

$$P = \frac{E}{R} \times E = \frac{E^2}{R} \qquad \text{(Equation 4-3)}$$

Another way to write this equation is given as Equation 4-4. Here we show the peak envelope voltage times 0.707, and multiply this quantity by itself. This is the same as squaring the term. Divide by the load resistance, often written as R_L, to obtain the peak envelope power.

$$PEP = \frac{(0.707 \, PEV)(0.707 \, PEV)}{R_L}$$ (Equation 4-4)

Suppose we measured the output of a transmitter using an oscilloscope and found 200-V peak-to-peak across a 50-Ω dummy load. This represents a peak value of 100 V, and an RMS value of 70.7 V:

$$\text{Peak Envelope Voltage} = \frac{\text{Peak - to - Peak Voltage}}{2}$$ (Equation 4-5)

$V_{RMS} = PEV \times 0.707$
$V_{RMS} = 100 \, V \times 0.707 = 70.7 \, V$

Using this value for E in Equation 4-3, we can calculate the PEP of our transmitter.

$$P = \frac{E^2}{R} = \frac{(70.7 \, V)^2}{50 \, \Omega}$$

$$P = \frac{5000 \, V^2}{50 \, \Omega} = 100 \, W$$

Notice here that the units of volts squared divided by ohms give watts for the answer.

As another example, suppose we measure 500 V peak-to-peak across a 50-Ω load resistor at the output of a transmitter. What is the PEP output?

First we will find the peak envelope voltage, which is 250 V. Then we will find the RMS value by multiplying by 0.707:

$V_{RMS} = 250 \, V \times 0.707 = 176.75 \, V$

Next we use Equation 4-3 to calculate the peak envelope power:

$$P = \frac{E^2}{R} = \frac{(176.75 \, V)^2}{50 \, \Omega}$$

$$P = \frac{31240 \, V^2}{50 \, \Omega} = 625 \, W$$

Average power is found by using the value for RMS voltage as read on a voltmeter with an RF probe attached. (This reading will not be the same as the RMS value of the peak envelope voltage, since the meter reads an average over many RF cycles rather than just the voltage of a single RF cycle.) Substitute that value for E in Equation 4-3.

For a constant-amplitude sine wave, such as for a CW signal, PEP and average power are the same. If an average-reading wattmeter measures 1060 W for an unmodulated carrier from a transmitter, the PEP output of that signal is also 1060 W. This is not true for a modulated wave.

Another way to measure the power of an amateur transmitter is with a peak reading wattmeter. This type of power meter has capacitors in the metering circuit that charge to the peak voltage of the input wave. This voltage is then measured with a voltmeter and displayed on a scale calibrated in watts. To measure the actual transmitter power output, the wattmeter should be connected as close to the transmitter output terminal as possible. There are more details about alternating current and its measurement in Chapter 5.

Suppose you know the output power from your transmitter and want to know the RMS voltage across the antenna terminals. We can solve Equation 4-3 for voltage to answer this question.

$$P = \frac{E^2}{R}$$

Cross multiply the terms on each side of the equals sign:

$$E^2 = P \times R$$

Finally, take the square root of both sides of this equation to get a final working equation to calculate the voltage.

$$\sqrt{E^2} = \sqrt{P \times R} \quad \text{so}$$
$$E = \sqrt{P \times R}$$

Suppose you find that your transmitter is putting out 1200 W across a 50-Ω load, such as a dummy antenna.

$$E = \sqrt{P \times R} = \sqrt{1200 \text{ W} \times 50 \text{ }\Omega} = \sqrt{60,000 \text{ V}^2}$$
$$E = 245 \text{ V}$$

The transmitter will produce 245 V RMS across the 50-Ω load. If your are curious, you can multiply the RMS value by 1.414 to find the peak value, 346 V peak. Then if you want to know the peak-to-peak voltage this signal will produce, multiply the peak value by 2: 692 V peak-to-peak.

[Study the following questions in Chapter 13 before you go on to the next section: G4D03 through G4D05 and G4D10. Review this section as needed.]

STATION ACCESSORIES

There are a number of station accessories that make the difference between simply operating a radio and having a full awareness of the communications medium called Amateur Radio. In this section we will look at some of the accessories you may find useful when operating your station. Most amateur transceivers include these accessories.

The S Meter

Because reception is (or should be) the first aspect of station operation, we'll examine the **S meter** first. S meters are used to indicate the strength of a received signal. Signal-strength meters are useful when there is a need to make comparative readings.

S meters are calibrated in S units from S1 to S9; above S9 they are calibrated in decibels (dB). (You will learn more about decibels in Chapter 5.) An attempt was made by at least one manufacturer in the 1940s to establish some significant numbers for S meters. S9 was to be equal to a signal level of 50 microvolts, with each S unit equal to 6 dB. This means that for an increase of one S unit, the received signal power would have to increase by a factor of four. Such a scale is useful for a theoretical discussion, although real S-meter circuits fall far short of this ideal for a variety of reasons.

Suppose you were working someone who was using a 25-watt transmitter, and your S meter was reading S8. If the other operator increased power to 100 watts (an increase of four times the original power), your S meter should read S9. If the signal was S9, the operator would have to multiply the power by 10 to make your S meter read 10 dB over S9.

S meters on modern receivers may or may not respond in this manner. S-meter operation is based on the output of the automatic gain-control (AGC) circuitry; the S meter measures the AGC voltage. As a result, every receiver S meter responds differently; no two S meters will give the exact same reading. The S meter is useful for giving relative signal-strength indications, however. You can see changes in signal levels on an S meter that you may not be able to detect just by listening to the audio output level.

[Now turn to Chapter 13 and study question G4B09. Review the material in this section if you have difficulty with this question.]

TR SWITCHING

The **transmit-receive (TR) switch** connects the antenna to the transmitter during periods of sending, and to the receiver during periods of reception. With a separate receiver and transmitter, the TR switch may be a manually operated switch, an antenna relay that is switched by the transmitter or receiver, or an electronic circuit. When an electronic TR switch is used, the transmitter is continuously connected to the antenna. The receiver antenna connection is made through the TR switch. During transmit periods, the TR switch prevents large amounts of RF from reaching the receiver by electronically disconnecting the receiver from the antenna.

Since no mechanical device is involved in the antenna switching, an electronic TR switch can follow CW keying. An electronic TR switch operates faster than a mechanical switch. The receiver is left operative continuously, and if it can recover from a strong signal quickly enough, it will be sensitive to received signals between the transmitted character elements. The higher operating speed of an electronic TR switch can provide what is called "full-break-in (QSK) operation," in which an operator is able to hear signals between the dots and dashes of their code transmission.

The diode, transistor or tube that performs the switching function in a TR switch may be driven into nonlinearity by the strong transmitter signal. This is why some electronic TR switches can generate harmonics, which may cause television interference (TVI). A low-pass filter placed between the TR switch and the antenna, as shown in **Figure 4-10**, may eliminate the TVI, but the low-order harmonics (say, the second harmonic of 40 meters) may cause interference to other communications

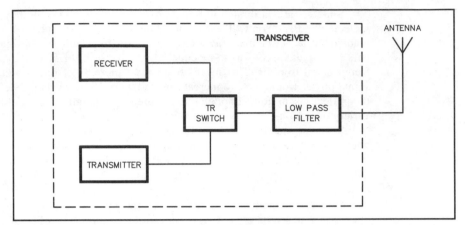

Figure 4-10 — A low-pass filter connected between the TR switch and the antenna will help filter out harmonics generated in the TR switch.

(on 20-meter SSB, for instance).

This discussion also describes the TR switching between the receiver and transmitter stages in a transceiver. Figure 4-10 can represent a transceiver block diagram as well as a separate transmitter and receiver. The TR switch is still followed by a low-pass filter to block any harmonics generated in the switch.

Many electronics circuits, including some TR switches, use diodes as switches because the current through the circuit can be controlled on or off depending on the diode bias conditions. When the positive side of a voltage supply is connected to the anode end of a diode, and the negative side of the supply is connected to the cathode end of the diode, there will be current through the diode. The diode is *forward biased* in this condition. When the positive side of a voltage supply is connected to the cathode end of a diode, and the negative side of the supply is connected to the anode end of a diode, there will be no current through the diode. The diode is *reverse biased* in this condition. PIN diodes are specially constructed diodes that will allow RF signals to pass through when a forward-bias DC voltage is applied across the diode.

[Now turn to Chapter 13 and study questions G4A04, G4A05 and G4A13. Review the material in this section if you have difficulty with either of these questions.]

Speech Processors

The human voice does not have a constant amplitude. When you speak, the speech amplitude is constantly changing, and only occasionally reaches a maximum, or peak, value. As can be seen in **Figure 4-11A**, there are large "valleys" between the voice peaks — periods where there is very little voice energy, hence little transmitter output. Because of the valleys in the speech waveform, the average power contained in a voice signal is small, as compared to the peak power. Peak-to-average ratios of 2 or 3 are quite common. Using a **speech processor**, the amplitude of the valleys can be increased while leaving the voice peaks unchanged. See

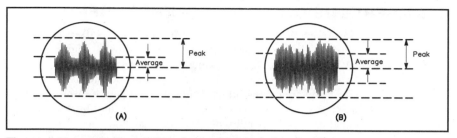

Figure 4-11 — A typical SSB voice-modulated signal might have an envelope of the general nature shown at A, where the RF amplitude (current or voltage) is plotted as a function of time, which increases to the right. The envelope pattern after speech processing to increase the average level of power output is shown at B.

Figure 4-11B. This serves to increase the *average* power contained in the waveform and improve the signal-to-noise ratio, but can lead to excessive background noise pickup. Properly used, speech processing can improve speech intelligibility at the receiver when propagation conditions are poor or the interference level is high.

The increased average power contained in a processed signal places a higher power demand on the transmitter. This can result in overheating of final amplifier and power supply components. If overheating becomes evident, you may need to reduce the processing level or transmitter power.

Speech processing, by its very nature, is a distortion of the audio signal, so processed signals sound less natural at the receiving end than unprocessed ones. For these reasons, processing should be employed only when necessary. If the frequency is clear and propagation is good, keep your speech processor turned off. If you do need the processor, use the minimum amount of processing necessary to increase intelligibility for the receiving station. Over processed, distorted signals are more difficult to copy than unprocessed ones. Distortion products generated during speech processing can increase the bandwidth of your signal and cause "splatter" (interference to stations on nearby frequencies). Keep in mind that speech processing cannot increase the peak power of the signal, even if the transmitter is 100% modulated. A speech processor adds nothing to the transmitter output PEP.

One problem with some single sideband (SSB) signals on the amateur bands is "splatter" or audio distortion caused by overdriving the transmitter's modulator stage. Such a transmitted signal has a wider bandwidth than necessary, and can cause interference to other stations on nearby frequencies. The worst cases are when amateur have their microphone gain set too high, overdriving the modulator stage and then they add excessive speech processing. The best way to prevent this, of course, is to adjust the microphone gain properly. A simple speech clipping circuit in the the microphone audio circuit will limit the maximum amplitude of the audio signal. Speech clipping can prevent overdriving the transmitter's modulator stage, reducing the chance of causing splatter interference.

[Turn to Chapter 13 and study questions G4D01, G4D02 and G4D09 at this time. If you have difficulty with any of these questions, review the material in this section.]

RADIO-FREQUENCY INTERFERENCE (RFI)

At one time or another, many amateurs have suffered the wrath of neighbors or family as their transmission was intercepted by a radio or television set. If you've never suffered problems like these, consider yourself lucky. An important means of survival in this age of electric and electronic devices is the ability to recognize and cure **radio-frequency interference**. Even if your transmitter puts out no harmonics or other spurious signals, interference is still an unfortunate possibility. Your RF signal can result in audio interference to another electronic device, such as a radio, television, stereo, electronic organ or intercom. For the purpose of this discussion, we'll call these audio devices.

RFI to audio devices is usually caused by **audio rectification**. Audio rectification happens when the RF signal gets into the audio device and is rectified (detected). The resulting audio signal is then amplified by the audio device, resulting in interference.

You might think that there would have to be a diode or semiconductor device in a circuit to rectify the RF signal, but that is not true. Any place where there are electrical conductors with a poor electrical connection between them, you have the possibility of an RF signal being rectified by this "natural" diode. Weathered joints between sections of TV mast, wire fence and fence stakes, sections of metal rain gutter can all result in a poor electrical connection between those joints. The lengths of metal conductors on either side of the joint can act as antenna elements, picking up RF energy and feeding it to the rectifier. The resulting nonlinear current is rich in harmonic energy, which can be reradiated and picked up by nearby receivers. This type of interference can be particularly difficult to locate and cure.

When the amateur operator is using double-sideband AM voice, the interference to the electronic device will be heard like any normal signal applied to the amplifier. The amateur's voice will be heard through the electronic equipment, and may be only slightly distorted. Audio rectification from an amateur SSB transmitter will sound like an unintelligible, garbled noise. If the amateur is using an FM transmitter, no sound will usually be heard: Instead, a decrease in volume will occur when the transmitter is on. A humming or clicking sound may be heard when the transmitter is keyed on and off. Similar humming or clicking will be heard if the interference is from a CW transmitter.

The first step in curing audio rectification involves finding out how the RF

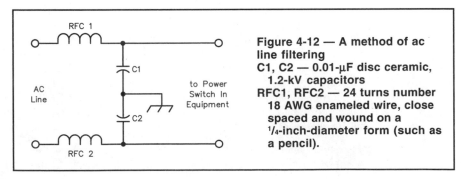

Figure 4-12 — A method of ac line filtering
C1, C2 — 0.01-µF disc ceramic, 1.2-kV capacitors
RFC1, RFC2 — 24 turns number 18 AWG enameled wire, close spaced and wound on a ¼-inch-diameter form (such as a pencil).

energy is entering the device, and "closing the door" with bypassing and filtering to keep it out. Several entry points are possible: the ac line cord, the antenna connection (if any), speaker wires or interconnecting cables (in a component stereo system). Another possibility is direct radiation into the circuitry of the audio device itself (usually through a wooden or plastic case).

Speaker wires and the ac power cord on an audio device form good entry points for RF, especially for signals on the higher amateur bands. On the higher bands, these lines can become a sizable fraction of a wavelength, and act as an antenna. A simple ac line filter, shown in **Figure 4-12**, can be built from readily available parts and installed at the audio device. The preferred location for the filter is inside the audio device, where the ac line enters. Notice that this filter uses inductors (often called RF chokes) in series with the lines, along with capacitors to *bypass* the unwanted signals to ground.

If RF is entering through external speaker wires, a simple filter using an RF choke and bypass capacitors at the speaker terminals can be effective. You should not install bypass capacitors directly across the output of a transistor amplifier. This can produce unwanted oscillations on the amplifier, even leading to destruction of the transistors. For transistor amplifiers you should place an RF choke in series with the amplified output, and then connect the bypass capacitors after the chokes, as shown in **Figure 4-13**.

Another common victim of audio rectification is the telephone. If you or your neighbor are experiencing telephone interference, first determine who owns the phone and who maintains the phone lines inside the home. If the phones are rented from the phone company and the phone company maintains the lines inside the home, contact your local telephone company. They have inductors and other filter components to help eliminate the problem. With so many consumer-owned electronic telephones, and customers maintaining the phone wires inside their own homes, you may have to install filters (or have the owner install them) yourself. **Figure 4-14** shows a K-Com filter, available at many consumer-electronics stores. Simply plug this filter into the phone, and then plug the phone line into the filter.

Installation of any of the devices mentioned in this section should be performed only by a qualified repair technician. If you install a filter in your neighbor's equipment, you may be blamed for any problems that occur with the equipment later,

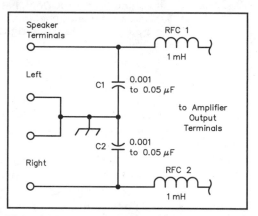

Figure 4-13 — A speaker filter for installation at the output of a transistor audio amplifier. The capacitors bypass the undesired signals to ground.
C1, C2 — 0.001 to 0.05-µF disc ceramic, 1.4-kV capacitors.
RFC1, RFC2 — 24 turns number 18 AWG enameled wire, close spaced and wound on a ¼-inch diameter form (such as a pencil).

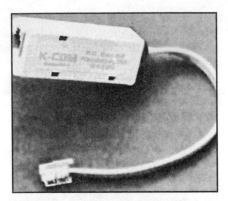

Figure 4-14 A typical telephone interference filter. This model, manufactured by K-Com, has modular phone connectors to simplify the installation right at the telephone.

whether they relate to the work you did or not!

Station Grounding

Make sure your station equipment is operating properly, and that it does not cause interference to audio equipment in your own home. Your transmitter and all station equipment should be effectively connected to a rod that is at least 8 feet long, driven entirely into the ground. The ground lead must be at least number 10 wire or heavier conductor such as copper ribbon. The greater the surface area of the ground lead, the more effective it will be. The outer shield braid from a piece of RG-213 or other large-diameter coaxial cable can be used as a conductor to connect your station equipment to a station ground.

The National Electrical Code forms the basis for electrical safety regulations in most building codes. You must be familiar with the requirements of the National Electrical Code or have a licensed electrician install the station ground for you. The Code specifies that there be only one grounding *system* for a building. That means your RF station ground must be connected to the electrical safety ground.

Keep the ground lead as short as possible. Proper equipment grounding is one of the most effective steps you can take to minimize the possibility of audio rectification of your transmitter's signals in a stereo or other electronics equipment.

An effective station ground connection should be short; not more than a few feet in length. If your ground wire is a resonant length at your operating frequency, it will act more like an antenna than an RF ground connection. A resonant "ground lead" won't act like a ground at all, and may lead to stray RF energy in your amateur station equipment. It may also lead to stray RF energy getting into home entertainment equipment, causing audio rectification or other types of RFI.

Suppose your amateur station is located in a third-floor apartment, or on the third floor of your house. The shortest ground lead you can have is 33 feet to the ground. This length will be resonant on several HF bands, and will bring stray RF into your "shack." In fact, you may even get an *RF burn* if you touch the front panel of your transceiver while you are transmitting. Such *RF hot spots* are quite common in a station located above the ground floor when the equipment ground includes a long ground wire.

A ground loop is an undesirable condition in which electrical currents flow to ground through unwanted paths. Avoid ground loops by always providing a single direct path from the equipment chassis to ground. If you connected a second wire to a different ground from another point in the circuit, you could create a condition in which current circulates through the wires and the two ground points. Don't connect the ground wire from one piece of equipement to the next in a series chain, eventually going to the Earth ground connection, because this is a good way to create a ground

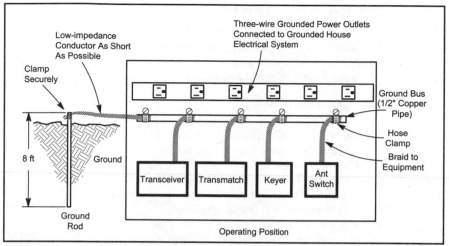

Figure 4-15 — You can make an effective station ground by connecting all equipment to a ground bus. A length of ½ inch copper pipe along the back of your operating desk or table makes a good ground bus. Heavy copper braid, such as the outer braid from RG-8 coaxial cable makes a good, flexible strap to connect each piece of equipment to the ground bus. The whole system then connects to a good earth ground — an 8-foot ground rod located as close to the station as possible is the minimum ground you should use. More than one ground rod may be needed in some locations.

loop. **Figure 4-15** shows one way to install a station ground system.

The National Electrical Code requires that all Earth grounds be connected by conductors. It is a good idea to use star washers or lock washers with any bolts you use to make ground connections, because the teeth on the washers will dig into the conductors and create a good electrical connection. The proper way to create an RF ground for your station is to connect the ground conductors from each piece of equipment to a single point and then connect that point to a good Earth ground.

If the RF ground in your station had an intermittent connection you might hear a severe broadband radio-frequency noise in your receiver. The intermittent connection forms a diode that can rectify any RF signals. The rectified RF is a nonlinear current that is rich in harmonic energy, producing the broadband noise.

An effective RF ground will reduce electrical noise, and reduce the possibility of interference. Even more important, however, a good ground connection will reduce the possibility of electric shock.

Remember that amateurs may use only the amount of power necessary to establish the desired communication. Operating with excessive power is more likely to cause audio-rectification problems than with low power.

[Before going to the next chapter, turn to Chapter 13 and study questions G4C01 through G4C09. Also study questions G4C11, G4C12 and G4C13. Review the material in this section if you have difficulty with any of these questions.]

AMATEUR RADIO SAFETY

Your station equipment not only makes use of ac line voltage, which can be dangerous in itself — it also generates additional potentially lethal voltages of its own. For your own safety, and that of others who may come in contact with your equipment, you should be familiar with some basic precautions.

Power-Line Connections

In most residential electrical power systems, three wires are brought in from the outside to the distribution board. Older systems may use only two wires. In the three-wire system, the voltage between the two "hot" wires is normally 240. The third wire is neutral and is grounded. Half of the total voltage appears between each of the hot wires and neutral, as indicated in **Figure 4-16**. In systems of this type, the 120-V outlets, lights and appliances are divided as evenly as possible between the two sides of the circuit. Half of the load is connected between one hot wire and the neutral, and the other half of the load is connected between the other hot wire and neutral.

Heavy appliances, such as electric stoves and most high-power amateur amplifiers, are designed for 240-V operation and are connected across the two hot wires. While *both* ungrounded wires should be fused, a fuse or switch should *never* be used in the neutral wire. If the appliance is connected to the supply voltage with a four-conductor power cord, the black and red wires should be connected to the fused hot wires. The white and green (or bare) wires should not have a fuse or switch in the line. The reason for this is that opening the neutral wire does not disconnect the equipment from the household voltage. It simply leaves the equipment on one side of the 240-V line in series with whatever load may be across the other side of the circuit. Furthermore, with the neutral wire open, the voltage will then

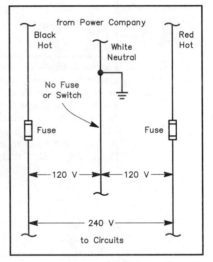

Figure 4-16 — **In home electrical systems, two wires carrying 240 V are split between the house circuits carrying 120 V each. Some heavy appliances use the full 240 V. As shown, a fuse should be placed in each hot wire, but no fuse or switch should be placed in the neutral wire.**

be divided between the two sides in inverse proportion to the load resistance. The voltage will drop below normal on one side and soar on the other, unless both loads happen to be equal.

High-power amateur amplifiers are designed to use the full 240 V because only half as much current is required to supply the same power as would be needed with a 120-V supply. The power-supply circuit efficiency improves with the higher voltage and lower current from a 240-V supply. Having a separate 240-V line for the amplifier also ensures that the capacity of the house circuit supplying 120 V to the shack is not exceeded. If your amplifier draws more current than the house wiring was designed to handle, it could cause the house lights to dim when the amplifier is operating. This is because the heavy current causes the power-line voltage to drop in value.

Three-Wire 120-V Power Cords

To meet the requirements of state and national electrical-safety codes, electrical tools, appliances and many items of electronic equipment now being manufactured to operate from the 120-V line must be equipped with a three-conductor power cord. The National Electrical Code, published by the National Fire Protection Association, is concerned with all aspects of electrical safety. The Code specifies the size and composition of electrical conductors, including grounding conductors. It also specifies proper wiring practices for all electrical wiring, and this helps ensure electrical safety inside your ham shack. The Code even specifies minimum conductor sizes for different lengths of amateur antennas. The code does not mention RF exposure limits to the human body, however. The FCC Rules establish safe RF exposure limits from all types of RF transmitters. These Rules are based on guidelines from the Institute of Electrical and Electronics Engineers (IEEE) and the American National Standards Institute (ANSI). See Chapter 10 of this book for further information about RF radiation exposure and the FCC Rules regarding this exposure.

In a three-wire power cord, two of the conductors (the "hot" and "neutral" wires) carry power to the device, while the third conductor (the ground wire) is connected to the metal frame of the device. See **Figure 4-17**. The "hot" wire is usually covered with black or red insulation, the "neutral" wire has white insulation, and the frame/ground wire has green insulation, or is bare.

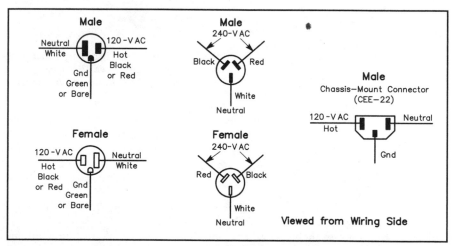

Figure 4-17 — Correct wiring technique for 120-V and 240-V power cords and receptacles. The white wire is neutral, the green wire is ground and the black or red wire is the hot lead.

When plugged into a properly wired mating receptacle, a three-contact plug connects the third conductor to an earth ground, thereby grounding the chassis or frame of the appliance and preventing the possibility of electric shock to the user. A defective power cord that shorts to the case of the appliance will simply blow a fuse. Without the ground connection, the case could carry the full line voltage, presenting a severe shock hazard. All commercially manufactured electronic test equipment and most ac-operated amateur equipment is supplied with these three-wire cords. Adapters are available for use where older electrical installations do not have mating receptacles. For proper grounding, the lug protruding from the adapter must be attached underneath the screw securing the cover plate of the outlet box. The outlet itself must be grounded for this to be effective. If the power wires coming into the electric box are inside a flexible metal covering, the outlet should be grounded through the metal covering. This type of wire is commonly referred to as armored cable.

A "polarized" two-wire plug and mating receptacle ensures that the hot wire and the neutral wire in the appliance are connected to the appropriate wires in the house electrical system. A polarized plug has one blade that is wider than the other. Without this polarized plug and receptacle, the power switch in the equipment may be in the hot wire when the plug is inserted one way, and in the neutral wire when inserted the other way. This can present a dangerous situation. It is possible for the equipment to be "hot" even with the switch off. With the switch in the neutral line, the hot line may be connected to the equipment chassis. An unsuspecting operator could form a path to ground by touching the case, and might receive a nasty shock!

Sometimes you may find it necessary to install a new plug on the end of an electrical

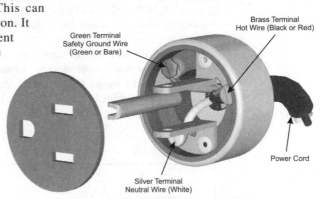

Figure 4-18 — This drawing shows the correct wiring of a 120-V ac line cord to a new plug. Be sure to observe correct color coding and grounding practices.

power cord. This could occur because the cord is worn or the insulation broken, for example. Be sure to follow accepted wiring practices when you install the new plug. If the wires in the power cord are stranded, be sure to twist the wires tightly and wrap them around the proper screw terminals of the new plug. Loose strands that aren't secured neatly under the screw head may come in contact with the other terminals, causing a short circuit.

Be sure to follow the proper color coding. The hot (black or red) lead should connect to the brass-colored screw terminal. The neutral (white) lead should connect to the silver-colored terminal. Also be sure to follow proper grounding techniques for the wiring. The safety ground (green or bare) lead should connect to the green-colored terminal, or the terminal that connects to the longer, round pin of the plug. See **Figure 4-18** for an illustration of this wiring.

Table 4-1

Current-Carrying Capability of Some Common Wire Sizes

Wire Size (AWG)	Continuous-Duty Current (A)*
8	46
10	33
12	23
14	17
16	13
18	10
20	7.5
22	5

* Wires or cables in conduits or bundles

Current Capacity

Another factor that must be taken into account when you are wiring an electrical circuit is the current-handling capability of the wire. **Table 4-1** shows the current-handling capability of some common wire sizes. From the table we can see that number 14 wire could be used for a circuit carrying 15 A, while number 12 (or larger) would have to be used for a circuit carrying 20 A.

To remain safe, don't overload the ac circuits in your home. The circuit breaker or fuse rating indicates the maximum load that may be placed on the line at any one time. Simple mathematics can be used to calculate the current your ham radio equipment will draw. Most equipment has the power requirements printed on the back. If not, the owner's manual should contain such information. When calculating current requirements, make sure to include any other household appliances that may be on the same line — including lights! If you were to put a larger fuse in the circuit, too much current could be drawn, the wires would become hot and a fire could result. If the household wiring to the outlet is American Wire Gauge (AWG) number 12, the maximum fuse or circuit-breaker size is 20 amperes. If the household wiring uses AWG number 14 wire, the maximum fuse or circuit-breaker size is 15 amperes.

[Turn to Chapter 13 now and study question G4C10 and questions G4D06 through G4D08. Also study question G4E02. Review this section if you have difficulty with any of these questions.]

OTHER SOURCES OF ELECTRICAL POWER

Amateurs should always be concerned about alternate sources of electrical power for times when the commercial power may go off. Having a source of emergency power could be important for your communications ability during various types of natural or man-made disasters.

If you live in an area where the commercial power goes out frequently, you might want to consider an emergency generator. This could also be a good source of power for Field Day and other Amateur Radio portable situations.

Whether you plan a temporary or permanent installation for your generator, you should be sure to install it in a well-ventilated area. You don't want carbon monoxide and other exhaust fumes to accumulate in your garage, basement or other living area, so don't run the generator inside a building or right outside an open window. You should also ensure proper grounding for your generator. This can be through a connection to your electrical system ground, for example. Of course you will need a supply of fuel, oil and other maintenance products for your generator. Don't store fuel in the generator fuel tank. Keep a fresh supply of fuel in a properly labeled, approved container. (Use stored gasoline in your car and get a fresh container once a month.) Don't store the fuel in an inhabited area. A detached garage or shed are safer storage locations.

Always check the engine oil on a generator before you begin using it. Check the oil level again every time you refuel the generator to make sure the engine is well lubricated.

Some generators produce an ac waveform that is not a very clean sine wave. Some of them are more like square wave generators, or exhibit other waveform distortion. That may be okay for powering an electric drill or saw, but isn't the best waveform for most electronic equipment. Be sure your generator produces a nice clean sine wave output.

Some operators use their gasoline-fueled generator to provide power for other household uses when the commercial ac power goes off for some reason. The proper way to make use of a generator to supply power for household use is to have a qualified electrician install a transfer switch in the electrical power feed to the house. Either the commercial ac mains are connected to the house, or the generator is connected to the house, but never both at the same time.

As a quick way to power the house, some people have made a power cord that will plug the generator output into an ac wall outlet in the house, thus feeding power from the generator throughout the house. This is a dangerous practice for several reasons. First, it is very dangerous for the power company workers, who may suddenly find that a "dead" line has power applied! Small portable generators do not generally produce enough current to operate appliances, lights and other household loads. Trying to power your house with such a generator can lead to an overloaded generator, and damaged equipment. In addition, if the commercial ac power is restored to your house while your generator is connected, it may damage or destroy the generator.

Many modern rigs require a source of 12 V dc at 20 amps or so. A deep-discharge lead-acid storage battery is a good emergency power source for such radios. Of course the battery is only useful if it is kept charged and ready to go at all times. A lead-acid battery gives off hydrogen gas as it is being charged. Since hydrogen gas can explode if ignited by a spark, it is important to provide proper ventilation for the battery as you charge it.

Natural sources of energy are becoming more economical and more practical as a way to power your amateur station equipment. **Photovoltaic conversion** is the process by which a semiconductor PN junction converts sunlight into electricity. Each **photovoltaic cell** produces about 0.5 volt in full sunlight, if there is no load connected to the cell. This is called the *open-circuit voltage*. The size or surface area of the cell determines the maximum current that the cell can supply. A solar panel is an array of photovoltaic cells that convert light into electricity. Twenty eight such cells connected in series provide a 14-volt supply that can be regulated to 13.8 volts and used to charge a lead-acid storage battery. Solar panels are rated according to output voltage and maximum output current. When setting up a solar electric system you must select a solar panel that is designed to supply the proper voltage and maximum output current for your system. After the initial expense of building such a solar electric system, the sun will provide a ready supply of free energy. The storage batteries maintain the power supply for nighttime needs or for cloudy periods when the solar panel may not produce enough electricity to power your station.

Some amateurs have also built wind-powered generating systems. The design and construction of a suitable windmill is not trivial. These systems provide a source of free energy after the expense of the initial construction, however. Again, the greatest disadvantage is the need for a battery storage system to supply power when the wind is not blowing.

[Turn to Chapter 13 now and study questions G4E06 through G4E11 and G4E13 through G4E15. Review this section if you have difficulty with any of those questions.]

MOBILE OPERATING

Many amateurs enjoy operating their HF Amateur Radio stations from their cars. While you must be very careful not to allow your ham activity to interfere with your concentration while driving, you can have many enjoyable contacts while someone else is driving or while stopped. When you install a 100-watt HF radio in your car, you will have to pay attention to a few details that you may have overlooked for a low-power VHF FM radio installation. For example, you might be able to use the cigarette lighter as a power connection for a low-power hand-held radio, but not for a full-power HF rig. The cigarette lighter socket wiring probably is not adequate for the 20 amps or so your HF radio will draw when transmitting. Use heavy gauge wire and make a direct fused connection to the battery. (Both leads should have fuses, placed as close to the battery as possible.) See **Figure 4-19**.

HF mobile antenna systems are the most limiting factor for effective operation of your station. The best place to locate your mobile antenna is in the center of your metal car roof. Placing even an 8-foot vertical antenna on top of a small car makes a dangerously tall system, however. If you have a mid-sized or larger vehicle such an antenna system is almost out of the question. For the lower-frequency bands, like 75 meters, the antenna will always use some type of inductive loading to shorten the length that would be required.

There are several types of connectors commonly used for coaxial cable RF transmission lines. Many hand-held radios use BNC connectors because this connector requires only about a quarter turn to install or remove the connector. BNC

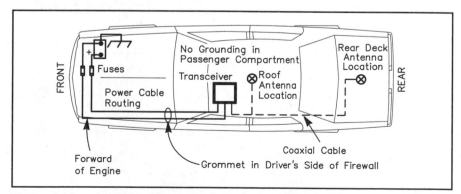

Figure 4-19 — A wiring diagram for a typical mobile transceiver installation.

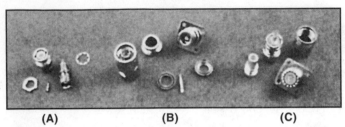

(A)   (B)   (C)

Figure 4-20 — Some common coaxial-cable connectors. Part A shows a BNC connector pair. Many hand-held radios use BNC connectors. They are popular when a weatherproof connector is needed for RG-58 sized cables. Part B shows a pair of N connectors. These are often used for UHF equipment because of their low loss. Type N connectors provide a weatherproof connector for use with RG-8 sized cables. Part C shows a PL-259 coaxial connector and its mating SO-239 chassis connector. Most HF equipment uses these connectors. They are designed for use with RG-8 sized cables although reducer bushings are available for the smaller-diameter cables.

connectors are designed for use with RG-58-type coaxial cables. Type N connectors are common on UHF and higher-frequency equipment. They are the best choice to use on the feed line between your 10-GHz equipment and antenna. Type N connectors are designed for use with RG-8 type coax. Most HF radios use the PL-259 connector, which is sometimes called a UHF connector (although it is generally considered to have too much loss for use at UHF). These connectors are designed for use with RG-8 type coaxial cable, although adapters are used with the smaller-diameter RG-58 cables. See **Figure 4-20**.

[This completes your study in Chapter 4. Before going on to Chapter 5, turn to Chapter 13 and study questions G4E01, G4E03, G4E04 and G4E05. Also study question G4E12. Review this section if you are uncertain of the answers to any of these questions.]

Alternating current (ac) — Electric current that flows first in one direction in a wire and then in the other direction. The applied voltage is changing polarity as the current direction changes. This direction reversal continues at a rate that depends on the frequency of the ac.

Capacitor — An electrical component composed of two or more conductive plates separated by an insulating material. A capacitor stores energy in an electric field.

Coil — A conductor wound into a series of loops. (Also see **inductor**.)

Current — A flow of electrons in an electric circuit.

Decibel (dB) — The smallest change in sound level that can be detected by the human ear. Power gains and losses are also expressed in decibels.

Direct current (dc) — Electric current that flows in one direction only.

Effective voltage — The value of a dc voltage that will heat a resistive component to the same temperature as the ac voltage that is being measured.

Electromotive force (EMF) — The force or pressure that pushes a current through a circuit.

Impedance — A term used to describe a combination of reactance and resistance in a circuit.

Inductor — An electrical component usually composed of a coil of wire wound on a central core. An inductor stores energy in a magnetic field.

Ohm — The basic unit of resistance, reactance and impedance.

Ohm's Law — A basic law of electronics, it gives a relationship between voltage, resistance and current ($E = IR$).

Power — The rate at which energy is consumed. In an electric circuit, power is found by multiplying the voltage applied to the circuit by the current through the circuit.

Reactance — The opposition to current that a capacitor or inductor creates in an ac circuit.

Resistance — The ability to oppose an electric current.

Root-mean-square (RMS) voltage — Another name for effective voltage. The term refers to the method of calculating the value.

Transformer — Two coils with mutual inductance used to change the voltage level of an ac power source to one more suited for a particular circuit.

Voltage — The EMF or pressure that causes electrons to move through an electric circuit.

Electrical Principles

To pass your Element 2 written test and receive a Technician license from the FCC, you had to learn some basic radio theory. For your Element 3 test, you will need to understand a few more electrical principles. Your General exam will include two questions about the electrical principles covered in this chapter. Those questions come from the following two syllabus groups:

G5A   Impedance, including matching; resistance, including ohm; reactance; inductance; capacitance and metric divisions of these values.

G5B   Decibel; Ohm's Law; current and voltage dividers; electrical power calculations and series and parallel components; transformers (either voltage or impedance); sine wave root-mean-square (RMS) value.

Be sure to turn to Chapter 13 and study the appropriate questions when the text tells you to study certain questions. This will help you check your progress, and decide if you need a little extra studying. We cannot tell you everything about a particular topic in one chapter — you may want to refer to some other reference books for additional information. *Understanding Basic Electronics*, published by the ARRL is a good place to start. *The ARRL Handbook* and other ARRL publications also contain lots of additional information about electrical principles.

E = VOLTAGE

Electrons need a push to get them moving; we call the force that pushes electrons through a circuit **electromotive force (EMF)**. EMF is similar to water pressure in a pipe. The more pressure, the more water that flows through the pipe. The greater the EMF, the greater the flow of electrons in a circuit.

EMF is measured in volts, so it is usually called **voltage**. A voltage-measuring instrument is called a voltmeter. The more voltage applied to a circuit, the more electrons will flow through the circuit.

I = CURRENT

The word **current** comes to us from a Latin word meaning "to run" and it always implies movement. When someone speaks of the current in a river, we think of a flow of water; a current of air can be a light breeze or a hurricane; and a current of electricity is a flow of electrons.

Current is indicated in equations and diagrams by the letter I (from the French word *intensité*). We measure current in *amperes,* and the measuring device is called an ammeter. An ampere is often too large a unit for convenient use, so we can measure current in milliamperes (one thousandth of an ampere) or microamperes (one millionth of an ampere). Amperes are abbreviated amp or A; milliamperes as mA and microamperes as μA.

In considering current, it is natural to think of a single, constant force causing the electrons to move. (Think of this as similar to the pressure of water down a river or through a pipe.) When this is so, the electrons always move in the same direction through a circuit made up of conductors connected together in a continuous loop. Such a current is called a **direct current**, abbreviated **dc**. This is the type of current furnished by batteries and by certain types of generators.

R = RESISTANCE

Resistance is something that opposes motion. All materials have electrical resistance; there is always some opposition to the flow of electrons through the material, although the resistance of different materials varies widely. Some materials have very high resistance; we call these materials good insulators, or poor conductors. Other materials have very low resistance; we call these materials poor insulators, or good conductors. The basic unit of resistance is the **ohm**, and a resistance-measuring instrument is called an ohmmeter. The symbol for an ohm is Ω, the Greek capital letter omega.

Between the extremes of a very good insulator and a very good conductor, we can make devices with a range of opposition to the flow of electrons in a circuit. We call such devices resistors. If we use the analogy of water flowing in a pipe to help us understand electron flow, these materials act like a pipe with a sponge in it. Some water will still flow in the pipe, but the flow will be reduced. **Figure 5-1** illustrates the concept of the water pressure in a pipe representing voltage, the water flowing through the pipe representing current and a valve that you can use to restrict the flow of water representing resistance.

The primary function of a resistor is to limit electron flow (current). The greater the resistance, the greater the reduction in current. If we want to control the current reduction over a certain range, we can use a variable resistor. A variable resistor, as its name implies, can be made to change resistance over a range. This allows us to control the electron flow between a maximum and a minimum value over the variable resistor's range.

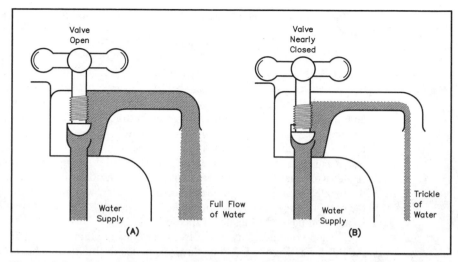

Figure 5-1 — Water pressure can represent the voltage in an electric circuit. The amount of water can represent electric current and a blockage in the water line (such as the spigot shown here) can represent the resistance in a circuit. Part A shows a water faucet with the valve open. There is little resistance and a lot of water flowing out of the faucet. Part B shows a faucet with the valve almost closed. In this case there is a lot of resistance, and only a small amount of water flows out of the faucet.

This current reduction has a price, however; the energy lost by the electrons as they flow through a resistor must be dissipated in some way. The energy is converted into heat, and the resistor gets warm. If the current is higher than the resistor can handle, the resistor will get very hot and may even be destroyed.

OHM'S LAW

Resistance, voltage and current are related in a very important way by what is known as **Ohm's Law**. Georg Simon Ohm, a German scientist, discovered that the voltage drop across a resistor is equal to the resistance multiplied by the current through the resistor. In symbolic terms, Ohm's Law can be written:

$$E = I \times R$$
(Equation 5-1)

By simply rearranging the terms, we can show that the current is equal to the voltage drop divided by the resistance:

$$I = \frac{E}{R}$$
(Equation 5-2)

and that the resistance equals the voltage drop divided by the current:

$$R = \frac{E}{I}$$

(Equation 5-3)

Ohm's Law is an important mathematical relationship; you should understand and memorize it. A simple way to remember the Ohm's Law equations is shown in **Figure 5-2**. If we draw the "Ohm's Law Circle" shown, we can find the equation for any of the three quantities in the relationship.

To use the circle, just cover the letter for which you want to solve the equation. If the two remaining letters are across from each other, you must multiply them. If the two letters are arranged one on top of the other, you must divide the top letter by the bottom. For example, if you cover I, you have "E over R" (Equation 5-2). Cover E, and you have "I times R" (Equation 5-1).

When you know any of the two quantities in Ohm's Law, you can always find the third. See **Figure 5-3**. The three examples in this figure show how to calculate voltage, current and resistance given the other two quantities in each case. You won't go wrong if you write the equation you want to use first, and then substitute the values you are given and solve for the unknown, as shown in these examples.

KIRCHHOFF'S LAWS

When we analyze a circuit, it is sometimes useful to calculate the current through, and the voltage drop across a circuit or portion of a circuit. We have two electrical principles, called Kirchhoff's Laws, that make it easy for us to find the current and voltage drop.

Kirchhoff's First Law (Kirchhoff's Current Law) states that when a circuit branches into two or more directions, the current entering the branch point, or node, is the same as the current leaving the node. Another way to say this is that the total current equals the sum of the branch currents through each resistor. We can use this law to find an unknown quantity in a parallel circuit.

For example, look at **Figure 5-4**. We know that the total current I_T entering point A is 1 A. The diagram also shows that the current through R_1 is 0.25 A and that the current through R_2 is 0.5 A. What is the current through R_3? We can use Kirchhoff's Current Law to answer that question.

$$I_1 + I_2 + I_3 = I_T = 1 \text{ A}$$

(Equation 5-4)

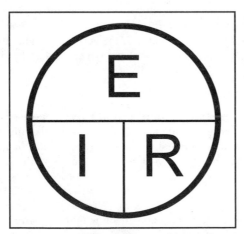

Figure 5-2 — The "Ohm's Law Circle." Cover the letter representing the unknown quality to find an equation to calculate the quantity. For example, if you cover the I, you are left with E / R.

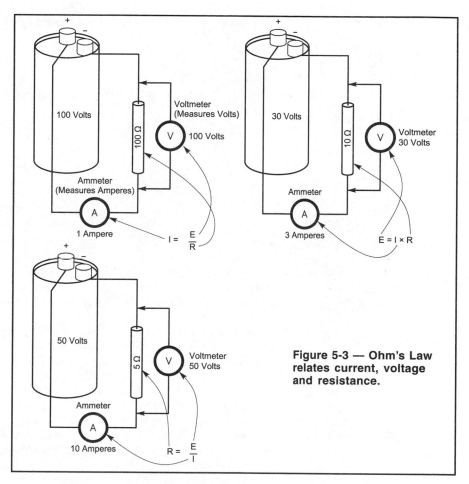

Figure 5-3 — Ohm's Law relates current, voltage and resistance.

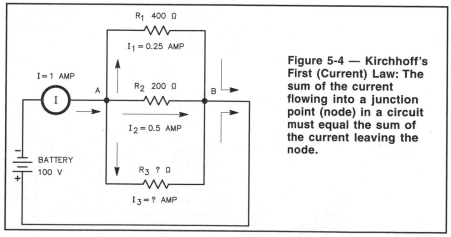

Figure 5-4 — Kirchhoff's First (Current) Law: The sum of the current flowing into a junction point (node) in a circuit must equal the sum of the current leaving the node.

Substituting our known values into this equation and solving for I_3, we have:

$$0.25 \text{ A} + 0.5 \text{ A} + I_3 = 1 \text{ A}$$

$$I_3 = 1 \text{ A} - 0.25 \text{ A} - 0.5 \text{ A} = 0.25 \text{ A}$$

Once we know the current through R_3 we can also find its resistance. This is a simple application of Ohm's Law. Since we know that 100 V is applied across points A and B, we can use Equation 5-3 to find the resistance:

$$R = \frac{E}{I}$$

$$R = \frac{100 \text{ V}}{0.25 \text{ A}} = 400 \, \Omega$$

We can draw some general conclusions from this discussion. First, the example shows that for a parallel circuit, the total circuit current is always equal to the sum of the currents in each branch of the circuit. We can also say each branch current is less than the total circuit current.

If the branch resistors have equal values, the current divides equally between the branches. For example, if two equal-value resistors are connected in parallel, each resistor carries half the total current. We don't even need to know the resistor values or the applied voltage in such a case. If the total circuit current is 1 A with two equal-value resistors connected in parallel, each resistor will carry 0.5 A.

Kirchhoff's Second Law (Kirchhoff's Voltage Law) tells us that the sum of the voltage drops must equal the sum of the voltage rises around any closed loop in a circuit (such as in a series circuit). So in the series circuit in **Figure 5-5**, the sum of the voltage drops across the resistors must equal the applied voltage. The voltage

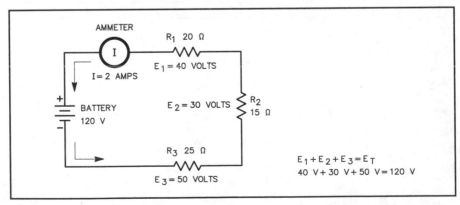

Figure 5-5 — Kirchhoff's Second (Voltage) Law: The sum of the individual voltage drops in a closed loop (such as a series circuit) equals the sum of the voltage rises.

drop across each resistor can be found by multiplying the current through the resistor by the resistance. (This is a simple Ohm's Law relationship, $E = I \times R$.) Since all the circuit current flows through each resistor, we can find the current value by dividing the applied voltage (120 V in this example) by the total resistance of the circuit, 60 Ω. (Did you remember that you simply add the values of resistors in series to find the total resistance?) From this, we find that 2 A flows through each resistor. We can then calculate the voltage drops.

$$E_{R1} = 2 \text{ A} \times 20 \text{ } \Omega = 40 \text{ V}$$

$$E_{R2} = 2 \text{ A} \times 15 \text{ } \Omega = 30 \text{ V}$$

$$E_{R3} = 2 \text{ A} \times 25 \text{ } \Omega = 50 \text{ V}$$

Total voltage drop = $40 \text{ V} + 30 \text{ V} + 50 \text{ V} = 120 \text{ V}$

Notice that the larger the resistance, the larger the voltage drop across that resistor.

We can use the principles of Kirchhoff's Laws and Ohm's Law to calculate information about various circuit conditions. Look at Figure 5-5 and suppose that the three resistors have some equal, but inknown value. Then suppose that we measure the battery voltage to be 13.5 V and the total circuit current to be 30 mA (0.030 A). Using Ohm's Law, Equation 5.3 we can calculate the total circuit resistance to be:

$$R = \frac{E}{I} = \frac{13.5 \text{ V}}{0.03 \text{ A}} = 450 \text{ } \Omega$$

When resistors are connected in series, the total combination is equal to the sum of the individual resistance values.

$$R_{\text{Total in Series}} = R_1 + R_2 + R_3 + \cdots + R_N \qquad \text{(Equation 5-5)}$$

Where N is the total number of resistors connected in series.

$$R_{\text{Total in Series}} = 450 \text{ } \Omega = R_1 + R_2 + R_3 = 3 \times R_1$$

$$R_1 = \frac{450 \text{ } \Omega}{3} = 150 \text{ } \Omega$$

So each of those unknown resistors have a value of 150 Ω. You may be wondering why we didn't just use an ohmmeter to measure the resistance. That certainly would be one way to find the resistance values. This example calculation shows how you could find the value if you had a voltmeter and an ammeter, but no ohmmeter.

Now take another look at Figure 5-4, and suppose that there were three equal-value unknown resistors connected in parallel. Again suppose we measure the battery voltage of this circuit to be 10 V and the total circuit current to be 200 mA (0.200 A). Again, we can use Equation 5-3 of Ohm's Law to calculate the total circuit resistance:

$$R = \frac{E}{I} = \frac{10 \text{ V}}{0.200 \text{ A}} = 50 \, \Omega$$

When resistors are connected in parallel, the total combination is given by the equation:

$$R_{\text{Total in Parallel}} = \frac{1}{\dfrac{1}{R_1} + \dfrac{1}{R_2} + \dfrac{1}{R_3} + \cdots + \dfrac{1}{R_N}} \qquad \text{(Equation 5-6)}$$

Once again, we can calculate the value of each resistor using the equation for resistors in parallel.

$$R_{\text{Total in Parallel}} = 50 \, \Omega = \frac{1}{\dfrac{1}{R_1} + \dfrac{1}{R_2} + \dfrac{1}{R_3}} = \frac{1}{\dfrac{3}{R_1}}$$

$$50 \, \Omega = \frac{1}{1} \times \frac{R_1}{3}$$

$$R_1 = 50 \, \Omega \times 3 = 150 \, \Omega$$

[Now turn to Chapter 13 and study questions G5B02 and G5B15. Review any material in this section that you have difficulty with.]

P = POWER

Suppose we want to know how fast energy is being consumed in a circuit. If you want to compare how bright two different light bulbs will be, or how much it will cost to run a new freezer for a month, you will have to know how fast they use electric energy. **Power** is defined as the time rate of energy consumption. The basic unit for measuring power is the watt, named for James Watt, who is probably best known for developing a working steam engine. The watt is usually abbreviated W.

Current is a measure of how many electrons are moving past a particular point in an electric circuit in a given time. Voltage is a measure of the force needed to move the electrons. There is a very simple way to calculate power using current and voltage. We use the equation:

$$P = I \times E \qquad \text{(Equation 5-7)}$$

where:
P = power, measured in watts
E = EMF in volts
I = current in amperes

To calculate the power in a circuit, multiply volts times amperes. For example, in **Figure 5-6**, a 12-V battery provides 0.2 A of current to a light bulb that is operating normally. Using Equation 5-7, we can find the power rating for the bulb by multiplying the voltage by the current:

$$P = I \times E = 0.2 \text{ A} \times 12 \text{ V} = 2.4 \text{ W}$$

If you know any two of the three quantities in Equation 5-7, you can find the unknown third quantity, just as you can find the unknown in an Ohm's Law problem. We can draw a "power circle" like the Ohm's Law Circle, as shown in **Figure 5-7**. Cover the unknown quantity to find the equation to use.

$$I = \frac{P}{E} \qquad \text{(Equation 5-8)}$$

$$E = \frac{P}{I} \qquad \text{(Equation 5-9)}$$

If 10 V is applied to a circuit consuming 20 W, how much current is flowing in the circuit? Use Equation 5-8 to find the current:

$$I = \frac{P}{E} = \frac{20 \text{ W}}{10 \text{ V}} = 2 \text{ A}$$

If a 5-A current flows in a 60-W circuit, how much voltage is applied to the circuit? Equation 5-9 will help us solve this problem:

$$E = \frac{P}{I} = \frac{60 \text{ W}}{5 \text{ A}} = 12 \text{ V}$$

Sometimes you may need to make a calculation involving power, but you will only know one of the quantities in these equations. In that case, you will probably know another quantity, such as a circuit resistance. Ohm's Law can help you find the additional information you need to

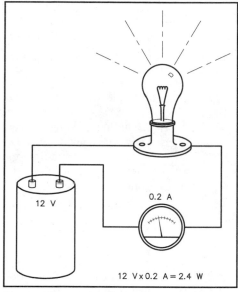

Figure 5-6 — Applied voltage multiplied by current equals power.

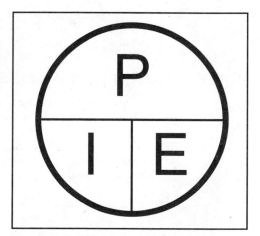

Figure 5-7 – The "Power Circle." Cover the letter representing the unknown quantity to find an equation to calculate that quantity. If you cover the P, you are left with I × E.

solve the problem. For example, suppose we apply 400 V to an 800-Ω resistor, and want to know how much power the resistor will have to dissipate? First, we use Ohm's Law (Equation 5-2) to find the circuit current:

$$I = \frac{E}{R} = \frac{400 \text{ V}}{800 \text{ }\Omega} = 0.5 \text{ A}$$

Then, we can use Equation 5-7 to find the power:

$$P = I \times E = 0.5 \text{ A} \times 400 \text{ V} = 200 \text{ W}$$

This will have to be a pretty large resistor, or it will burn up! In a practical example, this might be an electric heater element.

What if we want to know the power dissipated in a 1.25-kΩ resistor that has 7.0 milliamps (mA) of current flowing through it? To use Equation 5-7 we must know the applied voltage — but we can find that with Ohm's Law (Equation 5-1). First convert the current to amperes and the resistance to ohms: 7.0 mA = 0.007 A (move the decimal point three places left) and 1.25 kΩ = 1250 Ω (move the decimal point three places right).

$$E = I \times R = 0.007 \text{ A} \times 1250 \text{ }\Omega = 8.75 \text{ V}$$

Now we can use Equation 5-7:

$$P = I \times E = 0.007 \text{ A} \times 8.75 \text{ V} = 0.06125 \text{ W}$$

We can easily convert this value to milliwatts by moving the decimal point three places to the right:

$$0.06125 \text{ W} = 61.25 \text{ mW}.$$

[Study questions G5B03 through G5B05 in Chapter 13 before you go on to the next section. Review this section if you have difficulty with any of these questions.]

POWER GAIN, POWER LOSS AND THE DECIBEL

Power gains and losses in electric circuits are often spoken of in terms of **decibels** (one-tenth of a bel). Decibel is abbreviated **dB**. The bel was named for Alexander Graham Bell, and was first used in early audio telephone work to describe a change in loudness levels. The decibel is defined as the change in sound power level that is just perceptible to the human ear. In mathematical terms power gain or loss in decibels is given by the equation:

$$dB = 10 \times \log_{10}\left(\frac{P_2}{P_1}\right) \qquad \text{(Equation 5-10)}$$

where:
  P_1 = reference power
  P_2 = power being compared to the reference value

The $\log_{10}$ in this equation means "the logarithm to the base 10," also called the common logarithm. The common logarithm of a number is defined as the power to which you must raise 10 in order to obtain that number. If $X = 10^A$, then A is the $\log_{10}$ of X. Your scientific calculator probably has keys labeled "log" and "y^x" to help you with these calculations. You won't even need a calculator for a lot of decibel calculations if you learn a few basic relationships, though.

The $\log_{10}$ of 10 is 1, because 10 raised to the first power is 10 ($10^1 = 10$). The $\log_{10}$ of 100 is 2 ($10^2 = 100$); the $\log_{10}$ of 10,000 is 4, and so on. You may notice from these examples that it is very easy to find the common logarithm of a positive whole number consisting of a one and any number of zeros. Simply count the zeros to find the logarithm! The logarithm gives us the answer to the question "10 raised to what power will give us this number?"

Let's look at some sample problems using the power gain and loss relationship. If we increase our transmitter power level from 100 W to 1000 W, we can express this power gain in decibels:

$$dB - 10 \times \log_{10}\left(\frac{P_2}{P_1}\right)$$

$$dB = 10 \times \log_{10}\left(\frac{1000 \text{ W}}{100 \text{ W}}\right)$$

$$dB = 10 \times \log_{10}(10)$$

$$dB = 10 \times 1 = 10 \text{ dB}$$

A power decrease from 1500 W to 15 W would result in a loss of 20 dB:

$$dB = 10 \times \log_{10}\left(\frac{P_2}{P_1}\right)$$

$$dB = 10 \times \log_{10}\left(\frac{15 \text{ W}}{1500 \text{ W}}\right)$$

$$dB = 10 \times \log 10 (0.01)$$

$$dB = 10 \times -2 = -20 \text{ dB}$$

Notice that the decrease from 1500 W to 15 W can be thought of as one drop of 10 dB from 1500 W to 150 W, and another 10 dB drop from 150 W to 15 W so the total power loss is -10 dB + -10 dB = -20 dB. This is a very powerful property of logarithms. For example, if we double our transmitter power from 10 W to 20 W, we have a 3-dB gain:

$$dB = 10 \times \log_{10}\left(\frac{20 \text{ W}}{10 \text{ W}}\right)$$

$$dB = 10 \times \log 10 \ (2)$$

$$dB = 10 \times 0.301 = 3 \text{ dB}$$

If we then double the power again to 40 watts, we have another 3-dB gain:

$$dB = 10 \times \log_{10}\left(\frac{40 \text{ W}}{20 \text{ W}}\right)$$

$$dB = 10 \times \log 10 \ (2)$$

$$dB = 10 \times 0.301 = 3 \text{ dB}$$

So if we multiply our power level by a factor of four times, from 10 watts to 40 watts, we would have a total power gain of 3 dB + 3 dB = 6 dB. Checking our calculations:

$$dB = 10 \times \log_{10}\left(\frac{40 \text{ W}}{10 \text{ W}}\right)$$

$$dB = 10 \times \log 10 \ (4)$$

$$dB = 10 \times 0.602 = 6 \text{ dB}$$

Any time you double the power, there is a 3-dB increase. If you multiply the power by 4, there is a 6-dB increase. If you decrease the power by 3-dB, it means you have cut the power in half.

Receiver signal-strength meters (S meters) are usually marked in decibels above the strength 9 reading. If you receive a signal-strength report of "10 dB over S9," you could reduce your power by a factor of 10 (such as from 1500 W to 150 W) and still have an S9 signal. A report of "20 dB over S9" means you could reduce your power by a factor of 100 (such as from 1500 W to 15 W) and still have an S9 signal!

Power gains and losses expressed in decibels can be added (or subtracted) to find the total system gain or loss. We often specify amplifier gains, feed line losses and other system gains and losses in decibels.

Suppose you are using an antenna feed line that has a signal loss of 1 dB. We can use the decibel equation to calculate the power that reaches the antenna through this feed line, and from that we can calculate the power lost in the feed line. If we assume a transmitter power of 100 W, then the resulting answer will be the percentage of the original power that is lost in the feed line. First we will solve the decibel equation for P_2, the power that reaches the antenna.

$$dB = 10 \times \log_{10}\left(\frac{P_2}{P_1}\right)$$

(Equation 5-10)

$$\frac{dB}{10} = \log_{10}\left(\frac{P_2}{P_1}\right)$$

$$\log_{10}^{-1}\left(\frac{dB}{10}\right) = \left(\frac{P_2}{P_1}\right)$$

Note: The notation that shows the logarithm raised to the negative 1 power means the antilog, or inverse logarithm. On some scientific calculators this button is also labeled 10^x, which means "raise 10 to the power of this value." The decibel value is given as a loss of 1 dB, so we will write that as –1 dB.

$$P_1 \log_{10}^{-1}\left(\frac{dB}{10}\right) = P_2 = 100 \text{ W} \times \log_{10}^{-1}\left(\frac{-1 \text{ dB}}{10}\right) = 100 \text{ W} \times 0.794 = 79.4 \text{ W}$$

If 79.4 W is the power actually reaching the antenns, then we can calculate the lost power by subtracting this value from the original power: 100 W – 79.4 W = 20.6 W. Because we used 100 W as the reference power, this value is the percentage of the original power that is lost in our feed line, when it has a loss of 1 dB. The percentage of the power lost in this feed line is 20.6%.

[Now turn to Chapter 13 and study questions G5B01 and G5B14. Review this section if this question gives you any difficulty.]

ALTERNATING CURRENT

It is possible to have a voltage source that periodically reverses polarity. With this kind of voltage the current flows first in one direction and then in the opposite direction. Such a voltage is called an *alternating voltage*. The current produced by this alternating voltage is called an **alternating current** (abbreviated **ac**). The reversals (alternations) may occur at any rate from less than one per second up to several billion or more per second.

The difference between alternating current and direct current is shown in **Figure 5-8**. In these graphs, the horizontal axis indicates time, increasing toward the right. The vertical axis represents the amplitude or strength of the current, increasing either in the up or down direction away from the horizontal axis. If the graph is above the horizontal axis, the current is flowing in one direction through the circuit (indicated by the + sign). If the graph is below the horizontal axis the current is flowing in the opposite direction (indicated by the – sign).

In Figure 5-8A, assume that we close the circuit or make the path for the current complete at the time indicated by X. The current instantly takes the amplitude indicated by the height A. After that, the current continues at the same amplitude as

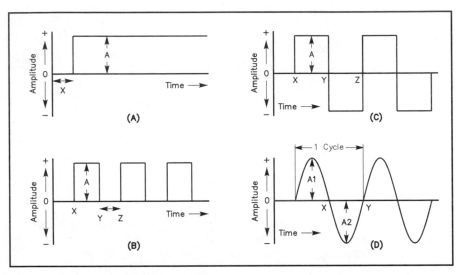

Figure 5-8 — Four types of current are shown. A — direct current; B — intermittent (pulsating) direct current; C — square-wave alternating current; D — sine-wave alternating current.

time goes on. This is an ordinary direct current.

In Figure 5-8B, the current starts flowing with the amplitude A at time X, continues at that amplitude until time Y and then instantly ceases. After an interval from Y to Z the current begins to flow again and the same sort of stop-and-start performance is repeated. This is an intermittent direct current (also called a pulsating direct current). We could produce this type of current by alternately opening and closing a switch in the circuit. It is a direct current because the direction of the current does not change; the amplitude is always on the + side of the horizontal axis. This series of current pulses is sometimes called square waves.

The waveform in Figure 5-8C is a little different. The current starts with amplitude A at time X just as the wave in Figure 5-8B, but at time Y the current instantly reverses direction. From time Y to time Z, the current flows in the opposite direction and the voltage has an equal, but opposite polarity. At time Z the current reverses direction again. This is called a square-wave alternating current.

In Figure 5-8D the current starts at zero, smoothly increases in amplitude until it reaches the amplitude A1 while flowing in the + direction, then decreases smoothly until it drops to zero amplitude once more. At that time (X) the direction of the current reverses. As time goes on the amplitude increases, with the current now flowing in the − direction, until it reaches amplitude A2. Then the amplitude decreases until finally it returns to zero (at point Y) and the direction reverses once more. The type of smoothly varying alternating current shown in Figure 5-8D is known as a sine wave.

Looking at the sine wave, the first thing we notice is that the waveform reaches a maximum value (peak) at two points in the ac cycle — once during the positive part and once during the negative part. For the remainder of the cycle, the waveform is constantly varying between these two values. Another interesting characteristic of a sine wave is that the average dc voltage contained in the wave is zero. The positive half of the cycle is a "mirror image" of the negative side — for every positive value, there is a corresponding negative value, and the positive and negative values cancel each other out. This doesn't mean that a sine wave contains no energy, however — it means only that the average dc potential is zero.

So how can we find a value for ac voltage and current? The wave is constantly changing. If we could measure the voltage at one point on the wave at some instant of time, in the next instant the value would be different. There is no one value we can point to and say: "That's the value of the voltage."

We could use the peak value the wave reaches on its positive half-cycle. Or we could measure the voltage from the positive peak to the negative peak, and use this to specify the current or voltage. This value is called the *peak-to-peak* value. We can also use a measuring scale that lets us compare the ac voltage to an equivalent dc voltage.

If we pass dc through a resistor, the resistor will get warm. If we pass ac through the same resistor, it will also get warm. The ac voltage that will heat a resistor to the same temperature as 1 V dc is said to have an **effective voltage** of 1 V ac. If we pass 10 V dc through a resistor and measure its temperature, we can then find an effective ac voltage of 10 V ac that will heat the resistor to the same temperature.

Most ac voltages are specified by their effective values. Your wall outlets provide 120 V effective ac. If we plug a directly heated soldering iron into a source of 120 V dc, it will get just as hot as it will if we plug it into the ac wall outlet.

The effective voltage is also called the **root-mean-square (RMS)** value of the ac wave. The phrase root-mean-square describes the actual process of calculating the effective voltage. The method involves squaring the instantaneous values for a large number of points along the waveform and then finding the average of the squared values. (Another mathematical term for average is *mean*.) We then take the square root of this average (the *root* of the average *[mean]* of the *squares*).

If we know the peak-to-peak value of a sine-wave ac waveform, how do we calculate the RMS value? The RMS value and the peak value are related by a factor

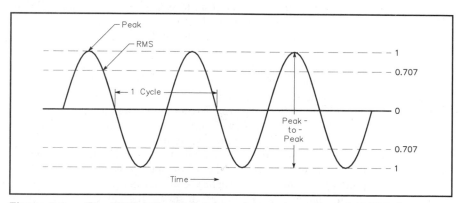

Figure 5-9 — The RMS (effective) value of an ac waveform is 0.707 times the peak value. The peak value is one-half the peak-to-peak value.

of the square root of two. For sine waves, the following relationships are useful:

$$V_{Peak} = V_{RMS} \times \sqrt{2} = V_{RMS} \times 1.414 \qquad \text{(Equation 5-11)}$$

$$V_{RMS} = \frac{V_{Peak}}{\sqrt{2}} = V_{Peak} \times 0.707 \qquad \text{(Equation 5-12)}$$

Remember that the peak value is one-half the peak-to-peak value. **Figure 5-9** shows the relationships between peak, peak-to-peak and RMS ac values.

For example, to find the peak-to-peak value of a 120-V RMS ac wave, we use Equation 5-11:

$$V_{Peak} = V_{RMS} \times \sqrt{2} = V_{RMS} \times 1.414$$

$$V_{Peak} = 120 \text{ V} \times 1.414 = 169.8 \text{ V}$$

Remember that the peak-to-peak voltage is peak voltage times 2, so:

$$V_{peak\text{-}to\text{-}peak} = 169.7 \text{ V} \times 2 = 339.4 \text{ V}$$

To find the RMS value of a 17 V_{Peak} ac wave, we use Equation 5-12:

$$V_{RMS} = \frac{V_{Peak}}{\sqrt{2}} = V_{Peak} \times 0.707$$

$$V_{RMS} = V_{Peak} \times 0.707 = 17 \text{ V} \times 0.707 = 12 \text{V}$$

If you were given the peak-to-peak value, you would divide that by 2 before substituting the value into Equation 5-12.

If you want to know the resulting RMS voltage that will be produced by combining several signal voltages, you can use the root-mean-square method to calculate it. The frequencies of the signals do not have to be the same, or even be related to each other to use this method. The signals do not even have to be sine waves, which means this method is an effective way of handling complex waveforms that you might display on an oscilloscope. While a voltmeter may have a scale calibrated in volts RMS, that scale will only be useful for sine waves. It will not give an accurate measurement of complex waveforms.

The method involves adding the values of the instantaneous voltages of the individual waveforms for a large number of points along each waveform. (The more points you compute, the better. Calculus provides a way to find the value for an infinite number of points.) Next, square the values for each point along the combined waveform, found by by that addition. Add the squared values, and find the average value. (*Mean* is a mathematical term for *average*.) Finally, take the square root of that average. This is the RMS value of the combined waveform.

[Turn to Chapter 13 now and study questions G5B08 through G5B10. Review this section as needed.]

X = REACTANCE

A **capacitor** is an electronics component made of two or more conductive plates separated by an insulating material. A capacitor stores electric energy in an electric field. A capacitor responds to an alternating current by storing and releasing electric energy in an electric field as the current passes through the component.

A **coil** (or **inductor**) is a component made by winding a conductor into a series of loops. An inductor stores electric energy in a magnetic field. An inductor responds to an alternating current by storing and releasing electric energy in a magnetic field as the current passes through it.

As the amount of stored energy changes, there is an opposition to the current with both capacitors and inductors. This opposition is called **reactance**, and it is measured in ohms. In equations, reactance is designated by the letter X. Capacitive reactance is designated X_C and inductive reactance is designated X_L. Unlike resistance, the reactance of a component will vary with the frequency of the applied alternating current.

Capacitive Reactance

A discharged or "empty" capacitor looks like a short circuit. This means that when the circuit is first energized, a lot of current will flow into the capacitor as it charges. Initially, the voltage across the capacitor is zero; as the voltage builds up, the current through the capacitor will drop toward zero. If direct current is applied to a capacitor, the current will drop to zero when the capacitor charges to the applied voltage. See **Figure 5-10**. A capacitor blocks dc.

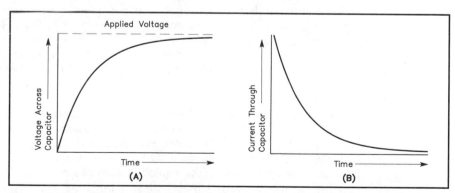

Figure 5-10 — When a circuit containing a capacitor is first energized, the voltage across the capacitor is zero, and the current is very large. As time passes, the voltage across the capacitor increases, as shown at A, and the current drops toward zero, as shown at B.

When an alternating current is applied to a capacitor, things are a little different. If the frequency of the ac is low, the capacitor has time to charge to the full
applied voltage during each half cycle. As the capacitor is charged, it opposes any
further increase in voltage. On the next half cycle the capacitor discharges and is
then recharged with the opposite polarity. A high-frequency ac wave will not fully
charge the capacitor during each half cycle, and the capacitor will pass the full
current. We can see from this that for a capacitor, as the frequency of the applied ac
increases, the opposition to current (the **reactance**) *decreases*.

The reactance of a capacitor is given by the equation:

$$X_C = \frac{1}{2\pi f\, C}$$
(Equation 5-13)

where:
  X_C = capacitive reactance in ohms
  f = frequency in hertz
  C = capacitance in farads
  π = 3.14

You should always use the proper units. If capacitance is given in microfarads,
you will have to convert to farads before using Equation 5-13. (To convert microfarads to farads, multiply microfarads by 10^{-6}. This means moving the decimal point 6
places to the left.) The equation shows the inverse relationship between capacitive
reactance and frequency; as the *frequency* goes *up*, the *reactance* will go *down*.

When capacitors are connected in series, the values combine the same way
resistors in parallel combine. If there are only two capacitors, the equation to calculate the combined series value is:

$$C_{\text{Total in Series}} = \frac{C_1 \times C_2}{C_1 + C_2}$$
(Equation 5-14)

For two equal value capacitors in series, the total capacitance will be one half
the value of one of the capacitors. Each capacitor in the combination will be able to
withstand a voltage equal to the rated voltage for that capacitor. That means the
total applied voltage across the series combination of capacitors can be the sum of
all the capacitors. For example, if each capacitor can withstand 150 V, then the
combination of two equal capacitors in series will be able to withstand 300 V.

Inductive Reactance

An inductor presents opposition to a change in *current*. This means that when
a circuit containing an inductor is first energized, no current will flow in the inductor. When a direct current is applied to an inductor, at first the inductor looks like an
open circuit. The current will gradually increase to the maximum value, which is
determined by the amount of the applied voltage and any circuit resistance. (The
word *gradually* is a relative term here. It may only take microseconds or milliseconds to reach the full current, depending on the value of inductance and other circuit conditions.) At first, all the applied voltage appears across the inductor. Then as
the current *increases*, the voltage across the inductor *decreases* toward zero. See

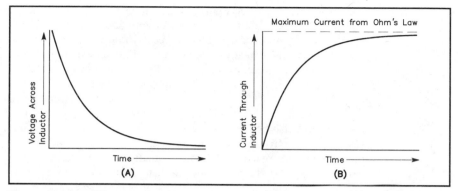

Figure 5-11 — When a circuit containing an inductor is first energized, the initial current is zero and the full applied voltage appears across the inductor. As time passes, the voltage drops toward zero as shown at A, and the current increases, as shown at B.

Figure 5-11. Inductors pass dc.

When an alternating current is applied to an inductor, as the frequency of the applied ac *increases*, the opposition to the current (the **reactance**) *increases*. Inductors oppose any change in current. The higher the frequency of the applied ac, the more rapidly the current is changing, and the greater the opposition to the current. The equation for inductive reactance is:

$$X_L = 2 \pi f L \qquad \text{(Equation 5-15)}$$

where:
X_L = inductive reactance in ohms
f = frequency in hertz
L = inductance in henrys
π = 3.14

Again, the correct units must be used. If the inductance is expressed in units other than henrys (such as millihenrys) you will have to change them to the basic unit. From the equation, we can see that as the frequency of the applied ac *increases*, inductive reactance *increases*.

Z = IMPEDANCE

Impedance is a number obtained by dividing the voltage applied to a circuit by the current flowing through it. It represents the opposition to the flow of ac in a circuit that may contain resis-

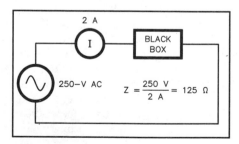

Figure 5-12 — In this circuit, we know that the box presents 125 Ω of opposition to the current, but we don't know what's in the box.

tance and reactance. See **Figure 5-12**. Suppose we measure the current, I, flowing into the black box and find it to be 2 A with 250 V ac applied. Dividing 250 V by 2 A results in an impedance of 125. We say that we have 125 ohms of impedance, since the ohm has already been established as representing the ratio of voltage to current in a dc circuit. So this circuit has an impedance of 125 Ω. The letter Z is used to represent impedance. The impedance does not tell us what is inside the black box, however.

We could get this result (125 Ω) if we had a 125-Ω resistor in the box. The result would also be the same if we used an inductor (or capacitor) with 125 Ω of reactance at the frequency of the applied voltage. We could also use a combination of resistance and reactance. Such combinations of resistance and reactance not only exist, but are actually more common than either pure resistance or reactance alone.

Impedance Matching

Impedance is an important concept in electronics. Remember that an antenna system operates most efficiently when the SWR (standing-wave ratio) on the feed line is low. The SWR is a measure of how well the impedance of the transmitter, feed line and antenna are matched; that is, the closer the impedances are to the same value, the better the antenna system will operate.

Maximum power transfer occurs when the impedance of the electric load is the same as the impedance of the power source. Some devices (like antennas) require a specific input impedance for proper operation. This input impedance may not be the same as the optimum output impedance for the power source (transmitter). In that case, we use an additional circuit between the two that will match or transform the output impedance of the source (the transmitter) to the input impedance of the load (the antenna). The impedance of the source must be matched to the impedance of the load to ensure that the source can deliver maximum power to the load. This process is called *impedance matching*.

[You should turn to Chapter 13 now and study question G5A01 through G5A10. Also study question G5B13. Review this section if you have trouble with any of those questions.]

TRANSFORMERS

When two coils are arranged so that a changing current in one induces a voltage in the other, the combination of windings is called a **transformer**. See **Figure 5-13**. Every transformer has a primary winding and a secondary winding. The primary winding is connected to the current source, and the secondary winding is connected to the load.

With a transformer, electric energy can be transferred from one circuit to another without direct connection. In the process, the signal voltage can be readily changed from one level to another. Mutual inductance between the two windings causes a voltage to appear across the secondary winding when a voltage source is connected across the primary winding.

Turns Ratio and Voltage Transformation

For a given magnetic field, the voltage induced in a coil will be proportional to the number of turns in the coil. It follows, therefore, that if the primary and second-

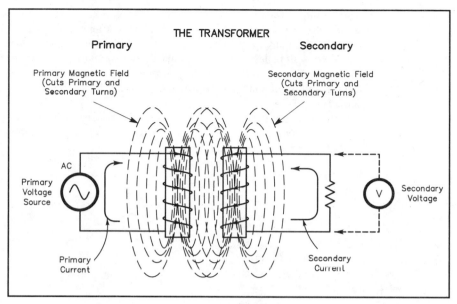

Figure 5-13 — The transformer is an application of mutual inductance between two or more coils or inductors placed close to each other. Both coils of most transformers are wound on the same core material.

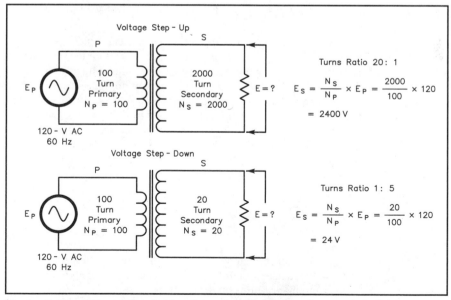

Figure 5-14 — The amount of voltage transformation is controlled by the ratio of turns in the primary to turns in the secondary.

ary windings of a transformer have the same number of turns and we ignore losses in the coils, we would expect the full primary voltage to be induced into the secondary. If the secondary has fewer turns than the primary, the voltage induced in the secondary will be less than the primary voltage, and the transformer is called a *step-down transformer*. If the secondary has more turns than the primary, the secondary voltage will be greater than the primary, and the transformer is a *step-up transformer*. See **Figure 5-14**.

If you know the number of turns on the primary and secondary windings of a transformer, and you know the applied voltage, it is easy to calculate the secondary voltage. We use the equation:

$$E_S = \frac{N_S}{N_P} \times E_P \qquad \qquad \text{(Equation 5-16)}$$

where:
    E = voltage
    N = number of turns
    S = secondary
    P = primary

For example, a certain transformer has 2250 turns on the primary winding and 500 turns on the secondary winding. If 120 V ac is applied to the primary winding, what will be the secondary voltage?

$$E_S = \frac{N_S}{N_P} \times E_P = \frac{500}{2250} \times 120 \text{ V} = 0.222 \times 120 \text{ V} = 26.7 \text{ V}$$

It is interesting to note that a transformer can't increase the power in a circuit. So if the transformer steps up the voltage, there will be less current in the secondary by the same ratio. If the transformer steps down the voltage, there can be more current in the secondary winding than in the primary winding. Since a transformer changes both the voltage and current between the primary and secondary windings, you can probably guess that we can apply Ohm's Law to find an impedance relationship.

In an ideal transformer — one without losses — the following relationship is true between the primary and secondary impedance and the number of turns in the transformer:

$$Z_P = Z_S \left(\frac{N_P}{N_S} \right)^2 \qquad \qquad \text{(Equation 5-17)}$$

where:
    Z_P = impedance seen by a device connected to the primary
    Z_S = impedance seen by a device connected to the secondary
    N_P = number of turns on the primary
    N_S = number of turns on the secondary

From this equation, we can derive an equation for the turns ratio necessary to

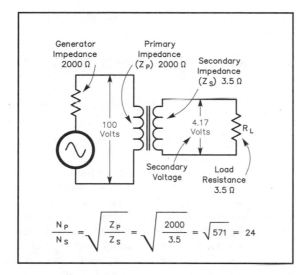

Figure 5-15 — To match a 2000-Ω primary impedance to a 4-Ω secondary impedance, we use a transformer with a turns ratio of 24 to 1.

match a given input and output impedance.

$$\frac{N_P}{N_S} = \sqrt{\frac{Z_P}{Z_S}}$$

(Equation 5-18)

In other words, the turns ratio of the transformer must equal the square root of the impedance ratio. So, if we have an audio amplifier with a 600-Ω output impedance to match to a speaker that has a 4-Ω input impedance, the turns ratio of the transformer (N_P to N_S) is:

$$\frac{N_P}{N_S} = \sqrt{\frac{Z_P}{Z_S}} = \sqrt{\frac{600\,\Omega}{4\,\Omega}} = \sqrt{150} = 12.2$$

Therefore, the primary must have 12.2 times the number of turns as the secondary. See **Figure 5-15** for another example of how to find the turns ratio for a matching transformer.

Suppose we have an audio amplifier with an output impedance of 2000 Ω, and the matching transformer used to drive a speaker has a turns ratio of 24 to 1. What is the impedance of the speaker for this circuit? To answer this question we must solve Equation 5-17 for the secondary impedance. We will cross multiply the turns ratio term to solve for Z_S. This process results in reversing the positions of the primary and secondary turns in the equation:

$$Z_P = Z_S \times \left(\frac{N_P}{N_S}\right)^2$$

(Equation 5-19)

Electrical Principles   5-23

Now we can substitute the values from our problem into Equation 5-19:

$$Z_S = Z_P \times \left(\frac{N_S}{N_P}\right)^2 = 2000\,\Omega \times \left(\frac{1}{24}\right)^2 = 2000\,\Omega \times (0.0147)^2$$

$Z_S = 2000\,\Omega \times 1.74 \times 10^{-3} = 3.47\,\Omega$

Round off this value to about 3.5 Ω. This transformer would be used to drive a 4-Ω speaker with this audio amplifier.

Suppose you wind some turns of wire on a magnetic core material, such as a ferrite or powdered iron core to form an impedance matching transformer. When you connect your transmformer into a circuit and apply a small signal, the transformer core will have a small magnetic flux density inside. If you increase the current, the flux density will increase. If you continue to increase the current through your transformer, the magnetic flux density in the core will continue to increase proportionally, up to a point. Eventually, the flux density won't change as much as the current increases. At this point, the core is becoming saturated with magnetic flux. Further increases in current will not increase the flux density, so the core permeability actually decreases. The core may overheat, and there may be permanent changes in the core permeability. When the core becomes saturated the transformer will operate in a nonlinear way. That can result in the creation of harmonics and other signal distortion. Select a core with a suitable material for the frequency range you want to cover with your impedance transformer. Select a core that is physically large enough (or stack multiple cores) to handle the power you expect to feed through the transformer. That will ensure that the core does not saturate.

[This completes the material for this chapter. To finish your study, however, you should turn to Chapter 13 and study question G5A11 and questions G5B06, G5B07 and G5B12. Review this section if you have difficulty with any of these questions.]

CHAPTER 6
KEYWORDS
KEYWORDS
KEYWORDS

Bipolar transistor — A transistor made of two PN semiconductor junctions, using two layers of similar-type material (N or P) with a third layer of the opposite type between them.

Capacitor — An electronic component composed of two or more conductive plates separated by an insulating material.

Junction diode — An electronic component formed by placing a layer of N-type semiconductor material next to a layer of P-type material. Diodes allow current to flow in one direction only.

Magnetizing current — A small current that flows in a transformer primary winding, even with no load connected to the secondary.

Maximum average forward current — The highest average forward current that can flow through a diode for a given junction temperature.

Peak-inverse-voltage (PIV) — The maximum voltage a diode can withstand when it is reverse biased (not conducting).

Primary winding — The coil in a transformer that is connected to the energy source.

Resistor — Any material that opposes a current in an electrical circuit. An electronic component specifically designed to oppose current.

Secondary winding — The coil in a transformer that is connected to the load.

Suppressor capacitor — A capacitor (often ceramic) connected across the transformer primary or secondary winding in a power supply. These capacitors are intended to suppress any transient voltage spikes, preventing them from getting through the power supply.

Temperature coefficient — A number used to show whether a component will increase or decrease in value as it gets warm.

Transformer — Mutually coupled coils used to change the voltage level of an ac power source to one more suitable for a particular circuit.

Circuit Components

Before you can understand the operation of most complex electronic circuits, you must know some basic information about the parts that make up those circuits. This chapter presents the information about circuit components that you need to know to pass your General written exam. You will find information about resistors, capacitors, inductors and transformers, along with some basic information about semiconductor diodes. These components are combined with other devices to build practical electronic circuits, some of which are described in Chapter 7. Your General exam will have one question from the circuit components group. There is only one syllabus group for this topic:

G6A   Resistors; capacitors; inductors; rectifiers and transistors; etc.

As you study the characteristics of the components described in this chapter, be sure to turn to the Element 3 questions in Chapter 13 when you are directed to do so. The questions will help you check on your progress, and will show where you need to do some extra studying. If you thoroughly understand how these components work, you should have no problem learning how they can be connected to make a circuit perform a specific task.

RESISTORS

Current is the flow of electrons from one point to another in a circuit. A perfect insulator would allow no electrons to flow (zero current), while a perfect conductor would allow an infinite number of electrons to flow (infinite current). In practice, however, there is no such thing as a perfect conductor or a perfect insulator. Partial opposition to electron flow occurs when the electrons collide with other electrons

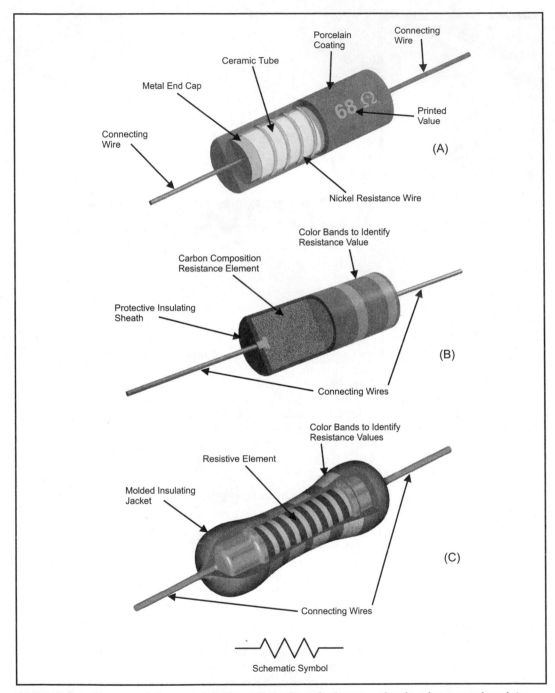

Figure 6-1 — Some resistor construction types. Part A shows a simple wire-wound resistor. Part B is a composition resistor. Part C shows the construction of a film resistor.

or atoms in the conductor. The result is a reduction in current, and heat is produced in the conductor. **Resistors** allow us to control the current in a circuit by controlling the opposition to electron flow. As they oppose the flow of electrons they dissipate electrical energy in the form of heat. The more energy a resistor dissipates, the hotter it will become.

There are several types of resistors, depending on the method used to construct them. For example, a *wire-wound resistor* is made by winding a specific length of wire onto a ceramic form. The length, type and size of the wire determines the resistance of a wire-wound resistor. Another way to form a resistor is to connect leads to both ends of a block or cylinder of material that is a mixture of carbon and clay. The proportions of carbon and clay determine the value of the resistor. This type of resistor is called a *carbon-composition resistor*. Another type of resistor uses a film of metal or carbon deposited onto a form. The conducting metal or carbon film is then trimmed (often into a spiral) to produce the desired resistance. These are called *metal-film* or *carbon-film resistors*. **Figure 6-1** shows the construction of several resistor types.

Each resistor type has advantages and disadvantages. Wire-wound resistors, for example, can get rid of more heat, and so can be made with higher power ratings than the other types. The coil-shape of the wire winding results in significant inductance. That can be quite a problem in ac circuits, especially at radio frequencies. The inductance associated with a wire-wound resistor can detune a resonant circuit. Some advantages of carbon-composition resistors are the wide range of available values, low inductance and capacitance, good surge-handling capability and the ability to withstand small power overloads without being completely destroyed. The main disadvantage is that the resistance of the composition resistor will vary widely as the operating temperature changes and as the resistor ages. Film-type resistors are replacing carbon-composition types in many applications. It is easier to control the exact resistance during manufacture of film resistors.

As the temperature of a carbon-composition resistor changes, its resistance will change. How the resistance will change is determined by the **temperature coefficient** of the resistor. With a positive temperature coefficient, the resistance will increase as the temperature increases. With a negative coefficient, the resistance will decrease as the temperature increases. Some types of resistors have positive coefficients, and some types have negative coefficients. One thing is certain — as the temperature changes, the value of the resistor will change.

[Now turn to Chapter 13 and study questions G6A01 and G6A12. Review this section if you are uncertain how to answer either of those questions.]

CAPACITORS

In Chapter 5 we learned that the basic property of a **capacitor** is the ability to store an electric charge. In an uncharged capacitor, the potential difference between the two plates is zero. As we charge the capacitor, this potential difference increases until it reaches the full applied voltage. The difference in potential creates an *electric field* between the two plates. See **Figure 6-2**. The electric field between the plates of our capacitor is invisible — a field is an invisible force of nature. We put energy into the capacitor by charging it, and until we discharge it or the charge leaks away somehow, the energy is stored in the electric field. When the field is not

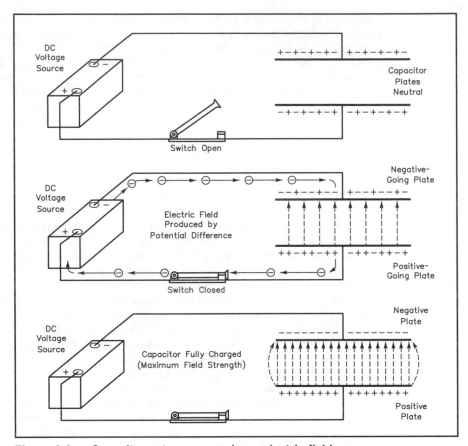

Figure 6-2 — Capacitors store energy in an electric field.

moving we sometimes call it an electrostatic field.

When a dc voltage is applied to a capacitor, current will flow in the circuit until the capacitor is fully charged. After the capacitor has charged to the full dc voltage, no more current will flow in the circuit. For this reason, capacitors can be used to block dc in a circuit. Because ac voltages are constantly reversing polarity, we can block dc with a capacitor while permitting ac to pass.

Practical Capacitors

Capacitors are described by the material used for their *dielectric*, which is the insulating material used between the conducting plates. Mica, ceramic, mylar, polystyrene, and electrolytic capacitors are in common use today. They each have properties that make them more or less suitable for a particular application.

Mica capacitors consist of strips of metal foil separated by thin strips of mica. (Mica is a mineral that easily separates into thin sheets that are translucent, which means it allows light through the thin sheet, but isn't quite clear, like glass.) Alter-

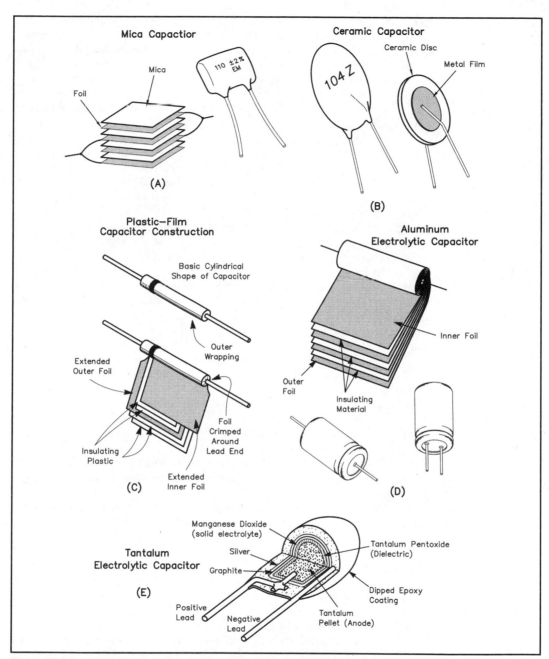

Figure 6-3 — Some capacitor construction types. Part A shows a mica capacitor, formed by interleaving metal foil with thin sheets of mica. Part B shows how the electrodes are deposited on both sides of a ceramic disc to form a ceramic capacitor. The construction of a plastic-film capacitor is shown in Part C. Part D shows an aluminum electrolytic capacitor and Part E shows a tantalum electrolytic capacitor.

nate plates are connected, and each set of plates is connected to an electrode. The entire unit is then encased in plastic or ceramic insulating material. An alternative to this form of construction is the "silvered-mica" capacitor. In the silvered-mica type, a thin layer of silver is deposited directly onto one side of the mica, and the plates are stacked so that alternate layers of mica have a layer of silver conductor between them. **Figure 6-3A** illustrates the construction of a mica capacitor.

Mica has a very high voltage breakdown rating. This means the mica dielectric will withstand several thousand volts on the capacitor plates without allowing a spark to jump through the insulator. Mica capacitors are frequently used in transmitters and high power amplifiers, where the ability to withstand high voltages is important. Mica capacitors also have good temperature stability — their capacitance does not change greatly as the temperature changes. Typical capacitance values for mica capacitors range from 1 picofarad to 0.1 microfarad, and voltage ratings as high as 35,000 are possible.

Ceramic capacitors are constructed by depositing a thin metal film on each side of a ceramic disc. Wire leads are then attached to the metal films, and the entire unit is covered with a protective plastic or ceramic coating. Ceramic capacitors are inexpensive and easy to construct, and they are in wide use today. **Figure 6-3B** illustrates the construction of a ceramic capacitor. The range of capacitance values available with ceramic capacitors is typically 1 pF to 0.1 µF, with working voltages up to 1000.

Ordinary ceramic capacitors cannot be used where temperature stability is important — their capacitance will change with a change in temperature. Special ceramic capacitors (called NP0 for negative-positive zero) are used for these applications. The capacitance of an NP0 unit will remain substantially the same over a wide range of temperatures.

Ceramic capacitors are often connected across the transformer primary or secondary winding in a power supply. These capacitors, referred to as **suppressor capacitors**, are intended to suppress any transient voltage spikes, preventing them from getting through the power supply.

Plastic-film capacitors use mylar or polystyrene insulation between thin sheets of foil. The unit is rolled into a cylinder, and a protective outer covering is usually added. **Figure 6-3C** shows this construction technique. The plastic material produces capacitors with a high voltage rating in a physically small package. Plastic-film capacitors also have good temperature stability. Typical values range from 5 pF to 0.47 µF.

In *electrolytic capacitors*, the dielectric is formed after the capacitor is manufactured. The construction of *aluminum electrolytic capacitors* is similar to that of plastic-film capacitors. Two sheets of aluminum foil separated by paper soaked in a chemical solution are rolled up and placed in a protective casing. After assembly, a voltage is applied to the capacitor, causing a thin layer of aluminum oxide to form on the surface of the positive plate next to the chemical. The aluminum oxide acts as the dielectric, and the foil plates act as the electrodes.

Because of the extremely thin layer of oxide dielectric, electrolytic capacitors can be made with a very high capacitance value in a small package. Electrolytic capacitors are polarized — dc voltages must be connected to the positive and negative capacitor terminals with the correct polarity. The positive and negative electrodes in an electrolytic capacitor are clearly marked. Connecting an electrolytic capacitor incorrectly causes gas to form inside the capacitor, and the capacitor may actually explode. This can be very dangerous. At the very least the capacitor will be destroyed by connecting it incorrectly.

Tantalum capacitors are another type of electrolytic capacitor, and they have several advantages over aluminum electrolytic capacitors. Tantalum capacitors can be made even smaller than aluminum electrolytic capacitors for a given capacitance value. They are manufactured in several forms, including small, water droplet-shaped solid-electrolyte capacitors. These are formed on a small tantalum pellet that serves as the anode, or positive capacitor plate. An oxide layer on the outside of the tantalum pellet serves as the dielectric, and a layer of manganese dioxide is the solid electrolyte. Layers of carbon and silver form the cathode, or negative capacitor plate. The entire unit is dipped in epoxy to form a protective coating on the capacitor. Their characteristic shape explains why these tantalum capacitors are often called "tear drop" capacitors.

Electrolytic capacitors are available in voltage ratings of greater than 400 V, and capacitance values from 1 μF to 100,000 μF (0.1 F). Electrolytic capacitors are used in power-supply filters, where large values of capacitance are necessary for good smoothing of the pulsating dc from the rectifier.

[Turn to Chapter 13 now and study questions G6A02 and G6A03. Review this material as necessary.]

INDUCTORS

Just as a capacitor stores energy in an electric field, an inductor stores energy in a magnetic field. Remember from Chapter 5 that inductors oppose any change in current. When a dc voltage is applied across an inductor, there is no current at first. A voltage is induced in the coil that opposes the applied voltage, and tries to prevent a current. Gradually the current will build up to a value that is limited only by any resistance in the circuit (the resistance in the wire of the coil will be very small). In the process of getting this current to flow, energy is stored in the coil in the form of a magnetic field around the wire. While the current is increasing and the magnetic field strength is also increasing, a voltage forms that opposes the changing current. When the current reaches a steady value the inductor no longer opposes the current. An inductor passes a steady direct current with no opposition.

When the applied voltage is shut off, the magnetic field collapses and returns the stored energy to the circuit as a current flowing in the same direction as the original current. This current is produced by the EMF generated by the collapsing magnetic field. Again, this voltage opposes any change in the current already flowing in the coil. The amount of opposition to changes in current, called reactance, depends on the inductance of the coil.

If an alternating voltage is applied across the inductor, energy is stored in the magnetic field as the voltage increases. That energy is returned to the circuit as the voltage decreases to zero. As the voltage reverses direction, the direction of the magnetic field also reverses (the North and South poles reverse positions) and energy is stored again. Then as the voltage decreases to zero again the magnetic field returns the energy to the circuit again. An inductor continually opposes alternating current. The higher the frequency of the applied voltage, the greater is the inductor's opposition to this current. (This opposition is called *reactance*.) Inductors can be used to pass direct current and block alternating current. Inductors and capacitors are helpful in directing ac and dc to various parts of a circuit, so the proper type of signal reaches the appropriate circuit sections.

Inductors are made by winding turns of wire into a coil. If the form used to hold the wire is made from a material that contains a mixture of iron compounds, it will either be a ferrite core or a powdered iron core, depending on the exact mix of materials. The Latin word for iron is *ferrum*, so the ferrite materials get their name from this Latin word. The inductance of a coil depends on the *permeability* of the core material. Permeability refers to the strength of a magnetic field in the core as compared to the strength of the field if the core was air. (Air has a permeability of 1.) Ferrite cores make it possible to obtain large values of inductance in a relatively small package, as compared to the number of turns that would be required with an air-core inductor.

A core material becomes saturated when the increasing the current through the coil does not result in a linear increase in magnetic field strength. For most inductor applications, you do not want to saturate the core material in your operating circuit. For some switched-mode power supply designs, however, the transformer is designed so the core saturates and creates an output pulse when the input voltage goes above a certain level. Special transformers, known as *peaking transformers*, produce brief voltage pulses near the peaks of the source-voltage waveform. This type of design uses ferrite core material.

An inductor that is formed by winding a coil of wire on a straight coil form is called a solenoid inductor. Compare that to an inductor wound on a donut-shaped toroidal inductor. See **Figure 6-4.** A solenoid inductor may be wound with an air core or it may be wound on a magnetic material, such as iron, powdered iron material or ferrite material. While the magnetic field associated with a toroidal inductor is all contained inside the toroid, the magnetic field associated with a solenoid core extends through the center of the core and then around the outside of the coil. To minimize any interaction between the magnetic fields around solenoid inductors in a circuit, the coils should be placed at right angles to each other. (To ensure mini-

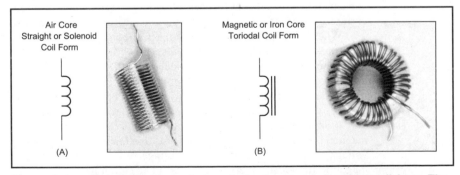

Figure 6-4 — Part A shows an air-core inductor on a solenoidal coil form. The magnetic field goes through the center of the solenoid and around the outside to return to the other end of the solenoid. Part B shows a toroidal inductor wound on a ferrite or powdered-iron toroid core. All of the magnetic field is contained within the torid core.

mum interaction between solenoid coils you might also have to place each inductor inside its own shielded enclosure.)

Because all of the magnetic field is contained inside the core of toroidal inductors, they have a self-shielding property that makes them ideal for use in RF circuits. You do not usually want interaction between nearby inductors in RF circuits.

TRANSFORMERS

When coils are positioned close to each other, so the magnetic field of one coil goes through another coil, we say the coils have *mutual inductance*. A changing current in one coil can induce a voltage in the other. Therefore, two coils having mutual inductance form a **transformer**. The coil connected to the source of energy is called the **primary winding**, and the coil connected to the load is called the **secondary winding**. See **Figure 6-5**.

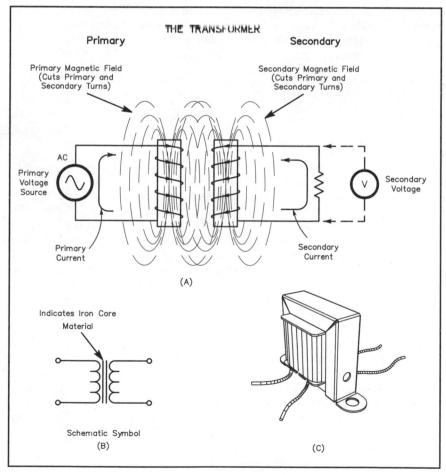

Figure 6-5 — Part A illustrates the operation of a transformer. Part B shows the schematic symbol used to represent an iron-core transformer. Part C shows what a typical small transformer might look like.

With a transformer, electrical energy can be transferred from one circuit to another without direct connection. In the process, the signal voltage can be readily changed from one level to another. Common house current is 120 V ac. If you need 240 V ac, you would use a transformer that has twice as many turns on the secondary coil as it does on its primary coil. The induced voltage in the secondary would then be double the primary voltage, or 240 V. A transformer can only be used with alternating current, because no voltages will be induced in the secondary if the magnetic field is not changing.

The primary and secondary coils of a transformer may be wound on a *core* of a magnetic material, such as iron. This increases the coupling of the coils so that practically all of the field set up by the current in the primary coil will cut the turns of the secondary coil. A small amount of current, called the **magnetizing current**, flows in a transformer primary winding, even with no load connected to the secondary. The amount of this magnetizing current depends on the transformer construction and the inductance of the transformer primary.

Transformers have power ratings, just as other components do. Usually, the primary and secondary voltages are specified, along with the maximum current that can be drawn from the secondary winding. Sometimes the secondary voltage and current are combined into a "volt-ampere" (VA) rating. A transformer with a 12-V, 1.5-A secondary would have a volt-ampere rating of 18 VA.

[At this point, you should turn to Chapter 13 and study questions G6A04, G6A05, G6A13, G6A15 and G6A16. Review this section if you have any difficulty with those questions.]

DIODES

Semiconductor material exhibits properties of both metallic and nonmetallic substances. Generally, the layers of semiconductor material are made from germanium or silicon. To control the conductive properties of the material, impurities are added as the semiconductor crystal is "grown." This process is called doping. If a material with an excess of free electrons is added to the structure, N-type material is produced. This type of material is sometimes referred to as *donor material*. When an impurity that results in an absence of electrons is introduced into the crystal structure, you get P-type material. The result in this case is a material with positive charge carriers, called holes. P-type material is also called *acceptor material*.

The **junction diode**, also called the PN-junction diode, is made from two layers of semiconductor material joined together. One layer is made from P-type material and the other layer is made from N-type material. The name PN junction comes from the way the P and N layers are joined to form a semiconductor diode. **Figure 6-6** illustrates the basic concept of a junction diode.

The P-type side of the diode is called the anode, while the N-type side is called the cathode. (The anode connects to the positive supply lead and the cathode connects to the negative supply lead.) Current flows from the cathode to the anode. The excess electrons from the N-type material flow to the P-type material, which has holes (an electron deficiency). Electrons and holes are called charge carriers because they carry current from one side of the junction to the other. When no voltage is applied to a diode, the junction between the N-type and P-type material acts as a

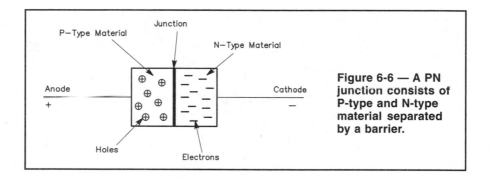

Figure 6-6 — A PN junction consists of P-type and N-type material separated by a barrier.

barrier that prevents carriers from flowing between the layers.

When voltage is applied to a junction diode as shown at A in **Figure 6-7**, however, carriers will flow across the barrier and the diode will conduct (electrons will flow through it). When the diode anode is positive with respect to the cathode, electrons are attracted across the barrier from the N-type material, through the P-type material and onto the positive battery terminal. As the electrons move through the P-type material, they fill some holes and leave other holes behind as they move, so the holes appear to move in the opposite direction, toward the negative battery terminal. When the diode is connected in this manner it is said to be forward biased.

If the battery polarity is reversed, as shown in Figure 6-7B, the excess electrons in the N-type material are attracted away from the junction by the positive battery terminal. Electrons in the P-type material are forced toward the junction by the negative charge from the battery, and the electrons leave holes that appear to move toward the negative battery terminal. Because there are no holes in the N-type material to accept electrons, no conduction takes place across the junction. When the anode is connected to a negative voltage source and the cathode is connected to a positive source, the diode does not conduct, and it is said to be reverse biased.

Junction diodes are used as rectifiers because they allow current to flow in one direction only. (We will discuss rectifier circuits in Chapter 7.) When an ac signal is applied to a diode, the diode will be forward biased during one half of the cycle, so it will conduct, and current will flow to the load. During the other half of the cycle, the diode is reverse biased, and there is no current. The diode output is pulsed dc, and the current always flows in the same direction.

Diode Ratings

Junction diodes have maximum voltage and current ratings that must be observed, or damage to the junction will result. The voltage rating is called the **peak inverse voltage (PIV)** and the current rating is called the **maximum average forward current**. PIV is sometimes called reverse breakdown voltage or peak reverse voltage (PRV). With present technology, diodes are commonly available with ratings up to 1000 PIV and 100 A forward current.

Peak inverse voltage is the voltage a diode must withstand when it isn't conducting. Although a diode is normally used in the forward direction, it will conduct in the reverse direction if enough voltage is applied. A few hole/electron pairs are

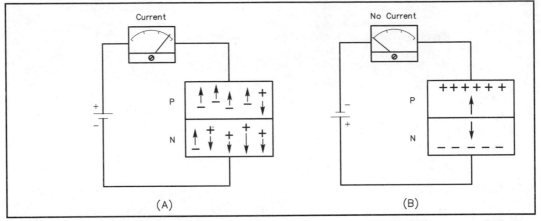

Figure 6-7 — At A, the PN junction is forward biased and is conducting. At B, the junction is reversed biased and it does not conduct.

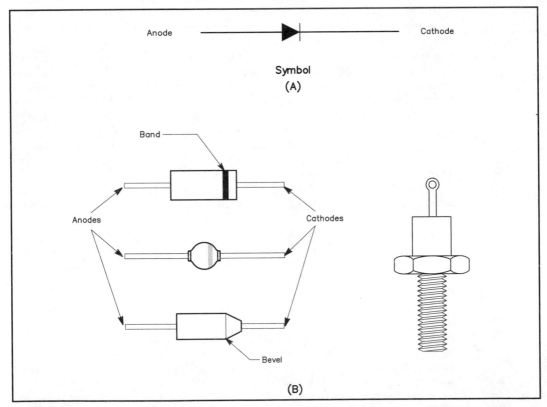

Figure 6-8 — The schematic symbol for a diode is shown at A. Diodes are typically packaged in one of the case styles shown at B.

thermally generated at the junction when a diode is reverse biased. These pairs cause a very small reverse current, called leakage current. Semiconductor diodes can withstand some leakage current, but if the inverse voltage reaches a high enough value the leakage current will rise abruptly, resulting in a heavy reverse current. The point at which the leakage current rises abruptly is called the *avalanche point*. A large reverse current usually damages or destroys the diode.

The maximum average forward current is the highest average current that can flow through the diode in the forward direction for a specified junction temperature. This specification varies from device to device, and it depends on the maximum allowable junction temperature and on the amount of heat the device can dissipate. As the forward current increases, the junction temperature will increase. If the diode is allowed to get too hot, it will be destroyed.

Impurities at the PN junction cause some resistance to current. This resistance results in a voltage drop across the junction. For silicon diodes, this drop is approximately 0.6 to 0.7 V; for germanium diodes, the drop is 0.2 to 0.3 V. When current flows through the junction, some power is dissipated in the form of heat. The amount of power dissipated depends on the current through the diode. For example, it would be approximately 6 W for a silicon rectifier with 10 A flowing through it:

$$P = I \times E \hspace{6cm} \text{(Equation 6-1)}$$

$$P = 10 \text{ A} \times 0.6 \text{ V} = 6 \text{ W}$$

If the junction temperature exceeds the safe level specified by the manufacturer, the diode is likely to be damaged or destroyed.

Diodes designed to safely handle forward currents in excess of 6 A generally are packaged so they may be mounted on a heat sink. These diodes are often referred to as stud-mount devices, because they are made with a threaded screw attached to the body of the device. This screw is attached to the heat sink so that the heat from the device will be effectively transferred to the heat sink. The heat sink helps the diode package dissipate heat more rapidly, thereby keeping the junction temperature at a safe level. The metal case of a stud-mount diode is usually one of the contact points, so it must be insulated from ground.

Figure 6-8 shows some of the more common diode-case styles, and the general schematic symbol for a diode. The line or spot on a diode case indicates the cathode lead. On a high-power stud-mount diode, the stud may either be the anode or the cathode. Check the case or the manufacturer's data sheet for the correct polarity.

TRANSISTORS

Transistors come in many shapes and sizes. They have applications almost too numerous to mention. **Bipolar transistors** use three layers of semiconductor material, arranged either with two layers of N-type material sandwiched around a layer of P-type material (an NPN transistor) or with two layers of P-type material sandwiched around a layer of N-type material (a PNP transistor). The middle layer is called the base and the two outside layers are the collector and the emitter. A small signal applied between the emitter and base can control a much larger output signal from the emitter, across the base layer and out the collector layer.

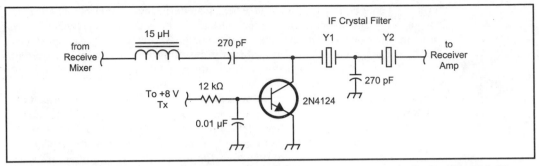

Figure 6-9 — This transistor switch circuit is taken from the Wilderness Radio NorCAL Sierra QRP transceiver. In receive, there is no signal applied to the transistor base, and the transistor is cut off. There is no collector-emitter current. When the radio switches to transmit mode an 8-V bias voltage is applied to the transistor base. That places the transistor in saturation, creating a short circuit across the transistor collector to emitter. The receiver input signal is shorted to ground through the transistor, protecting the sensitive receiver circuitry from the transmitted signal.

Field-effect transistors (FET) use a channel of one type of semiconductor material (source and drain) between sections of the opposite type of semiconductor material (gate). A voltage applied across the gate terminals controls the signal current through the source/gain channel.

Either type of transistors can be used as amplifiers, oscillators and in many other applications. On interesting application of a transistor is as a switch that can be used to turn an RF circuit on or off.

When a bipolar transistor is used as a switch in a logic circuit, it is important that the switch either be "all the way on" or "all the way off." This is achieved by operating the transistor either in its saturation region to turn the switch on, or in its cut-off region to turn the switch off. **Figure 6-9** shows the transmit/receive switching circuit from the Wilderness Radio NorCAL Sierra QRP transceiver. When the radio is receiving, there is no signal applied to the base of the 2N4124 transistor. In the case, there is no base current and the transistor is cut off, so there is no collector current. When the radio switches to transmit, there is an 8-V bias voltage applied to the transistor base, putting the transistor into saturation. The collector-emitter path becomes a short circuit, shorting the receiver input to ground. This protects the sensitive receive circuitry from the strong transmitter signal.

[Before proceeding to Chapter 7, turn to Chapter 13 and study questions G6A06, G6A07 and G6A14. Review this section as needed. The remaining questions in the G6A section will be covered by material in Chapter 7.]

CHAPTER 7 KEYWORDS KEYWORDS KEYWORDS

Balanced modulator — A mixer circuit that combines an audio input signal with a carrier-oscillator signal. The output signal contains the two sidebands produced by this mixing, but does not include the original carrier-oscillator signal or the pure audio signal. A modulated RF signal contains some information to be transmitted.

Band-pass filter — A circuit that allows signals to go through it only if they are within a certain range of frequencies, and attenuates signals above and below this range.

Bleeder resistor — A large-value resistor connected to the filter capacitor in a power supply to discharge the filter capacitors when the supply is switched off.

Detector — A mixer circuit used to recover the information signal from a modulated RF signal.

Filter — A circuit that will allow some signals to pass through it but will greatly reduce the strength of others. In a power-supply circuit, a filter smoothes the ac ripple.

Full-wave bridge rectifier — A full-wave rectifier circuit that uses four diodes and does not require a center-tapped transformer.

Full-wave rectifier — A circuit basically composed of two half-wave rectifiers. The full wave rectifier allows the full ac waveform to pass through; one half of the cycle is reversed in polarity. This circuit requires a center-tapped transformer.

Half-wave rectifier — A circuit that allows only half of the applied ac waveform to pass through it.

High-pass filter — A filter that allows signals above the cutoff frequency to pass through, and attenuates signals below the cutoff frequency.

Local oscillator (LO) — A receiver circuit that generates a stable, pure signal used to mix with the received RF to produce a signal at the receiver intermediate frequency (IF).

Low-pass filter — A filter that allows signals below the cutoff frequency to pass through and attenuates signals above the cutoff frequency.

Mixer — A circuit that takes two or more input signals, and produces an output that is the sum or difference of those signal frequencies.

Power supply — A device used to convert the available voltage and current source (often the 120-V ac household supply) to a form that is required for a specific circuit requirement. This will often be a higher or lower dc voltage.

Pulsating dc — The output from a rectifier before it is filtered. The polarity of a pulsating dc source does not change, but the amplitude of the voltage changes with time.

Rectifier — An electronic component that allows current to pass through it in only one direction.

Ripple — The amount of change between the maximum voltage and the minimum voltage in a pulsating dc waveform.

Single-sideband, suppressed-carrier signal — A radio signal in which only one of the two sidebands generated by amplitude modulation is transmitted. The other sideband and the RF carrier signal are removed before the signal is transmitted.

Superheterodyne receiver — A receiver that converts RF signals to an intermediate frequency before detecting them (converting the signals to audio).

Practical Circuits

Now that you have studied some basic electrical principles, and have learned about the properties of some simple components, you are ready to learn how to apply those ideas to practical Amateur Radio circuits. This chapter will lead you through examples and explanations to help you gain that knowledge.

In this chapter, we will look at how power supplies operate, and we will discuss band-pass, high-pass and low-pass filter circuits. You will be directed to turn to Chapter 13 at appropriate points, and to use the General question pool as a study aid to review your understanding of the material.

Keep in mind that there have been entire books written on each topic covered in this chapter. If you do not understand some of the circuits from our brief discussion, it would be a good idea to consult some other reference books. ARRL's *Understanding Basic Electronics* is a good starting point. *The ARRL Handbook* contains much more detailed information, but even that won't tell you everything about a topic. The discussion in this chapter will help you understand the circuits well enough to pass your General license exam, however. There will be one question from the practical circuits subelement. The syllabus point for that topic is:

G7A   Power supplies and filters; single-sideband transmitters and receivers.

POWER SUPPLIES

One circuit that is common to just about every type of radio circuit is the **power supply**. The 120 V ac available at your wall outlet isn't exactly what we mean by a power supply. The current at the outlet is an alternating current, and most electronic components require direct current. Additionally, circuits may require voltage levels other than the 120 V available at the wall outlet. A power supply is used to increase or decrease the value of the voltage and to convert ac to dc.

An ac power supply consists of three sections: the transformer, the diode **rectifier** and the **filter**. A power transformer is used to raise or lower the voltage. It transforms the input-voltage level (120 or 240 V) to the level required by the equipment. Power transformers normally operate at 60 hertz, the household-current frequency.

Many power transformers have multiple secondary windings to provide several different output voltages. The power transformer used in a vacuum-tube amateur transmitter or amplifier might have three secondaries: one delivering 6.3 V for the tube filaments, one 12-V winding for the solid-state circuits and one supplying high voltage for the plates of the final-amplifier tubes.

Rectifiers

The diode rectifier section of a power supply converts ac to dc. (Remember that diodes conduct only in one direction.) **Figures 7-1** and **7-2** show several common rectifier circuits. Let's examine the operation of the simplest type, the **half-wave rectifier**.

To understand diode-rectifier action, we'll consider a diode connected to a secondary winding of a power transformer, as shown in Figure 7-1A. During half of the ac cycle, the applied voltage makes the anode positive with respect to the cath-

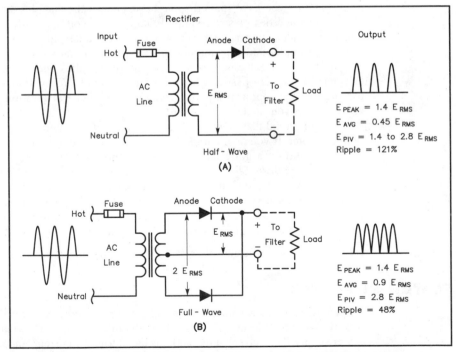

Figure 7-1 — Fundamental rectifier circuits are shown. (A) Half-wave. (B) Full-wave center-tap. These rectifier circuits are discussed in the text.

ode. The diode is *forward biased* and will conduct during this half cycle. Current flows through the diode and any load attached to the rectifier output. The current passing through the load will vary with time, because the diode allows current to pass only during one half of the cycle. During the other half of the cycle, the anode is not positive with respect to the cathode, so no current passes through the diode.

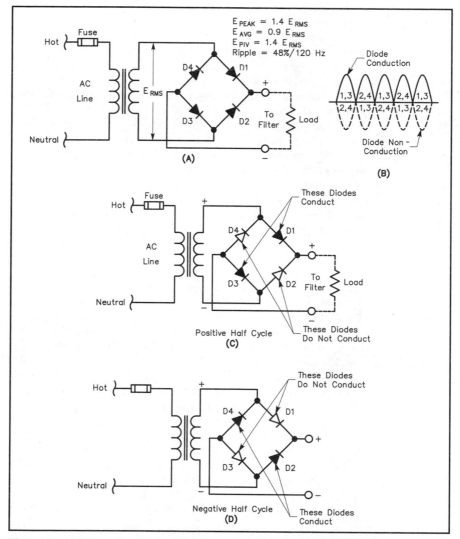

Figure 7-2 — The operation of a full-wave, bridge-rectifier circuit is shown. The basic circuit is illustrated at A. Diodes D_1 and D_3 conduct on one-half of the input cycle, while diodes D_2 and D_4 conduct on the other half. The nonconducting diodes have the full transformer output voltage across them in the reverse direction.

In this case the diode is *reverse biased*. The action of the diode in permitting the current to flow only in one direction is called *rectification*.

The output of the rectifier is also shown in Figure 7-1A. The half-wave rectifier circuit takes its name from the fact that only half the ac wave passes through the circuit. One complete cycle of an ac wave goes from zero to 360° and the half-wave rectifier uses only 180° (one half) of the wave. The half-wave rectifier conducts during 180° of each cycle. There is one output pulse for every complete cycle of the input voltage. The output waveform of a half-wave rectifier has a frequency equal to the ac input waveform.

If another diode is connected to the opposite end of the transformer secondary, the second diode will conduct during the portions of the ac cycle that the first diode cannot. A connection (called a center tap) is made in the center of the transformer winding and connected through the load to the cathodes of the diodes. The rectifier circuit that has diodes working on both halves of the cycle is called a **full-wave rectifier**. This particular full-wave rectifier is known as a *full-wave, center-tap rectifier*. The advantage of the full-wave rectifier can be seen from the output waveform in Figure 7-1B — output is produced during the entire 360° of the ac cycle.

There are two output pulses for every full cycle of the input voltage. This means the output waveform from a full-wave rectifier has a frequency equal to twice the ac input frequency.

A third type of rectifier circuit is the **full-wave bridge rectifier** shown in Figure 7-2. The output current or voltage waveshape is the same as for the full-wave center-tap rectifier. The important difference is in the output voltage from the rectifier compared to the transformer-secondary output voltage. In the center-tap circuit, each side of the center-tapped secondary must develop a peak voltage high enough to supply the desired dc output voltage, so the total transformer peak output voltage must be twice the power-supply output voltage (for a 6.3-V rectifier output we must use a center-tapped transformer with a peak output of 12.6 V). In the bridge circuit, no center tap is necessary — we can get 12.6-V output from a 12.6-V transformer. We can't get something for nothing, however — the elimination of the center tap requires the use of two additional rectifiers in the bridge-rectifier circuit. The bridge is another full-wave rectifier, and it uses the full 360° of the ac wave.

In the bridge circuit, two rectifiers are connected to the transformer secondary in the same way as the two rectifiers in the center-tap circuit shown in Figure 7-1B. The second pair of rectifiers is also connected in series between the ends of the transformer winding, but in reverse. See Figure 7-2. When the upper end of the transformer secondary is positive, Diode D_1 conducts. The current then flows through the load and back to D_3, which also conducts. This completes the circuit from one side of the transformer, through the load and back to the other side of the transformer. Diodes D_2 and D_4 are reverse biased during this time, so they do not conduct. During the second half of the ac cycle, diodes D_2 and D_4 are forward biased so they conduct. Diodes D_1 and D_3 are reverse biased, so they do not conduct.

Diode Ratings

Remember that diodes are rated with regard to their current-handling capability and the maximum voltage that may be applied in the reverse direction. These ratings must not be exceeded. For example, when one diode in a bridge rectifier is conducting, the two diodes connected to it are not. These diodes must each withstand the full transformer secondary voltage in the reverse direction. Since the peak value of the voltage is 1.41 times the RMS value, the diodes must withstand a peak inverse voltage (PIV) of 1.41 times the RMS value of the transformer secondary. A safe design practice is to use diodes rated for at least twice the transformer output voltage.

A full-wave center-tap rectifier has the full transformer output voltage across the reverse-biased diode. The diodes must be able to withstand a PIV equal to twice the peak output voltage. (The output voltage is only half the full transformer voltage because of the center-tapped transformer. See Figure 7-1B.) It is also a good design practice to select diodes with a peak inverse voltage rating (PIV) of two or more times the normal peak output voltage of the supply for full-wave rectifier circuits.

The diode in a half-wave rectifier circuit must withstand the full transformer peak output voltage. It is good design practice to select a diode with a PIV rating of one to two times the normal peak output voltage of the supply.

In full-wave rectifier circuits, each diode carries one-half the average output current, so the diodes must be rated to pass at least this much current without becoming too warm. Using diodes able to handle the full average current (instead of half) is a safer design practice. In addition, when a capacitor-input filter is used, the capacitor appears as a dead short to the rectifier when the supply is first switched on. This means that the diodes may pass as much as 10 to 100 times the average current in one brief burst. The diodes used with a capacitor-input filter must be able to handle this surge current.

Diodes in Series and Parallel

Suppose we're building a power supply, and we need rectifiers rated at 1000 PIV, but we only have 500-PIV diodes on hand. How can we use the 500-PIV diodes? If we connect two 500-PIV diodes in series, the pair will withstand 1000 PIV (2 × 500 = 1000). With modern silicon rectifier diodes, you can safely wire them in series to increase the effective PIV rating.

Diodes can also be placed in parallel to increase total forward current-handling capability. When diodes are in parallel, a resistor should be placed in series with each diode to ensure that each takes the same current. This is illustrated in **Figure 7-3**. Without a resistor in series with each diode, one diode may take most of the current, and that could destroy the diode. The resistor value should be chosen to provide a few tenths of a volt drop at the expected forward current. Be sure to select resistors with sufficient power rating.

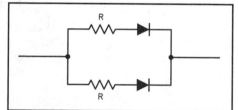

Figure 7-3 — Use equalizing resistors when connecting diodes in parallel to increase the forward current-handling capability. The resistor value should be selected to provide a few tenths of a volt drop at the expected forward current.

[Turn to Chapter 13 now and study questions G6A09 through G6A11. Also study questions G7A03 and G7A04. Review this section if you have trouble with any of these questions.]

Power-Supply Filters

The output from a rectifier is not like dc from a battery. It's neither pure nor constant. It's actually what's known as **pulsating dc** — flowing only in one direction but not at a steady value. **Figure 7-4A** shows a full-wave rectifier connected to a resistive load. The output waveform is a series of pulses at twice the frequency of the ac input, because there are two output pulses for every ac cycle. Using this type of current as a replacement for a battery in a transmitter or receiver would introduce a strong hum on all signals. To smooth out the pulsations (called **ripple**) and eliminate the hum and its effects, a **filter** is used between the rectifier and the load.

The simplest type of filter is a large-value capacitor across a load (**Figure 7-4B**). If the capacitance is high enough and the current drawn by the load is not excessive, the capacitor will charge during voltage peaks. The *filter capacitor* discharges as the pulse decreases, smoothing out the ripple. During the peaks in the pulsating dc waveform, the capacitor stores energy. As the value of the dc waveform decreases, the capacitor releases some of its energy, and supplies current to the load. This cycle occurs over and over, resulting in a smooth dc output waveform, with minimum ripple.

We can design a more effective filter if we place an inductor, or *filter choke*, between the rectifier and the load, as shown in **Figure 7-5A**. The series inductor

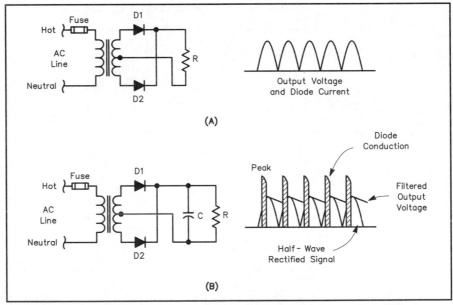

Figure 7-4 — The circuit shown at A is a simple full-wave rectifier with a resistive load. The waveform shown to the right shows the output voltage and diode current for this circuit. B illustrates how diode current and output voltage change with the addition of a capacitor filter. The diodes conduct only when the rectified voltage is greater than the voltage stored in the capacitor. Since this time is usually only a short portion of a cycle, the peak current through each diode will be quite high.

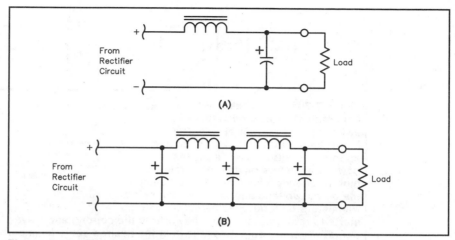

Figure 7-5 — Shown at A is a choke-input power-supply filter circuit. B shows a capacitor-input, multisection filter.

opposes changes in the circuit current the way the capacitor opposes changes in circuit voltage. We can connect several series-choke/parallel-capacitor sections one after another to improve the filtering action even more. The two basic filters are called *choke-input filters* and *capacitor-input filters*, depending on which component comes immediately after the rectifier. A capacitor-input multisection filter is shown in Figure 7-5B. In either case, capacitors and inductors are used to form the power-supply filter network.

The Bleeder Resistor

It's a good design practice to connect a high-value resistor across the filter capacitor in a power supply, as shown in **Figure 7-6**. Such a resistor, called a **bleeder resistor**, is used to discharge the capacitor when the

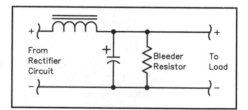

Figure 7-6 — A bleeder resistor connected to the output of a power supply will discharge the filter capacitor(s) when the supply is switched off.

power-supply primary voltage is turned off. It drains the charge from the capacitors and decreases the chances for accidental electrical shock when the power supply is turned off.

[Turn to Chapter 13 now and study question G6A08 and questions G7A01 and G7A02. Review this section as needed.]

Switched-Mode Power Supplies

Switched-mode power supplies normally take the 120 or 240 V ac input signal and rectify it with a full-wave or bridge rectifier. The rectified signal is filtered, and

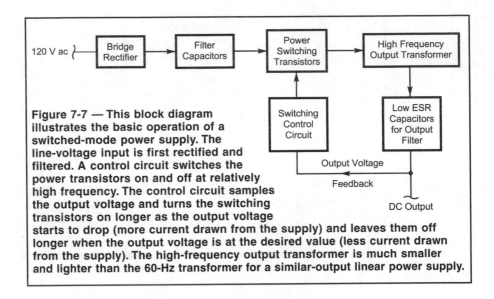

Figure 7-7 — This block diagram illustrates the basic operation of a switched-mode power supply. The line-voltage input is first rectified and filtered. A control circuit switches the power transistors on and off at relatively high frequency. The control circuit samples the output voltage and turns the switching transistors on longer as the output voltage starts to drop (more current drawn from the supply) and leaves them off longer when the output voltage is at the desired value (less current drawn from the supply). The high-frequency output transformer is much smaller and lighter than the 60-Hz transformer for a similar-output linear power supply.

a power oscillator then controls the current through a high-frequency transformer. There are several techniques used to connect the dc voltage to the high frequency transformer in the process of creating the output voltage.

Switched-mode power supplies use an oscillator operating at a frequency much higher than the 60 Hz ac line current. (Oscillator frequencies of 50 kHz or more are commonly used.) This allows the use of small, lightweight transformers. While the transformer in a 20 amp linear power supply might weigh 15 or 20 pounds, the transformer for a switched-mode power supply with a similar current rating might weigh 1 or 2 pounds (or less)!

Either switched-mode or linear power supplies can be designed to have just about any desired output voltage. Switched-mode power supplies generally require significantly more components than a simple linear supply, though. **Figure 7-7** shows a simple block diagram for the operation of a switched-mode power supply.

It is important to use filter capacitors that have a low equivalent series resistance (ESR) rating for the output filter of a switching power supply. The capacitors will have to

Power Supply Overvoltage Protection — The Crowbar Circuit

If the output voltage control circuit of a power supply should fail, a high voltage could be applied to your radio or other device connected to the power supply. To prevent damage to the equipment being powered by the supply, a "crowbar" circuit is often used at the supply output. The crowbar circuit is usually a silicon-controlled rectifier (SCR) with a voltage-sensing trigger circuit connected to the SCR gate. If the output voltage from the supply reaches the set point of the trigger circuit, the SCR turns on, shorting the output to ground. This will quickly blow the power supply fuses, preventing damage to your radio equipment. **Figure 7-10** is the schematic diagram for a simple crowbar circuit.

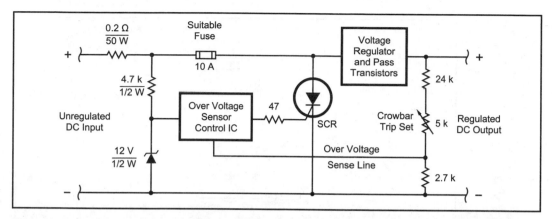

Figure 7-8 — This schematic diagram shows one implementation of a "crowbar" over-voltage protection circuit for a power supply. If the over-voltage control IC senses an output voltage higher than the set point, it will turn on the silicon-controlled rectifier (SCR), which will short the unregulated input to ground. The short circuit will blow the fuse or trip an in-line circuit breaker, quickly cutting the output voltage to zero. A crowbar circuit will protect sensitive electronic equipment from being damaged or destroyed by a dc operating voltage that is too high.

[Now turn to Chapter 13 and study questions G7A10 through G7A13. Review this section if you have difficulty answering any of those questions.]

FILTERS

We've already discussed one type of filter used in Amateur Radio: the power-supply filter. You should also be familiar with three others: the low-pass filter, the high-pass filter and the band-pass filter.

A **low-pass filter** is one that passes all frequencies below a certain frequency, called the *cutoff frequency*. Frequencies above the cutoff frequency are *attenuated*, or significantly reduced in amplitude. See **Figure 7-9A**. The cutoff frequency depends on the design of the low-pass filter. One common use for a low-pass filter in an amateur station is to connect one between the transmitter and the antenna, preferably as close to the transmitter as possible. It is important that the filter have the same source and load impedance as the feed line into which it is inserted. This ensures that the filter won't introduce an impedance mismatch that could increase the SWR on the feed line.

The low-pass filter cutoff frequency has to be higher than the highest frequency used for transmitting. A filter with a 45-MHz cutoff frequency would be fine for the "low bands" (1.8 through 29.7 MHz), but the filter would significantly attenuate 6-meter (50 MHz) signals.

A **high-pass filter** passes all frequencies above the cutoff frequency, and attenuates those below it. See **Figure 7-9B**. A high-pass filter is often connected to a television set, stereo receiver or other home-entertainment device that is receiving interference. The filter will attenuate the lower-frequency signals from an amateur HF station while allowing the higher-frequency television or FM broadcast signals

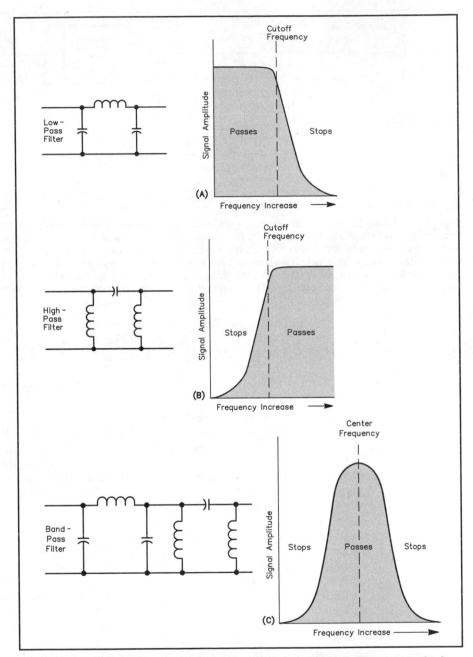

Figure 7-9 — Output-versus-frequency curves for different filter types. At A, the low-pass filter passes signals below the cutoff frequency, and attenuates signals above the cutoff. At B, the high-pass filter attenuates only those signals below the cutoff. The band-pass filter has a response curve like (C), and passes only those signals within the passband.

to pass through to the receiver. This is useful in reducing the amount of lower-frequency radiation (at the ham's operating frequency) that might overload a television set, causing disruption of reception. For best effect, the filter should be connected as close to the television set's tuner as possible — inside the TV if it is your own set and you don't mind opening it up, or directly to the antenna terminals on the back of the television. If the set belongs to your neighbors, have them contact a qualified service technician to install the filter, so you are not blamed if something goes wrong with the set later on. Again, it is important that the filter have the same source and load impedance as the feed line into which it is inserted.

A **band-pass filter** is a combination of a high-pass and low-pass filter. It passes a desired range of frequencies while rejecting signals above and below the pass-band. This is shown in **Figure 7-9C**. Band-pass filters are commonly used in receivers to provide different degrees of rejection — very narrow filters are used for CW reception, wider filters for single-sideband voice (SSB) and AM double-sideband reception.

[Now it is time to turn to Chapter 13 again. Study question G7A05. Review the material in this section if needed.]

SINGLE-SIDEBAND VOICE TRANSMITTERS

When you speak into the microphone of your single-sideband voice transmitter, the magic of radio converts your voice to radio signals. These radio signals come out of the transmitter and travel to the antenna. Then they radiate off into the air as they make their way to who knows where. This process of changing your voice into a radio signal is not as complicated as it may seem. **Figure 7-10** shows the block diagram of a basic transmitter that produces a **single-sideband, suppressed-carrier signal**. When we refer to this type of signal, we usually abbreviate it to single sideband, or SSB. You can compare this type of signal to those used by

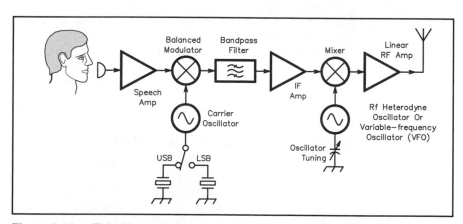

Figure 7-10 — This block diagram shows a basic single-sideband, suppressed-carrier (SSB) transmitter. The text explains the operation of the various stages in this circuit.

stations on the AM, Standard Broadcast band. There the transmitters use *double-sideband, full-carrier signals*. From the names, you can tell that SSB is just a standard AM signal with one sideband and the carrier removed. Let's follow the signals as they move through the various stages in our SSB transmitter.

When you speak, your vocal chords cause air molecules to vibrate. The vibrating air molecules bump into the microphone, which changes the air-pressure variations of your voice into a changing electrical signal. This electrical signal is usually quite weak, so it is normally amplified before being passed on to the **balanced modulator**. The balanced modulator has a second input, this one from the carrier oscillator. The carrier oscillator produces a signal at the intermediate frequency, or IF of the transmitter circuit. The system shown in Figure 7-10 uses two crystals to switch the carrier-oscillator frequency to select the upper sideband or the lower sideband. It is also possible to use a single-frequency oscillator and switch band-pass filters.

The balanced modulator processes signals from the speech amplifier and the carrier oscillator, and sends them on to a band-pass filter. The modulator is a mixer stage, which adds the two input signals and passes the sum along to the output. It also subtracts the audio from the carrier, and passes that difference along to the output. Because the modulator is *balanced*, however, it does not pass either the audio or the carrier signals to the output. The balanced modulator takes two input signals and produces two different output signals. The output signals are the two s*idebands*, but the carrier is suppressed.

The next stage of our transmitter completes the process of generating an SSB signal by removing one of the sidebands. This is the job of the band-pass filter. The filter must have a bandwidth that is narrow enough to pass one of the sidebands but reject the opposite one. After the band-pass filter there is often an amplifier stage to boost the strength of the signal before it is processed further.

Next is the **mixer** stage, which is similar to the balanced modulator. The mixer has two inputs and one output. The second input is from the RF heterodyne oscillator, which is set to a frequency that will add to or subtract from the signal coming from the modulator, to produce an output at the desired transmit frequency.

The output signal from the mixer is ready to be transmitted. The signal is often amplified by one or more amplifier stages before it is passed to the antenna and transmitted. In most transmitters, the modulation and mixing processes are carried out at low power levels. Any amplifier stages after the balanced modulator must be linear, so the signal does not become distorted.

[Turn to Chapter 13 now and study questions G7A06 and G7A07. Review this section if you do not understand either of those questions.]

SINGLE-SIDEBAND VOICE RECEIVERS

An SSB receiver reverses the processes carried out in the transmitter. The received radio signal is processed, and the speaker reproduces the air-molecule vibrations necessary to recreate the sound of the transmitting operator's voice. **Figure 711** illustrates the primary stages of a **superheterodyne receiver** suitable for SSB reception.

The received signal coming in from the antenna is usually quite weak. Many receivers have an amplifier stage before any further signal processing takes place.

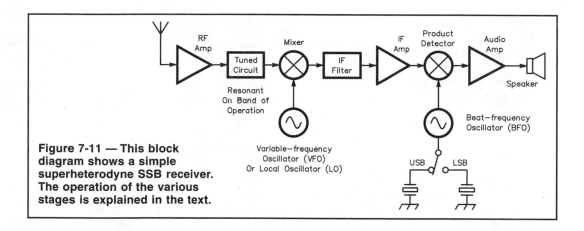

Figure 7-11 — This block diagram shows a simple superheterodyne SSB receiver. The operation of the various stages is explained in the text.

There may be one or more resonant networks at the input also. This is a tuned circuit used to select the range of frequencies that the receiver will process. If the receiver is to cover more than one band, it will have a bandswitch to select the proper tuned circuits for the desired operating band.

The mixer stage has two inputs and one output, just like the modulator and mixer in our transmitter circuit. The first input signal comes from the input filter and RF amplifier stage just discussed. The second input for the receive mixer is from the **local oscillator (LO)**, which generates a signal to mix with the received signal to produce an output at the receiver intermediate frequency (IF). The LO frequency is variable, so this oscillator is often called a variable-frequency oscillator, or VFO. As the frequency changes, different RF signals are converted to the IF. This frequency-conversion technique allows the IF stages to operate over a relatively narrow frequency range. The IF filters can be designed to provide a high degree of selectivity without the need to adjust the filter operating frequency.

The IF filter sets the overall receiver selectivity. Many receivers use several selectable filters for different modes. A narrow filter provides the needed selectivity for CW and RTTY reception, and a 2.4kHz filter for SSB operation. After filtering, there are usually one or more stages of amplification before the signal is passed on to the product **detector**. This amplifier operates at the intermediate frequency (IF) so it is usually called the IF amplifier.

The detector is another mixer stage, so it has two inputs and one output. The first input is from the IF amplifier. The second input this time is from the *beat-frequency oscillator*, or *BFO*. Notice that like the transmitter carrier oscillator, the BFO can change frequency to select either the upper sideband or the lower sideband for reception. The IF and BFO signals mix in the detector, giving an output signal that matches the original transmitted audio. After one or more amplification stages, this audio signal is ready to be fed to a speaker or headphones, where the electrical signals produce variations in air pressure again, creating sound.

[This completes the material for Chapter 7. Now turn to Chapter 13 and study questions G7A08 and G7A09. Review this section as needed before proceeding to Chapter 8.]

CHAPTER 8
KEYWORDS
KEYWORDS
KEYWORDS

Amplitude modulation (AM) — A method of combining an information signal and an RF carrier wave in which the amplitude of the RF envelope (carrier and sidebands) is varied in relation to the information signal strength.

Audio-frequency shift keying (AFSK) — A method of transmitting radioteletype information by switching between two audio tones fed into an FM transmitter microphone input. This is the RTTY mode most often used on VHF and UHF.

Balanced modulator — A circuit used in a single-sideband suppressed-carrier transmitter to combine a voice signal and the RF signal. The balanced modulator isolates the input signals from each other and the output, so that only the difference of the two input signals reaches the output.

Bandwidth — The frequency range (measured in hertz) over which a signal is stronger than some specified amount below the peak signal level. For example, if a certain signal is at least half as strong as the peak power level over a range of +3 kHz, the signal has a 3-dB bandwidth of 6 kHz.

Deviation ratio — The ratio between the maximum change in RF-carrier frequency and the highest modulating frequency used in an FM transmitter.

Flattopping — A distorted audio signal produced by an SSB transmitter with the microphone gain set too high. The peaks of the voice waveform are cut off in the transmitter because of overmodulation. Also called *clipping*.

Frequency deviation — The amount the carrier frequency in an FM transmitter changes as it is modulated.

Frequency modulation (FM) — The process of varying the frequency of an RF carrier in response to the instantaneous changes in an audible signal.

Frequency-shift keying (FSK) — A method of transmitting radioteletype information by switching an RF carrier between two separate frequencies. This is the RTTY mode most often used on HF.

Image response — A form of interference to received signals that is produced when a mixer responds to a signal frequency equal to the LO – the IF when the desired signal frequency is equal to the LO + the IF. Also when the mixer responds to a signal frequency equal to the LO + IF when the desired signal frequency is the LO – IF.

Mixer — A circuit that takes two or more input signals, and produces an output that includes the sum and difference of those signal frequencies.

Modulation — The process of varying some characteristic (amplitude, frequency or phase) of an RF carrier for the purpose of conveying information.

Modulation Index — The ratio between the maximum carrier frequency deviation and the audio modulating frequency at a given instant in an FM transmitter.

Peak envelope power (PEP) — The average power of the RF cycle having the greatest amplitude. (This occurs during a modulation peak.)

Phase — The time interval between one event and another in a regularly recurring cycle.

Phase modulation — Varying the phase of an RF carrier in response to the instantaneous changes in an audio signal.

Reactance modulator — An electronic circuit whose capacitance or inductance changes in response to an audio input signal.

RF envelope — The shape of an RF signal as viewed on an oscilloscope.

Sideband — The sum or difference frequencies generated when an RF carrier is mixed with an audio signal.

Single-sideband, suppressed-carrier, amplitude modulation (SSB) — A technique used to transmit voice information in which the amplitude of the RF carrier is modulated by the audio input, and the carrier and one sideband are suppressed.

Splatter — Interference to adjacent signals caused by overmodulation of a transmitter.

Varactor diode — A component whose capacitance varies as the reverse bias voltage is changed.

Signals and Emissions

One of the most exciting aspects of Amateur Radio is that it offers so many different ways to participate. This chapter will introduce the modes and emission types you will need to understand for your General exam. There will be two questions from this subelement on your exam, and these will come from two groups of questions:

G8A Signal information; AM; FM; single and double sideband and carrier; bandwidth; modulation envelope; deviation; overmodulation

G8B Frequency mixing; multiplication; bandwidths; HF data communications

SIGNAL QUALITY

The quality of the signal your station produces reflects on you! As a radio amateur and control operator, you are responsible (legally and otherwise) for all the signals emanating from your station. Your signals are the only way an amateur on the other side of the country or the world will get to know you.

Communication is the transfer of information from one place to another. We can derive information from a radio signal just by its presence and its strength, as in the case of a beacon. By switching the signal on and off in some prescribed manner we can transmit more information. Finally, we can modify the signal in other ways to convey still more information. An important consideration *and* limitation, however, is the amount of space in the radio-frequency spectrum that a signal occupies. This space is called **bandwidth**. The bandwidth of a transmission is determined by the information rate. Thus, a pure, continuous, unmodulated carrier has a very small bandwidth, while a television transmission, which contains a great deal of information, is several megahertz wide.

MODULATION

Modulation is the process of varying some characteristic (amplitude, frequency or phase) of a radio frequency (RF) wave in accordance with the instantaneous variations of some external signal, for the purpose of conveying information. Almost every amateur will think of phone transmission when the word modulation is mentioned. The subject of modulation, however, is much broader than that. The principles are the same whether the radio transmission is modulated by speech, audio tones or the on-off keying that modulates a CW transmitter with Morse code.

The sound of your voice consists of physical vibrations in the air, called sound waves. The range of audio frequencies (AF) generated by a person's voice may be quite large; the normal range of human hearing is from about 20 to 20,000 Hz. All the information necessary to make your voice understood, however, is contained in a narrower band of frequencies. In amateur communications, your audio signals are usually filtered so that only frequencies between about 300 and 3000 Hz are transmitted. This filtering is done to reduce the bandwidth of the phone signal.

Amplitude Modulation

Amplitude modulation is a method of combining an information signal and an RF carrier so the amplitude of the **RF envelope** (carrier and sidebands) varies in relation to the information signal strength. An amplitude modulation (AM) system changes the amplitude of an RF wave for the purpose of conveying information. For phone transmissions, the audio (information) signal is constantly changing in frequency and amplitude, or strength. The instantaneous amplitude of the RF signal envelope varies in accordance with the modulating audio.

On the high-frequency amateur bands, **single-sideband, suppressed-carrier, amplitude modulation** (usually called **SSB**) is the most widely used radiotelephone mode. Since SSB is a sophisticated form of amplitude modulation, it is worthwhile to take a look at some fundamentals of amplitude modulation.

Modulation is a mixing process. When RF and AF signals combine in a standard double-sideband, full-carrier, AM transmitter (such as one used for commercial broadcasting) four output signals are produced: the original RF signal, the original AF signal, and two **sidebands**, whose frequencies are the sum and difference of the original RF and AF signals. See **Figure 8-1**. The amplitude of the sidebands varies with the amplitude of the audio signal.

For higher-frequency modulating tones, the sideband frequencies will be further

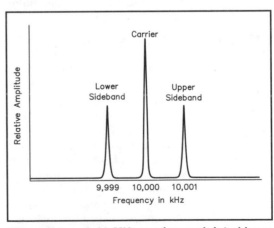

Figure 8-1 — A 10-MHz carrier modulated by a 1-kHz sine wave has sidebands at 9.999 MHz and 10.001 MHz.

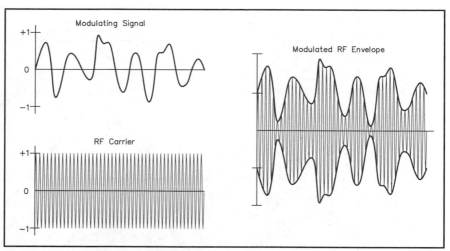

Figure 8-2 — This drawing illustrates the relationship between the modulating audio waveform, the carrier and the resulting RF envelope in a double-sideband, full-carrier AM signal.

from the carrier frequency, and for lower-frequency tones the sideband will be closer to the carrier frequency. The sum component is called the upper sideband, and increasing the frequency of the modulating audio causes a corresponding increase in the frequency of the RF output signal. The difference frequency is called the lower sideband, and is inverted, meaning that an increase in the frequency of the modulating AF causes a decrease in the frequency of the output signal. The amplitude and frequency of the carrier are unchanged by the modulation process. The resulting RF envelope, as viewed on an oscilloscope, has the shape of the modulating waveform, as **Figure 8-2** shows. The modulating waveform is mirrored above and below the zero axis of the RF envelope.

The amount of spectrum space occupied by the sidebands in a properly operating double-sideband phone transmitter depends on the bandwidth of the audio modulating signal. This is why most amateur transmitters limit the maximum audio-frequency component of your voice that is transmitted to about 3000 Hz. There is one emission type that does not produce sideband frequencies. That emission is an unmodulated carrier signal, which the FCC calls a test emission. [See §97.3(c)(9) of the FCC Rules.]

Modulation Percentage

One of the concepts that is easy to visualize by looking at the RF envelope is percentage of modulation. **Figure 8-3A** is a sine-wave signal used to modulate an RF carrier. In Figure 8-3B, we see that the modulating signal has just the right value to make the RF-envelope amplitude go to zero on the audio downswing and to twice the unmodulated value at the peak of the upswing. This is called 100% modulation. The percent of modulation is the ratio of the peak-to-peak voltage of the

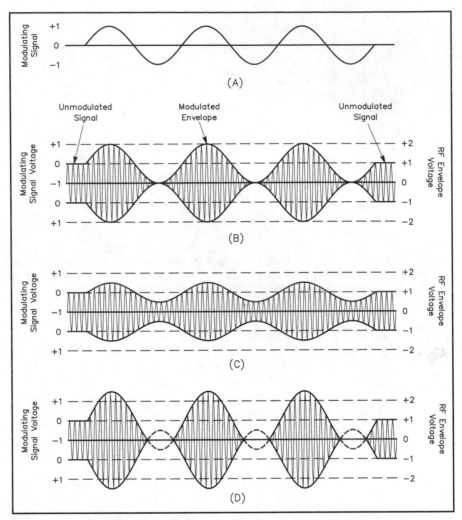

Figure 8-3 — The audio-frequency waveform at A is superimposed on an RF signal, and the resulting RF envelope is shown by the outline at B. The RF carrier is 100% modulated — the envelope just barely reaches zero between peaks. The RF envelope shown at C is less than 100% modulated because the maximum amplitude of the audio signal was decreased from the amplitude shown at A. At D, the audio-signal strength was increased, and the RF envelope is modulated more than 100%.

modulating signal to the peak-to-peak voltage of the unmodulated RF envelope. With voice modulation, the percentage is continually varying, because speech contains many variations in volume. If the modulating signal were only half as strong, so its peak-to-peak voltage was 1 instead of 2 as shown in Figure 8-3A, the modulated envelope would look like the one shown in part C. This represents a condition of 50% modulation.

Occasionally the envelope may be held at zero for longer than an instant, as shown in Figure 8-3D. This produces a gap in the RF output that distorts the shape of the modulation envelope so it differs greatly from the modulating signal. This situation is called overmodulation, since the carrier is modulated more than 100%. The distorted signal introduces harmonics and can cause **splatter** — a signal with excessive bandwidth that results in interference to nearby signals. The signal from an overmodulated single-sideband or double-sideband phone transmitter is distorted and occupies more bandwidth than it should.

[Study the following questions in Chapter 13 before you go on to the next section: G8A01, G8A05 and G8A09. Review this section if you have difficulty with any of these questions.]

SINGLE-SIDEBAND, SUPPRESSED-CARRIER SIGNALS

When the carrier signal is modulated, sidebands are generated in proportion to the strength of the modulating signal, as shown in **Figure 8-4A**. The carrier amplitude is not affected by the modulation. Since the carrier is never changed, it contains no information. All the information is contained in the sidebands, which are mirror images of each other. The only purpose of the carrier is to provide a reference signal for demodulation by an envelope detector.

Suppose we remove or suppress the carrier and amplify only the sidebands, as shown in Figure 8-4B. Removing the carrier means we can increase the power in the sidebands by using the power that was required to amplify the carrier in the final stages. The total power dissipation in the transmitter final amplifier remains the same. Taking this even further, since both sidebands contain the same information, we really need only one. If we remove one of the sidebands we can use the power that would have amplified that sideband to increase the power in the remaining sideband. See Figure 8-4C.

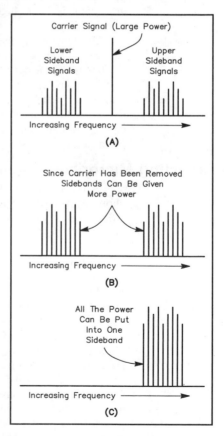

Figure 8-4 — **This drawing shows the frequency spectrum of amplitude-modulated radio signals. Part A shows a double-sideband, full-carrier signal. If the carrier is removed, as shown at B, the sidebands can be given more power without increasing the total transmitter power. Going further, if one sideband is suppressed, as shown at C, the full transmitter power can be concentrated in one sideband. The suppressed parts of the signal are not needed for intelligibility.**

Single-sideband phone is the most popular form of voice modulation in use on the amateur HF bands. FCC rules require that the carrier be reduced to a level at least 40 dB below the **peak envelope power (PEP)** output of the desired sideband in a single-sideband phone transmitter.

As you learned earlier in this chapter, the process of mixing two ac signals creates two new signals. One of the new signals has a frequency equal to the sum of the two input-signal frequencies and the other new signal has a frequency equal to the difference between the two input-signal frequencies. The output signal from a simple mixer will also include some signal energy at the original input signals. A balanced mixer includes some additional circuitry beyond the basic mixer. (A mixer used to modulate an oscillator signal in a transmitter is called a *modulator*.) When a **balanced modulator** is properly adjusted, neither of the original input signals reach the modulator output. In that case, only the sum and difference frequency signals reach the output. These signals are the upper (sum) and lower (difference) sidebands of an AM signal. (This is sometimes called a *double-balanced mixer*, because both input signals are *balanced* out. With a *single-balanced mixer*, one of the input signals is balanced out, but there will still be some energy from the other input signal.) A balanced modulator is used to suppress the carrier in a single-sideband phone transmitter. A band-pass filter placed after the modulator removes the unwanted sideband. In a typical single-sideband phone transmitter a 2.5-kHz band-pass filter is used.

By now, you probably realize that by removing one of the sidebands and the carrier, the range of radio frequencies is reduced quite a bit as compared to a double-sideband, full-carrier transmission. In fact, SSB signals have the narrowest band-width of any of the popular phone emissions. A properly adjusted SSB phone transmitter will have a transmitted-signal bandwidth of about 2.5 to 2.7 kHz. On the Amateur 60-meter band, where hams are only allowed to transmit upper side-band SSB signals on five specific channels, the maximum transmitted bandwidth permitted by FCC Rules is 2.8 kHz.

Adjusting Microphone Gain

Always make sure your transmitter is operating properly and adjust your micro-phone gain correctly. Generally speaking, if your rig has an automatic level control (ALC) meter, you should adjust the microphone gain until you see a slight move-ment of the ALC meter on modulation peaks. (Most transceivers have one meter, with several scales to indicate received signal strength, output power and some-times SWR in addition to ALC. With an ALC indicator light, adjust the microphone gain until the indicator just lights on voice peaks.)

Start with the MIKE GAIN control set to minimum. Speak into the mike and begin increasing the control until the meter moves slightly, but does not exceed the "ALC" zone on the meter scale, even on voice peaks. If you have a monitor oscilloscope, you can observe the effects of adjusting the MIKE GAIN, and see when it is set too high. Proper voice waveforms have rounded peaks. Peaks with flat tops indicate **flattopping** (also called *clipping*), caused by an excessive microphone gain control setting. See **Figure 8-5**. When the audio input is too high, this excessive drive causes a distorted output. Consult the owner's manual for your rig for instructions on proper operation and gain control settings.

Why is proper operation important? If your microphone gain is set too high,

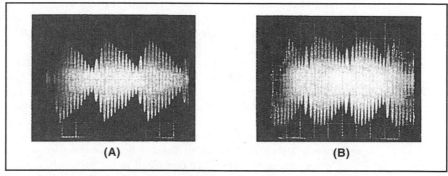

Figure 8-5 — The waveform of a properly adjusted SSB transmitter is shown at A. B shows a severely clipped signal.

your transmitted signal will be distorted. Distortion can cause splatter, resulting in interference to stations on other frequencies. Speak clearly into the microphone, while remaining a constant, close distance away from it. This will reduce background noise, raising the signal-to-background-noise ratio. If you use a speech processor, use the minimum necessary processing level to limit distortion. An over-processed, distorted signal is more difficult to copy than an unprocessed one.

Speech is the primary element of radiotelephone operation. For effective voice communication, your speech should be clearly enunciated, delivered directly into the microphone at a speed suitable for the purpose and conditions at hand. To be efficient, the operator must use the microphone and other speech equipment properly.

Microphones vary in design and purpose, but generally it is best to speak close to the microphone, holding it a few inches away from you. This will improve the signal-to-noise ratio by making your voice louder than any background noises, such as the kids, the TV set or traffic noise when you are in your car.

Pay attention to your mike technique to keep average speech levels high without overdoing it. Automatic level controls (ALC) help, but operator care is also needed to maintain relatively constant speech levels.

[Now turn to Chapter 13 and study questions: G8A06, G8A07, G8A08, G8A10, G8A11 and G8A13. Also study question G9B12. Review this section as needed.]

Mixer Operation

The **mixer** stages in a transmitter and receiver change the frequency of a signal that is being processed. A **mixer** takes two or more input signals, and produces an output that includes the sum and difference of those signal frequencies. One of the input signals is the signal being processed. This could be the audio from a transmitter microphone, an intermediate-frequency (IF) signal being changed to the final transmitter frequency, a received RF signal or an IF signal in a receiver. (The IF stages in transmitters and receivers provide a convenient way to filter and process the signals. All such filtering and processing can be done at a single frequency — or narrow range of frequencies — while the transmitter or receiver can be set to operate on one of several bands.

The second mixer input signal comes from an oscillator in the transmitter or receiver. This oscillator may be called the *local oscillator* (LO), *variable-frequency oscillator* (VFO), *beat-frequency oscillator* (BFO) or several other names. The frequency of this signal is selected to change the processed signal to the desired output frequency.

This process of mixing signals is sometimes called *heterodyning*. A receiver that mixes a local oscillator (LO) signal with a received RF signal to produce a signal at some intermediate frequency (IF) that is higher in frequency than the baseband audio signal is called a *superheterodyne receiver*, or a *superhet*. The IF signal is processed further by filtering and amplifying it, and then converting the signal to baseband audio. The audio signal is further amplified and fed to a speaker, or is connected to headphones so you can hear the received signal.

When the input signals are combined in the mixer they add and subtract. We will have four output signals with two input signals. Two of the output signals are the same as the original input signals. The other two output signals consist of the sum of the two input signals and the difference between the input signals. Usually there will be a filter at the output to select only the desired signal. An example will make this easier to understand.

Suppose you have a transmitter that processes signals at an IF of 5.3 MHz. You want to convert this signal to 14.3 MHz to transmit on the 20-meter band. A 9.0-MHz oscillator signal mixed with the 5.3-MHz IF will produce the desired output.

5.3 MHz + 9.0 MHz = 14.3 MHz

You could also use a 19.6-MHz oscillator to produce the desired output signal.

The process of mixing two ac signals creates two new signals. One of the new signals has a frequency equal to the sum of the two input-signal frequencies and the other new signal has a frequency equal to the difference between the two input-signal frequencies. Suppose you want to receive a 14.25-MHz signal. What frequency will the VFO in your receiver be set to if the receiver uses an IF of 455 kHz? We need a VFO frequency that is either the sum or difference of these two frequencies. (First convert 455 kHz to 0.455 MHz, so both frequencies are expressed in the same units.)

14.25 MHz + 0.455 MHz = 14.705 MHz

The oscillator could also be set to the difference frequency:

14.25 MHz − 0.455 MHz = 13.795 MHz

For any given oscillator signal, there are two possible signal frequencies to produce the same output frequency. For example, you might have a receiver that is designed to use a 13.800-MHz LO signal to mix with a 14.255-MHz received signal, producing a 455-kHz IF. If a 13.345-MHz signal finds its way into the receiver, the same LO signal will again produce a 455-kHz signal, which will be processed by the receiver. You might be listening to the 13.345-MHz signal even though your receiver indicates it is tuned to 14.255 MHz! The 13.345-MHz signal would produce interference to any 14.225-MHz signals you were receiving. This type of interference is known as **image response**, and is a result of the mixing process. If the desired signal has a frequency of LO + IF, the *image frequency* is at LO − IF, and vice versa. See **Figure 8-6**.

Mixer circuits can be designed to prevent one or more of the input signals from appearing at the output. If both input signals appear at the output, along with the

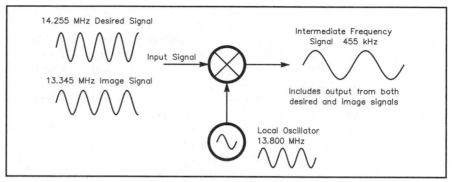

Figure 8-6 — A mixer combines two signals (usually an input signal being processed and an oscillator signal) and produces an output signal that includes the sum and difference of the input signals.

sum and difference signals, the mixer is *unbalanced*. A *single-balanced mixer* suppresses one or the other input signal at the output, and a *double-balanced mixer* suppresses both input signals at the output. Single or double-balanced mixers can simplify the filtering task at the output. They also improve other receiver-performance characteristics.

[Now turn to Chapter 13 and study questions G8B01 through G8B03 and question G8B13. Review this section if these questions give you difficulty.]

FREQUENCY MODULATION (FM)

We can transmit information by modulating any property of a carrier. We have already studied amplitude modulation. We can also modulate the *frequency* or the *phase* of the carrier. **Frequency modulation** and **phase modulation** are closely related, since the phase of a signal cannot be varied without also varying the frequency, and vice versa. Phase modulation and frequency modulation are especially suited for channelized local UHF and VHF communication because they feature good

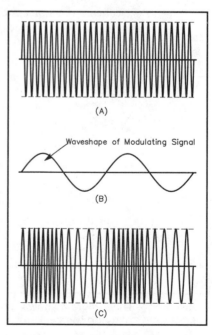

Figure 8-7 — This drawing shows a graphical representation of frequency modulation. In the unmodulated carrier at A, each RF cycle takes the same amount of time to complete. When the modulating signal at B is applied, the carrier frequency is increased or decreased according to the amplitude and polarity of the modulating signal, as shown by Part C.

audio fidelity and high signal-to-noise ratio, as long as the signals are stronger than a minimum, or threshold, level.

In FM systems, when a modulating signal is applied, the carrier frequency is increased during one half of the audio cycle and decreased during the other half of the cycle. The change in the carrier frequency, or **frequency deviation**, is proportional to the instantaneous amplitude of the modulating signal. **Figure 8-7** shows a representation of a frequency-modulated signal. The deviation is slight when the amplitude of the modulating signal is small and greatest when the modulating signal reaches its peak amplitude. The amplitude of the envelope does not change with modulation. This means that an FM receiver is not sensitive to amplitude variations caused by impulse-type noise, and this feature makes it popular for mobile communication.

Practical FM Transmitters

The simplest type of FM transmitter we could design is shown in **Figure 8-8**. Remember that a circuit containing capacitance and inductance will be resonant at some frequency. If we use a resonant circuit in the feedback path of an oscillator, we can control the oscillator frequency by changing the resonant frequency of the tuned circuit. A capacitor microphone is nothing more than a capacitor with one movable plate (the diaphragm). When you speak into the microphone, the diaphragm vibrates and the spacing between the two plates of the capacitor changes. When the spacing changes, the capacitance value changes. If we connect the microphone to the resonant circuit in our oscillator, we can vary the oscillator frequency by speaking into the microphone. If we add frequency multipliers to bring the oscillator frequency up to our operating frequency, and amplifiers to bring up the power, we have a simple FM transmitter.

Another way to shift the frequency of the oscillator would be to use a circuit called a **reactance modulator**. This is a vacuum tube or transistor wired in a circuit so that it changes either the capacitance or inductance of the oscillator reso-

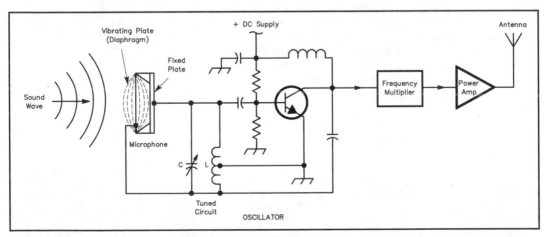

Figure 8-8 — This diagram shows a simple FM transmitter. The frequency of the oscillator is changed by changing the capacitance in the resonant circuit.

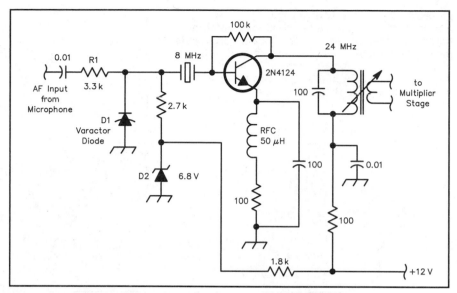

Figure 8-9 — A varactor diode can also be used to frequency-modulate an oscillator. The capacitance of the varactor diode changes when the bias voltage is varied.

nant circuit in response to an audio input signal.

Modern FM transmitters may use a **varactor diode** to modulate the oscillator. A varactor diode is a special diode that changes capacitance when its bias voltage is changed. Connected to a crystal oscillator as shown in **Figure 8-9**, the varactor diode will "pull" the oscillator frequency slightly when the audio input changes.

Phase Modulation (PM)

One problem with early direct-modulated FM transmitters was frequency stability. Using a capacitive microphone or a reactance modulator means that a crystal oscillator cannot be used. Frequency multiplication also multiplies any drift or other instability problems in the oscillator. With phase modulation, however, the modulation takes place after the oscillator, so a crystal oscillator can be used. For this reason phase modulation is used in most FM transmitters built for mobile applications, where stability may be a serious problem because of shocks and temperature variations. Phase modulation produces what is called indirect-modulated FM. The most common method of generating a phase-modulated telephony signal is to use a reactance modulator connected to the RF power amplifier.

The term **phase** essentially means "time," or the time interval between the instant when one thing occurs and the instant when a second related thing takes place. The later event is said to *lag* the earlier, while the one that occurs first is said to *lead*. When two waves of the same frequency start their cycles at slightly different times, the time difference (or phase difference) is measured in degrees. This is shown in **Figure 8-10**. In phase modulation, we shift the phase of the output wave

Figure 8-10 — When two waves of the same frequency start their cycles at slightly different times, the time difference (or phase difference) is measured in degrees. In this diagram wave B starts 45 degrees (one-eighth cycle) later than A, and so lags wave A by 45°. (We can also say that wave A leads wave B by 45°.)

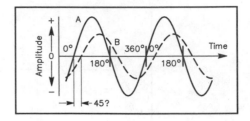

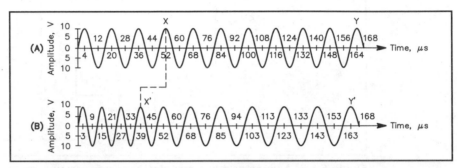

Figure 8-11 — This diagram shows a graphical representation of phase modulation. The unmodulated wave is shown at A. After modulation, cycle X lags cycle X', and all the cycles to the left of cycle X' are compressed. To the right, the cycles are spread out.

in response to the audio-input-signal amplitude.

The output from an RF oscillator is an unmodulated wave, as shown in **Figure 8-11A**. The phase-modulated wave is shown in Figure 8-11B. At time 0 and again 168 microseconds later the two waves are in phase. The phase modulation has shifted cycle X' so that it reaches its peak value at time 39 microseconds as compared with 52 microseconds for the unmodulated wave. (Cycle X' leads cycle X.)

The number of cycles in both waveforms is the same, so the center frequency of the wave has not been changed. All the cycles to the left of cycle X' have been compressed, and all the cycles to the right of cycle X' have been spread out. The frequency of the wave to the left is greater than the center frequency, and the frequency to the right is less — the wave has been *frequency modulated*.

Frequency Deviation

With direct FM, frequency deviation is proportional to the amplitude of the audio modulating signal. With phase modulation, the frequency deviation is proportional to both the amplitude and the frequency of the modulating signal. Frequency deviation is greater for higher audio frequencies. This is actually a benefit, rather than a drawback, of phase modulation. In radiocommunication applications, speech frequencies between 300 and 3000 Hz need to be reproduced for good intelligibility. In the human voice, however, the natural amplitude of speech sounds between 2000 and 3000 Hz is low. Something must be done to increase the amplitude of these frequencies when a direct FM transmitter is used. A circuit called a *preemphasis network* amplifies the sounds between 2000 and 3000 Hz. With phase modulation, the preemphasis network is not required, because the deviation already increases with increasing audio input frequency.

Modulation Index

The sidebands that occur from FM or PM differ from those resulting from AM. With AM, only a single set of sidebands is produced for each modulating frequency. FM and PM sidebands occur at integral *multiples* of the modulating frequency on either side of the carrier, as shown in **Figure 8-12**. Because of these multiple sidebands, FM or PM signals inherently occupy a greater bandwidth. The additional sidebands depend on the relationship between the modulating frequency and the frequency deviation. The ratio between the peak carrier frequency deviation and the audio modulating frequency is called the **modulation index**.

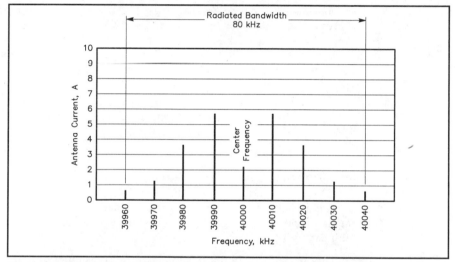

Figure 8-12 — When an RF-carrier is frequency-modulated by an audio signal, multiple sidebands are produced. This diagram shows a 40-MHz carrier modulated by a 10-kHz sinusoidal audio signal. Additional sidebands exist above 40.04 MHz and below 39.96 MHz, but their amplitude is too low to be significant.

Given a constant input level to the modulator, in phase modulation, the modulation index is constant regardless of the modulating frequency. For an FM signal, the modulation index varies with the modulating frequency. In an FM system, the ratio of the maximum carrier-frequency deviation to the highest modulating frequency used is called the **deviation ratio**. Whereas modulation index is a variable that depends on a set of operating conditions, deviation ratio is a constant. The deviation ratio for narrow-band FM is 5000 Hz (maximum deviation) divided by 3000 Hz (maximum modulation frequency) or 1.67.

[Now turn to Chapter 13 and study questions G8A02, G8A03, G8A04 and G8A12. Review this section as needed.]

Frequency Multiplication and Amplification

Since there is no change in the amplitude with modulation, an FM or PM signal can be amplified in a nonlinear amplifier without producing distortion. This has the advantage that modulation can be introduced in a low-level stage and the signal can then be amplified efficiently.

It is not usually practical to have an oscillator operating in the VHF range or higher, because such an oscillator would not be as stable as one operating at lower frequencies. Common practice is to use an oscillator operating at some submultiple of the desired operating frequency, and then to use *frequency multiplication* to obtain the proper output frequency. A frequency multiplier is an amplifier that produces harmonics of the input signal. A filter selects the desired harmonic for the output frequency.

When the modulation is applied to the oscillator in such a system, care must be taken to set the proper deviation ratio. When the modulated signal is multiplied to the output frequency, the deviation ratio will also be multiplied by the same factor.

For example, suppose we are using a 12.21-MHz reactance-modulated oscillator in our FM transmitter operating at 146.52 MHz. We want the maximum frequency deviation to be 5 kHz for the transmitted signal. What is the maximum deviation of the oscillator frequency? First we need to know how much multiplication the transmitter is using. We find this by dividing the output frequency by the oscillator frequency:

$$\text{Multiplication Factor} = \frac{\text{Transmitter Output Frequency}}{\text{Oscillator Frequency}} \qquad \text{(Equation 8-1)}$$

$$\text{Multiplication Factor} = \frac{146.52\,\text{MHz}}{12.21\,\text{MHz}} = 12$$

This means we are multiplying the oscillator frequency by a factor of 12 to bring it up to the operating frequency. So the deviation in the oscillator will also be multiplied by 12.

$$\text{Oscillator deviation} \times \text{Multiplication factor} = \text{Transmitter deviation}$$

$$\text{(Equation 8-2)}$$

$$\text{Oscillator deviation} \times 12 = 5\ \text{kHz}$$

$$\text{Oscillator Deviation} = \frac{5\ \text{kHz}}{12} = 0.4167\ \text{kHz} = 416.7\ \text{Hz}$$

Frequency multiplication offers a means for obtaining any desired amount of frequency deviation, whether or not the modulator is capable of that much deviation. Overdeviation of an FM transmitter causes **splatter** — out-of-channel emissions that can cause interference to adjacent frequencies.

A frequency-multiplier stage is an amplifier that produces harmonics of the input signal. By using a filter to select the desired harmonic, the proper output frequency can be obtained. A frequency multiplier is not the same as a mixer stage, which is used to obtain the desired output frequency in an SSB or CW rig. A mixer simply mixes two input signals to provide an output at some new frequency.

Bandwidth in FM and PM

With frequency deviation defined as the instantaneous change in frequency for a given signal, we can now define the total bandwidth of the signal. Since the frequency actually swings just as far in both directions, the total frequency swing is equal to twice the deviation. In addition, there are sidebands that increase the bandwidth still further. A good estimate of the bandwidth is twice the maximum frequency deviation plus the maximum modulating audio frequency:

$$Bw = 2 \times (D + M)$$ (Equation 8-3)

where:

Bw = bandwidth
D = maximum frequency deviation
M = maximum modulating audio frequency

With a transmitter using 5-kHz deviation and a maximum audio frequency of 3 kHz, the total bandwidth is approximately 16 kHz.

$$Bw = 2 \times (5 \text{ kHz} + 3 \text{ kHz}) = 2 \times 8 \text{ kHz}$$
$$Bw = 16 \text{ kHz}$$

The actual bandwidth is somewhat greater than this, but it is a good approximation.

Frequency modulated phone is not used below 29.5 MHz because the bandwidth of such signals would exceed FCC limits. Between 29.5 and 29.7 MHz on the 10-meter band, and on the VHF bands and higher frequencies, the allowable bandwidth is greater, so FM (and PM) phone are permitted.

[Now study these questions in Chapter 13: G8B04 through G8B07. Review this section as needed.]

DIGITAL COMMUNICATIONS

Digital communications refers to the type of coding used to transmit information. Any signal that has separate, distinct conditions, and switches between those conditions to form the pattern of a code, is a digital communications method. Usually there are only two different conditions used to form the code. The on-off keying of a Morse code signal is one simple example of digital communications. Morse code is designed to be copied by ear, but other digital codes are designed to be copied by machine. Baudot radioteletype (RTTY), AMTOR, PACTOR, packet radio and PSK31 are examples of codes designed to be copied by machine. They are all *digital communications*.

Instead of switching a carrier on and off to send information as in CW, the carrier can be left on continuously and switched between two different frequencies. This is called **frequency-shift keying (FSK)**. This is the type of keying used to produce digital communications signals below 50 MHz. FSK has the advantage over on-off keying that it gives definite signals for both the on and off states used for digital communication. These *on* and *off* states are referred to as "mark" and "space," respectively.

Many amateurs use the term RTTY or radioteletype to refer only to communications using the Baudot code. Radioteletype can refer to any system that sends a signal designed to be copied by machine and printed directly at the receiving station. Originally it referred to mechanical teletypewriters that automatically typed the message directly on paper at the receiving station. Modern equipment makes use of computers to create the signals that modulate the transmitter, and the receiving station displays the received message on a computer monitor. The computer can also save the text as a file, which can be printed at any time.

FSK can be produced in two ways. The first method is to change the oscillator frequency back and forth between two frequencies so the carrier also shifts between two frequencies. The second method makes use of the fact that there is no output from an SSB transmitter when there is no modulation. Only when an audio tone is applied to the microphone input is there a corresponding RF signal emitted from the transmitter. By switching between two fixed audio tones, the RF signal is shifted between the mark and space frequencies. The difference in the two RF signals is equal to the difference in frequency of the audio tones.

When using FSK emissions below 50 MHz, no more than 1000 Hz is permitted between the mark and space signals. Above 50 MHz, the frequency shift, in hertz, must not exceed the sending speed, in bauds, of the transmission, or 1000 Hz, whichever is greater. As the sending speed of a digital communications transmission is increased, the frequency shift between the mark and space tones must also be increased.

The electronic device that goes between the RTTY terminal (keyboard and display unit) and the transmitter is called a *modem*, which is short for *mo*dulator/*dem*odulator. As the sending speed is increased, it becomes more difficult for the modem to distinguish between the two frequencies used to represent mark and space. Increasing the shift between mark and space helps to solve this problem. This also increases the bandwidth of the signal, so the FCC bandwidth restrictions impose practical limits on the keying speed of an RTTY signal. At higher frequencies, where wider bandwidths are permitted, higher keying speeds can be used. Below 28 MHz, the sending speed may not exceed 300 bauds. Between 28 and 50 MHz you may use speeds up to 1200 bauds, between 50 and 220 MHz the speed limit is 19,600 bauds (19.6 kilobauds) and above 220 MHz you may use speeds up to 56 kilobauds. A baud is the unit used to describe transmission speeds for digital signals. For a single-channel transmission, a baud is equivalent to one digital bit of information transmitted per second. So a 300-baud signaling rate represents a transmission rate of 300 bits of digital information per second in a single-channel transmission.

Another way to transmit digital communication is to feed audio tones into the microphone input of an FM transmitter. These emissions are called **audio-frequency shift keying (AFSK)** emissions.

The audio frequencies used for AFSK have been standardized in the US at 2125 Hz and 2295 Hz for 170-Hz shift and 2125 Hz and 2975 Hz for 850-Hz shift. Most amateur operation uses 170-Hz shift. These tones were selected because they are close to the upper limit of the audio passband of most transmitters. This is important because harmonics produced in the tone generator will fall outside the audio passband and, therefore, will be attenuated in the transmitter. When lower-frequency audio tones are used, care must be taken to ensure that harmonics are eliminated from the signal produced by the tone generator. If the audio input to the transmitter contains harmonics, the RF output from the transmitter may contain signals at frequencies other than the desired fundamental. These extra signals can cause interference to other amateur stations, as well as producing RFI and TVI to consumer electronics devices.

It is important to consider the duty cycle of your transmitter when transmitting digital communications. For Morse code, your transmitter is producing the full output power every time you close the key. Each dot or dash is short, however, so your transmitter is actually on only a short time, and then turned off for a short time while you are "transmitting."

When transmitting Baudot RTTY and most other digital communications, your transmitter is producing full output power the entire time, because it just switches between the two transmitted frequencies. This is called 100% duty cycle. If your transmitter is not rated for 100% duty cycle it may overheat and be destroyed. It is common practice to reduce the transmitter power to about 50% of maximum for digital operation, to protect the transmitter.

Mode A AMTOR transmission has a duty cycle of about 50%, because your transmitter sends three characters, then turns off to await an acknowledgment. Your transmitter has a 100% duty cycle to transmit Mode B AMTOR, however. Packet radio has a duty cycle less than 100% because it stops transmitting after each "packet" to await an acknowledgment. PACTOR also has a duty cycle of less than 100% because the station transmits a block of data and then waits for an acknowledgment.

In order to minimize interference between various modes, concerned amateurs follow an agreement referred to as a band plan. The band plan specifies the location of various emissions on the band. In general, most amateurs operate digital modes in a band segment near the high-frequency end of the range designated by the FCC for CW, RTTY and Data operation. (CW is allowed on the entire band, so the actual FCC designation is for RTTY and Data operation.) Most hams refer to the low-frequency end of each band as a CW band, with the digital modes in what is commonly known as the RTTY segment. On the 20-meter band, most of the PSK31 activity will be near the low-frequency end of the RTTY segment, near 14.070 MHz. Tune your radio to 14.07015 MHz and use the "waterfall display" of your PSK31 computer program to select signals across the audio passband of your receiver.

Most AMTOR and PACTOR operation takes place just above the low-frequency end ("bottom") of the RTTY segment on each band, above the PSK31 frequencies. On 20 meters, most AMTOR and PACTOR activity is found near 14.075 MHz. HF packet radio operation is usually found at the high-frequency end ("top") of the RTTY segment on each band. There are many different digital modes that you can use in these band segments, so the goal is to find the operators using modes similar to the one you want to use.

[This completes the material for this chapter. You should now turn to Chapter 13 and study questions G8B08 through G8B11. Review any material in the chapter that you had difficulty with.]

**CHAPTER 9
KEYWORDS
KEYWORDS
KEYWORDS**

Balanced line — Feed line with two conductors having equal but opposite voltages, with neither conductor at ground potential.

Balun — A transformer used between a *bal*anced and an *un*balanced system, such as for feeding a balanced antenna with an unbalanced feed line.

Coaxial cable — Feed line with a central conductor surrounded by plastic, foam or gaseous insulation, which in turn is covered by a shielding conductor and the entire cable is covered with vinyl insulation.

Cubical quad antenna — A full-wavelength loop antenna built with its elements in the shape of squares.

Delta loop antenna — A full-wavelength loop antenna built with its elements in the shape of equilateral triangles.

Director — A parasitic element in "front" of the driven element in a multielement antenna.

Driven element — The element connected directly to the feed line in a multielement antenna.

Feed line — The wire or cable used to connect an antenna to the transmitter and receiver. (Also called *transmission line*.)

Front-to-back ratio — The energy radiated from the front of a directive antenna divided by the energy radiated from the back of the antenna.

Gain — An increase in the effective power radiated by an antenna in a certain desired direction, or an increase in received signal strength from a certain direction. This is at the expense of power radiated in, or signal strength received from, other directions.

Ground-plane antenna — A vertical antenna built with a central radiating element one-quarter-wavelength long and several radials extending horizontally from the base. The radials are slightly longer than one-quarter wave, and may droop toward the ground.

Half-wavelength dipole antenna — A fundamental antenna one-half wavelength long at the desired operating frequency, and connected to the feed line at the center. This is a popular amateur antenna.

Horizontally polarized wave — An electromagnetic wave with its electric lines of force parallel to the ground.

Main lobe — The direction of maximum radiated field strength from an antenna. (Also called *major lobe*.)

Parallel-conductor feed line — Feed line constructed of two wires held a constant distance apart; either encased in plastic or constructed with insulating spacers placed at intervals along the line.

Parasitic element — Part of a directive antenna that derives energy from mutual coupling with the driven element. Parasitic elements are not connected directly to the feed line.

Polarization — The orientation of the electric lines of force in a radio wave, with respect to the surface of the Earth.

Quarter-wavelength vertical antenna — An antenna constructed of a quarter-wavelength-long radiating element placed perpendicular to the Earth. (See **Ground-plane antenna**.)

Random-length wire antenna — A multiband antenna consisting of any convenient length of wire, connected directly to a transmitter or impedance-matching network without the use of feed line.

Reflector — A parasitic element placed "behind" the driven element in a directive antenna.

Standing-wave ratio (SWR) — The ratio of maximum voltage to minimum voltage along a feed line. Also the ratio of antenna impedance to feed line impedance when the antenna is a purely resistive load.

Unbalanced line — Feed line with one conductor at ground potential, such as coaxial cable.

Vertically polarized wave — A radio wave that has its electric lines of force perpendicular to the surface of the Earth.

Yagi antenna — A directive antenna made with a half-wavelength driven element, and two or more parasitic elements arranged in the same horizontal plane.

Antennas and Feed Lines

An *antenna system* includes the actual antenna (the part that actually radiates the signal), the feed line, and any coupling devices or matching networks used for transferring power from the transmitter to the line and from the line to the antenna. In some simple systems the feed line may be part of the antenna, and coupling devices may not be used. An applicant for the General license should know the basic design and operation of commonly used Amateur Radio antennas, and the principles of feed lines and standing wave ratio (SWR). There are four questions from this material on the General exam, and these questions come from the four syllabus groups:

G9A   Yagi antennas — physical dimensions; impedance matching; radiation patterns; directivity and major lobes

G9B   Loop antennas — physical dimensions; impedance matching; radiation patterns; directivity and major lobes

G9C   Random wire antennas — physical dimensions; impedance matching; radiation patterns; directivity and major lobes; feed-point impedance of $1/2$-wavelength dipole and $1/4$-wavelength vertical antennas

G9D   Popular antenna feed lines — characteristic impedance and impedance matching; SWR calculations

WAVE CHARACTERISTICS AND POLARIZATION

An electromagnetic wave consists of moving electric and magnetic fields. Remember that a field is an invisible force of nature. We can't see radio waves — the best we can do is to show a representation of where the energy is in the electric and magnetic fields. We can visualize a traveling radio wave as looking something like **Figure 9-1**. The lines of electric and magnetic force are at right angles to each other, and are also perpendicular to the direction of travel. These fields can have any position with respect to the Earth.

Polarization is defined by the direction of the electric lines of force in a radio wave. If the electric lines of force are parallel to the surface of the Earth, the wave is called a **horizontally polarized wave**. Similarly, if the electric lines of force are perpendicular to the Earth, the wave is called a **vertically polarized wave**.

The polarization of a radio wave as it leaves the antenna is determined by the orientation of the antenna. For example, a half-wavelength dipole parallel to the surface of the Earth transmits a wave that is horizontally polarized. A whip antenna, mounted vertically on an automobile, transmits a wave that is vertically polarized.

After radio waves have traveled some distance from the transmitting antenna and have interacted with their surroundings and with the ionosphere, the polarization may change. Longer wavelengths will typically retain their polarization, while shorter wavelengths will often change polarization as they interact with the ionosphere. These changes can occur quite rapidly. Most man-made noise tends to be vertically polarized; thus, a horizontally polarized antenna will receive less noise of this type than will a vertical antenna.

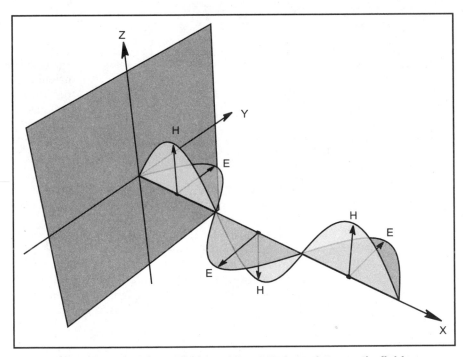

Figure 9-1 — This diagram illustrates the electric and magnetic field components of an electromagnetic wave. The polarization of a radio wave is the same as the plane of the electric field. In this example the electrical lines of force are parallel to the ground. This is a horizontally polarized electromagnetic wave.

RANDOM WIRE ANTENNA

One of the simplest antenna types is the **random-length wire antenna**, also known as the *random-wire antenna*. As the name implies, the length of this antenna is not related to the wavelength of the desired signals. The random-wire antenna is a multiband antenna. The antenna impedance will vary from band to band, and so requires the use of some type of impedance-matching network.

A random-wire antenna does not require a feed line. The antenna is a single wire, which connects directly to the impedance-matching network output. Because the antenna has no feed line, it comes right into the radio shack. This causes one of the greatest disadvantages of the random-wire antenna. You may experience RF feedback in your station if you use this type of antenna. RF feedback may result in radio-frequency interference of various types. Your radio or other station equipment may cause RF burns when you touch the equipment.

[You should turn to Chapter 13 now, and study questions G9C01, G9C02 and G9C03. Review this section if you have difficulty with any of these questions.]

VERTICAL ANTENNAS

Another popular antenna is the **quarter-wavelength vertical antenna**. It is often used to obtain low-angle radiation when a beam or dipole antenna cannot be placed far enough above ground. Low-angle radiation refers to signals that travel closer to the horizon, rather than signals that are high above the horizon as they leave the antenna. Low-angle radiation is usually advantageous when you are attempting to contact distant stations.

You might also want to use a vertical antenna if you don't have two supports separated by an appropriate distance to hold the ends of a dipole antenna. As shown in **Figure 9-2**, the quarter-wavelength vertical antenna is half of a half-wavelength antenna, mounted vertically. Conductivity in the ground produces an "image" of the missing quarter wavelength. An excellent ground connection is necessary for the most effective operation of a vertical antenna.

In most installations, ground conductivity is inadequate, so an artificial ground must be made from wires placed along the ground near the base of the antenna, as shown in Figure 9-2C. These wires, called *radials*, are usually one-quarter wavelength long or longer. Depending on ground conductivity, 8, 16, 32 or more radials may be required to form an effective ground. The radial wires of a ground-mounted vertical antenna should be placed on the ground surface or buried a few inches below the ground.

For best results, the vertical radiator should be located away from nearby conductive objects. Nearby conductive objects may distort, or *skew*, the antenna radiation pattern. Such objects may also cause the antenna feed-point impedance to change. You should ensure that no people can come near your antenna while you are transmitting because they may also skew the radiation pattern and cause the feed-point impedance to vary. Of course you should always take steps to keep people away from your antennas to avoid exposure to RF radiation. Radiation exposure is the most important reason to keep people away from your transmitting antenna.

Vertical antennas can be combined to form arrays in which each element is fed

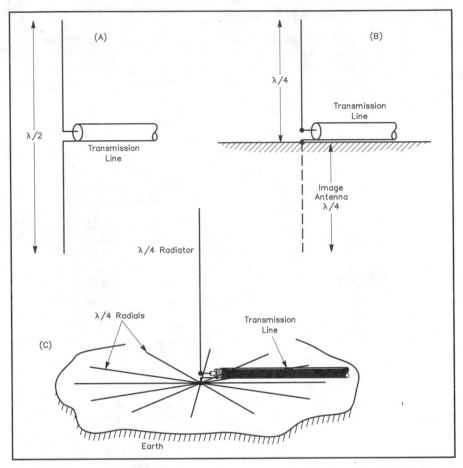

Figure 9-2 — A half-wavelength vertical antenna is shown at A. At B is its grounded quarter-wavelength counterpart. In ground with good conductivity, we consider the missing quarter wavelength to be supplied by the image. A practical quarter-wave vertical antenna installation will require the use of several ground radials, as shown at C.

with a signal from the transmitter. By controlling the phase relationships of the signals fed to each element, you can obtain a variety of radiation patterns.

The radiation resistance of a quarter-wavelength antenna over a perfect ground is about 36 Ω. The feed-point impedance of a real quarter-wavelength antenna may vary between 30 and 100 Ω, depending on ground resistance.

The Ground-Plane Antenna

The **ground-plane antenna** is a vertical antenna that uses an artificial metallic ground consisting of metal rods or wires either perpendicular to, or sloping from, the antenna base and extending outward, as shown in **Figure 9-3**. The entire antenna,

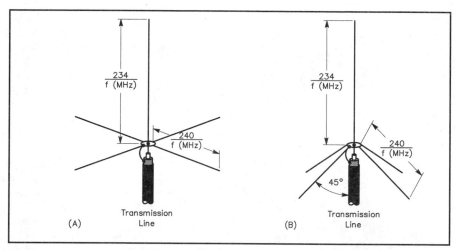

Figure 9-3 — This drawing illustrates a ground-plane antenna. Dropping the radials (as shown at B) raises the feed-point impedance closer to 50 Ω.

including the radials, is mounted above the ground. This permits the antenna to be mounted at a height clear of surrounding obstructions such as trees and buildings. The ground plane also requires only 3 or 4 radials to operate properly, as opposed to the large number required for some ground-mounted installations.

With the radials perpendicular to the antenna, the feed-point impedance is around 35 Ω. Sloping the radials downward raises the feed-point impedance of the antenna somewhat. If the radials are angled down at about a 45° angle, the feed-point impedance is around 50 Ω, which makes it easier to match the antenna to available coaxial-cable feed lines. Unlike other quarter-wavelength vertical antennas without an artificial ground, the ground-plane antenna will give low-angle radiation regardless of the height above actual ground. The radials should be slightly longer than the radiating element.

[Turn to Chapter 13 and study questions G9C04, G9C05 and G9C11 now. Review the material in this section if you have difficulty with any of these questions.]

THE HALF-WAVELENGTH DIPOLE ANTENNA

The radiation from an ideal dipole antenna in free space is not uniform in all directions. It is strongest in directions perpendicular to the wire, and nearly zero along the axis of the wire, as shown in **Figure 9-4**. The plane radiation pattern for an ideal **half-wavelength dipole antenna** is a figure-eight at right angles to the antenna.

When placed over real ground, this ideal pattern is modified. The antenna height above ground, the conductivity of the ground and other nearby objects all affect the pattern. A half-wavelength dipole exhibits excellent directivity when

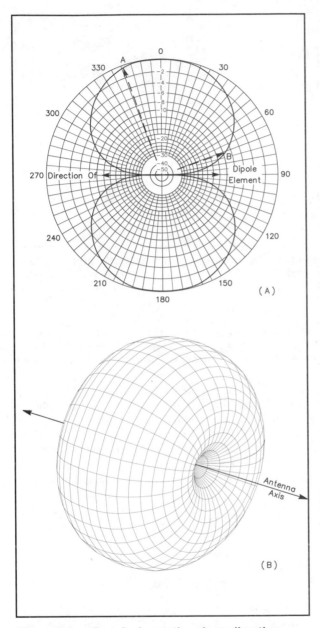

Figure 9-4 — Part A shows the plane directive diagram of a dipole. The solid line shows the direction of the wire. Note that the pattern is bidirectional; the ideal dipole radiates equally well in two directions. Part B shows the solid (three-dimensional) directive pattern of an ideal dipole in free space.

placed at least one-half wavelength above the ground, as shown by the pattern in **Figure 9-5A**. This is approximately a figure-eight pattern, with the maximum radiation at right angles to the antenna. A dipole cut for the 40-meter band must be mounted at least 20 meters (about 65 feet) above ground to exhibit such directivity. If the antenna is lower, the directivity effects are reduced. **Figure 9-5B** shows the radiation pattern for a dipole mounted only one-quarter wavelength above the ground. Notice that the pattern is more nearly circular. This pattern is nearly omnidirectional, meaning there is no directional advantage. Waves reflected from the ground significantly distort the ideal pattern.

Despite this apparent disadvantage, the dipole is the most common and practical antenna for the 80 and 40-meter bands. The impedance at the feed point of a half-wavelength dipole suspended horizontally at least one-quarter wavelength above ground is close to 70 Ω, so a dipole can be fed directly with 75-Ω coaxial cable. The feed-point impedance will be less than 70 Ω if the antenna is mounted less than one-quarter wavelength above the ground, as shown in **Figure 9-6**. Coaxial cable with a characteristic impedance of 50 Ω is often used as the feed line to dipole antennas.

[Now turn to Chapter 13 and study questions G9C06 and G9C07. Review this section as needed.]

MULTIPLE-ELEMENT ARRAYS

The radiation pattern of the ideal half-wavelength dipole antenna is described as bidirectional — it radiates equally well in two directions. A unidirectional pattern (maximum radiation in one direction) can be produced by coupling the half-wavelength antenna to additional elements. The

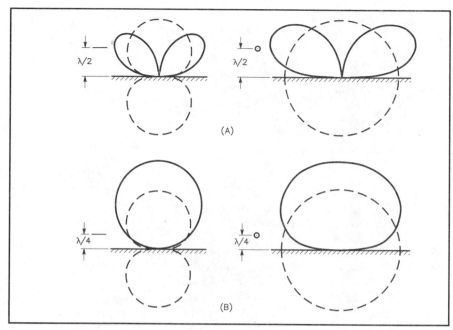

Figure 9-5 — The effect of ground on the radiation from a horizontal half-wavelength antenna, for heights of one-half (A) and one-quarter (B) wavelength is shown. The dashed lines show what the pattern would be if there were no reflection from the ground.

additional elements may be made of wire or metal tubing.

Parasitic Excitation

In most multiple-element antennas, the additional elements are not directly connected to the feed line. They receive power by mutual coupling from the **driven element** (the element connected to the feed line). They then reradiate it in the proper phase relationship to achieve gain or directivity over a simple half-wavelength dipole. These elements are called **parasitic elements**.

There are two types of parasitic elements. A **director** is generally shorter than the driven element and is located at the

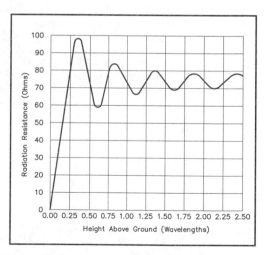

Figure 9-6 — The radiation resistance of a horizontal half-wavelength antenna varies depending on how high above ground the antenna is mounted.

"front" of the antenna. A **reflector** is generally longer than the driven element and is located at the "back" of the antenna. See **Figure 9-7**. The direction of maximum radiation from a parasitic antenna travels from the reflector through the driven element and the director. The term **main lobe** refers to the region of maximum radiation from a directional antenna. The main lobe is also sometimes referred to as the *major lobe*. If communication in different directions is desired, provision must be made to rotate the array in the *azimuth*, or horizontal plane. The minimum practical spacing between the elements of such an antenna is 0.1 wavelength.

A parasitic element that is slightly shorter than the driven element, placed 0.1 λ from a dipole antenna will produce a horizontal radiation pattern with a main lobe *toward* the parasitic element. A parasitic element that is slightly longer than the driven element, placed 0.1 λ from a dipole antenna will produce a horizontal radiation pattern with a main lobe *away* from the parasitic element.

[Turn to Chapter 13 and study questions G9C09 and G9C10 before going on to the next section. Review this text as needed.]

Yagi Arrays

An antenna that uses a half-wavelength dipole antenna and at least one parasitic element parallel to it is called a *Yagi-Uda antenna*, named for its developers. This name is normally shortened to **Yagi antenna**. The Yagi arrays in **Figure 9-8** are examples of antennas that make use of parasitic elements to produce a unidirectional radiation pattern. Though typical HF antennas of this

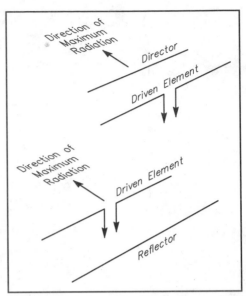

Figure 9-7 — In a directive antenna, the reflector element is placed "behind" the driven element. The director goes in "front" of the driven element.

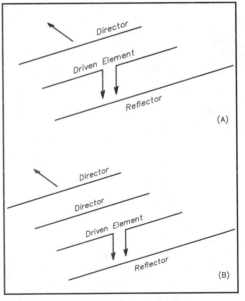

Figure 9-8 — Part A shows a three-element Yagi. Part B shows a four-element Yagi with two directors.

type have three elements, some may have six or more. Multiband Yagi antennas may have many elements. Some designs use several elements on one frequency range and other elements that are used for different frequencies. Other multiband designs use traps in the elements, so one set of elements will work on several bands.

The length of the driven element in the most common type of Yagi antenna is approximately one-half wavelength. This means that Yagi antennas are most often used for the 20-meter band and higher frequencies. For frequencies below 20 meters, the physical dimensions become so large that the antennas require special construction techniques and heavy-duty supporting towers and rotators. Yagis for 30 and 40 meters are fairly common, but 80-meter rotatable Yagis are few and far between. To overcome the mechanical difficulties, some amateurs build nonrotatable Yagi antennas for 40 and 80 meters, with elements made of wire supported on both ends.

In a three-element beam, the director will be approximately 5% (0.05) shorter than one-half wavelength, and the reflector will be approximately 5% longer than one-half wavelength. We can derive a set of equations to calculate these element lengths.

The driven element length is about the same as the length of a half-wavelength dipole antenna. The length is a bit shorter than a "free-space" half wavelength because of "end effects," the element diameter, and other factors. To calculate the length of a half-wavelength dipole antenna (or the driven element of a Yagi antenna) use:

$$(\text{Yagi})L_{\text{Driven}}(\text{in feet}) = \frac{468}{f(\text{in MHz})} \qquad \text{(Equation 9-1)}$$

To find the length of a director element for a Yagi antenna, we shorten this length by about 5%:

$$(\text{Yagi}) \ L_{\text{Director}} = L_{\text{Driven}} \ (1 - 0.05)$$

$$(\text{Yagi}) \ L_{\text{Director}} = L_{\text{Driven}} \times 0.95 \qquad \text{(Equation 9-2)}$$

To find the length of a reflector element for a Yagi antenna, we increase this length by about 5%:

$$(\text{Yagi}) \ L_{\text{Reflector}} = L_{\text{Driven}} \ (1 + 0.05)$$

$$(\text{Yagi}) \ L_{\text{Reflector}} = L_{\text{Driven}} \times 1.05 \qquad \text{(Equation 9-3)}$$

Suppose you want to find the approximate length of the driven element of a Yagi antenna for 14.0 MHz. Use Equation 9-1:

$$(\text{Yagi})L_{\text{Driven}}(\text{in feet}) = \frac{468}{f(\text{in MHz})}$$

$$(\text{Yagi})L_{\text{Driven}}(\text{in feet}) = \frac{468}{14.0 \text{ MHz}} = 33.4 \text{ ft}$$

You can round this value off to 33 ft as an approximate answer. Any values you calculate using the equations for antenna lengths will be subject to some adjustment when you build the antenna. Conditions like height, nearby objects and element diameter can all affect the optimum length.

Suppose you want to find the approximate length for the director element of a Yagi antenna for 21.1 MHz. First, use Equation 9-1 to calculate the driven-element length:

$$(\text{Yagi})L_{\text{Driven}}(\text{in feet}) = \frac{468}{f(\text{in MHz})} = \frac{468}{28.1\,\text{MHz}} = 16.7\,\text{ft}$$

Next, use Equation 9-2 to calculate the approximate director length:

(Yagi) $L_{\text{Reflector}} = L_{\text{Driven}} \times 1.05$

(Yagi) $L_{\text{Reflector}} = 16.7 \times 1.05 = 17.5$ ft   (Equation 9-3)

This is approximately 21 ft.

Now suppose you want to find the approximate length for the reflector element of a Yagi antenna for 28.1 MHz. First, use Equation 9-1 to calculate the driven-element length:

$$(\text{Yagi})L_{\text{Driven}}(\text{in feet}) = \frac{468}{f(\text{in MHz})} = \frac{468}{28.1\,\text{MHz}} = 16.7$$

Next, use Equation 9-3 to calculate the approximate reflector length:

(Yagi) $L_{\text{Reflector}} = L_{\text{Driven}} \times 1.05$

(Yagi) $L_{\text{Reflector}} = 16.7$ ft $\times 1.05 = 17.5$ ft

There can be considerable variation on these lengths, however. The actual lengths depend on factors, such as the spacing between elements, the diameter of the elements, and whether the elements are made from tapered or cylindrical tubing.

The director element is usually the shortest parasitic element on a three-element Yagi antenna. The reflector element is usually the longest parasitic element on a three-element Yagi antenna.

[Now turn to Chapter 13 and study examination questions G9A02 through G9A05. Review the material in this section as needed.]

Front-to-Back Ratio

A typical Yagi-antenna radiation pattern is shown in **Figure 9-9**. This pattern indicates that the antenna will reject signals coming from the sides and back, selecting mainly those signals from a desired direction. This helps explain why Yagi antennas are so popular for use on crowded HF bands like the 20-meter band. The antenna helps reduce interference from other stations in directions off to the side of the desired station, or from behind the receiving station. Transmitted signals are concentrated in the desired direction, reducing potential interference to stations in directions other than the desired one. The **main lobe** of a Yagi antenna radiation pattern is the direction of maximum radiated field strength from the antenna. If the antenna has higher **gain**, more of the transmitted signal energy is in the direction of the main lobe. One way to increase the gain of a Yagi antenna is by increasing the boom length and adding more directors to the antenna. In general, directors added in front of the driven element are shorter as you move away from the driven element. In other words, a second director added to the antenna will be shorter and farther from the driven element. Additional reflectors are not generally added to a Yagi antenna.

Front-to-back ratio is the ratio between the power radiated in the major radiation lobe (the direction of maximum radiation — the "front" of the antenna) compared to the power radiated in exactly the opposite direction (the "back" of the antenna). (Sometimes this is also called the *front-to-rear ratio* and is taken to mean the ratio of power radiated in the major lobe compared to power radiated generally behind the antenna rather than just 180° from the maximum radiation lobe.) If you point your antenna directly at a receiving station and transmit, and then turn the antenna 180° and transmit again, the difference in the received signal strength is the front-to-back ratio, which is usually expressed in decibels. An increase in the front-to-back ratio is brought about by adjusting the length of the parasitic elements or their position relative to the driven element and each other. The tuning condition that gives maximum attenuation to the rear is considerably more critical than the condition for maximum forward gain.

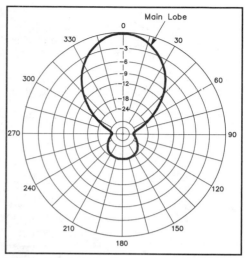

Figure 9-9 — This diagram represents the directive pattern for a typical, three-element Yagi. Note that this pattern is essentially unidirectional, with most of the radiation in the direction of the main lobe. There is also a small minor lobe at 180° from the direction of the main lobe, however.

In addition to the main radiation lobe, there will usually be one or more *minor* radiation lobes. For a Yagi antenna, there will be a minor lobe at about 180° from the direction of the main lobe, as shown in Figure 9-9. There may also be side lobes at various other angles. For example, suppose an antenna pattern shows a main lobe of radiation at 0° and a minor lobe at 180°. You might describe the radiation pattern by saying most of the signal would be radiated toward 0° and a smaller amount would be radiated toward 180°.

Gain versus Spacing

When we speak of antenna gain, we mean gain referenced to another antenna, used for comparison. One common reference antenna is an ideal half-wavelength dipole in free space. The gain of an antenna with parasitic elements varies with the boom length of the antenna, the spacing between elements, and the length and the number of parasitic elements. The maximum front-to-back ratio seldom, if ever, occurs at the same conditions that yield maximum forward gain, and it is frequently necessary to sacrifice some forward gain to get the greatest front-to-back ratio, or vice versa. To obtain maximum performance from a Yagi antenna the lengths and spacing of the elements must be optimized.

Changing the tuning and spacing of the elements also affects the impedance of the driven element. The impedance will vary as the operating frequency changes. The standing-wave ratio (SWR) will increase as you move higher and lower than the design operating frequency. The antenna SWR bandwidth is the frequency range over which the antenna will have less than some particular SWR, such as the 2:1 SWR bandwidth. The most effective method to increase the SWR bandwidth of a parasitic beam antenna is to use larger-diameter elements.

Theoretically, an optimized three-element (director, driven element and reflector) Yagi antenna has a gain of slightly more than 7 dB over a half-wavelength dipole in free space. Wide element spacing results in an antenna with high gain, wider bandwidth and less critical tuning.

The design of a Yagi antenna is generally optimized for forward gain, front-to-rear gain ratio and SWR bandwidth. You can use computer software to model the antenna and predict how changes to the various parameters will affect your design. One parameter that won't affect the design is whether you plan to orient the antenna elements either horizontally or vertically. (With the elements oriented horizontally the antenna will produce horizontally polarized radiation and with the elements oriented vertically the antenna will produce vertically polarized radiation.)

[Turn to Chapter 13 and study question G9A01 and questions G9A06 through G9A11. Also study question G9C08. Review this material as needed.]

Loop Antennas

There is another type of antenna that uses a parasitic element to produce a directional radiation pattern. The elements of this antenna are full-wavelength loops. The elements of a **cubical quad antenna** (or *quad*) are full-wavelength loops. A typical quad, shown in **Figure 9-10**, has two elements — a driven element and a reflector. A two-element quad could also use a driven element with a director. More elements, such as a reflector and one or more directors, can be added to increase the directivity of the antenna. By installing additional loops of the right dimensions, we can work different frequency bands with the same antenna. This is commonly done as a modification to a 20-meter quad to provide 15 and 10-meter coverage. The elements of the quad are arranged as a square, with each side a quarter-wavelength long. The total lengths of the elements are calculated as follows:

Circumference of driven element:

$$(\text{Quad})\,L_{\text{Driven}}(\text{in feet}) = \frac{1005}{f(\text{in MHz})} \qquad \text{(Equation 9-4)}$$

Circumference of director element:

$$(\text{Quad})\,L_{\text{Director}}(\text{in feet}) = \frac{975}{f(\text{in MHz})} \qquad \text{(Equation 9-5)}$$

Circumference of reflector element:

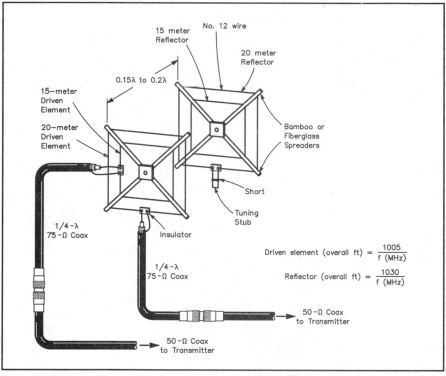

Figure 9-10 — This drawing shows the construction of a cubical quad antenna. The total length of the driven element can be found using Equation 9-4, and the antenna can be fed directly with coaxial cable.

Within the figure:

No. 12 wire

15 meter Reflector

20 meter Reflector

0.15λ to 0.2λ

15–meter Driven Element

20–meter Driven Element

Bamboo or Fiberglass Spreaders

Short

Tuning Stub

$1/4$-λ 75-Ω Coax

Insulator

$1/4$-λ 75-Ω Coax

50-Ω Coax to Transmitter

50-Ω Coax to Transmitter

$$\text{Driven element (overall ft)} = \frac{1005}{f\,(MHz)}$$

$$\text{Reflector (overall ft)} = \frac{1030}{f\,(MHz)}$$

$$(\text{Quad})L_{\text{Reflectorr}}(\text{in feet}) = \frac{1030}{f(\text{in MHz})} \qquad (\text{Equation 9-6})$$

So for a 21.4-MHz cubical quad, the element lengths would be:

$$(\text{Quad})L_{\text{Driven}}(\text{in feet}) = \frac{1005}{f(\text{in MHz})} = \frac{1005}{21.4\,\text{MHz}} = 47.0\,\text{ft}$$

$$(\text{Quad})L_{\text{Director}}(\text{in feet}) = \frac{975}{f(\text{in MHz})} = \frac{975}{21.4\,\text{MHz}} = 45.6\,\text{ft}$$

$$(\text{Quad})L_{\text{Reflector}}(\text{in feet}) = \frac{1030}{f(\text{in MHz})} = \frac{1030}{21.4\,\text{MHz}} = 48.1\,\text{ft}$$

These equations give the total length of the elements. To find the length of each side of the antenna, we must divide the total lengths by 4:

Antennas and Feed Lines   9-13

$(Quad)$ Driven Element Side $(in\ feet)=$

$$\frac{\text{Driven Element Length }(\text{in feet})}{4} = \frac{47.0\,\text{ft}}{4} = 11.7\,\text{ft}$$

$(Quad)$ Director Element Side $(in\ feet)=$

$$\frac{\text{Director Element Length }(\text{in feet})}{4} = \frac{45.6\,\text{ft}}{4} = 11.4\,\text{ft}$$

$(Quad)$ Reflector Element Side $(in\ feet)=$

$$\frac{\text{Reflector Element Length }(\text{in feet})}{4} = \frac{48.1\,\text{ft}}{4} = 12.0\,\text{ft}$$

Use Equations 9-4, 9-5 and 9-6 to calculate the length of wire needed for a quad antenna on any frequency. Divide these lengths by 4 to calculate the length of one side for each element.

The polarization of the signal from a quad antenna is determined by where the feed point is located on the driven element, as shown in **Figure 9-11**. With the feed point located in the center of a side parallel to the Earth's surface, the transmitted wave will be horizontally polarized; fed in the center of a perpendicular side, the transmitted wave will be vertically polarized. If we turn the antenna 45°, so it looks like a diamond, and feed it at the bottom corner, the transmitted wave will be horizontally polarized. If the antenna is fed at a side corner, the transmitted wave will be vertically polarized.

Polarization is especially important when building antennas for use at VHF and UHF. Propagation at these frequencies is generally line of sight, and the polarization of a signal does not change from transmitting antenna to receiving antenna. Best signal reception will occur when both transmitting and receiving stations use

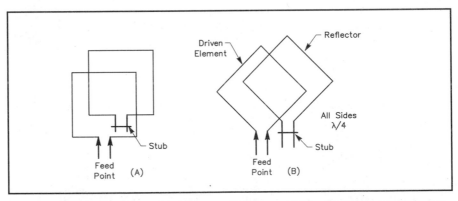

Figure 9-11 — The feed point of a quad antenna determines the polarization. Fed in the middle of the bottom side (as shown at A) or at the bottom corner (as shown at B), the antenna produces a horizontally polarized wave. Vertical polarization is produced by feeding the antenna at the side in either case.

the same polarization. On the HF bands, the polarization of the signal may change many times as it passes through the ionosphere, so antenna polarization is not as important.

A **delta loop antenna** also uses full-wavelength elements, as shown in **Figure 9-12**. In a delta loop, the elements are triangular rather than square. Equations 9-4, 9-5 and 9-6 also work to calculate the full element lengths for the delta loop. To find the length of each side of the delta loop antenna, the total length is divided by 3, because of the triangular shape of the elements.

For example, suppose you want to design a delta loop antenna for 24.9 MHz. Calculate the loop lengths and the lengths of each side as follows:

Circumference of driven element:

$$\text{(Delta Loop)} L_{\text{Driven}} \text{(in feet)} = \frac{1005}{f \text{(in MHz)}} \qquad \text{(Equation 9-4)}$$

Circumference of director element:

$$\text{(Delta Loop)} L_{\text{Director}} \text{(in feet)} = \frac{975}{f \text{(in MHz)}} \qquad \text{(Equation 9-5)}$$

Circumference of reflector element:

$$\text{(Delta Loop)} L_{\text{Reflectorr}} \text{(in feet)} = \frac{1030}{f \text{(in MHz)}} \qquad \text{(Equation 9-6)}$$

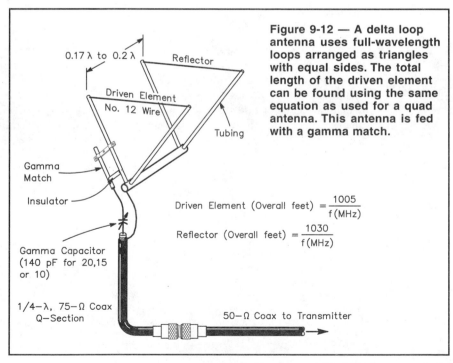

Figure 9-12 — A delta loop antenna uses full-wavelength loops arranged as triangles with equal sides. The total length of the driven element can be found using the same equation as used for a quad antenna. This antenna is fed with a gamma match.

$$\text{Driven Element (Overall feet)} = \frac{1005}{f \text{(MHz)}}$$

$$\text{Reflector (Overall feet)} = \frac{1030}{f \text{(MHz)}}$$

$$(\text{Delta Loop})\,L_{\text{Driven}}\,(\text{in feet}) = \frac{1005}{f\,(\text{in MHz})} = \frac{1005}{24.9\;\text{MHz}} = 40.36\;\text{ft}$$

$$(\text{Delta Loop})\,L_{\text{Director}}\,(\text{in feet}) = \frac{975}{f\,(\text{in MHz})} = \frac{975}{24.9\;\text{MHz}} = 39.2\;\text{ft}$$

$$(\text{Delta Loop})\,L_{\text{Reflectorr}}\,(\text{in feet}) = \frac{1030}{f\,(\text{in MHz})} = \frac{1030}{24.9\;\text{MHz}} = 41.4\;\text{ft}$$

These equations give the total length of the elements. To find the length of each side of the antenna, we must divide the total lengths by 3:

$$(\text{Delta Loop})\,\text{Driven Element Side}\,(\text{in feet}) =$$
$$\frac{\text{Driven Element Length}\,(\text{in feet})}{3} = \frac{40.36\;\text{ft}}{3} = 13.45\;\text{ft}$$

$$(\text{Delta Loop})\,\text{Director Element Side}\,(\text{in feet}) =$$
$$\frac{\text{Director Element Length}\,(\text{in feet})}{3} = 3 = 13.1\;\text{ft}$$

$$(\text{Delta Loop})\,\text{Reflector Element Side}\,(\text{in feet}) =$$
$$\frac{\text{Reflector Element Length}\,(\text{in feet})}{3} = \frac{41.4\;\text{ft}}{3} = 13.8\;\text{ft}$$

The radiation pattern of a typical quad is similar to that of the Yagi shown in **Figure 9-9**. A typical two-element quad or two-element delta loop has a gain of about 6.5 dB over a half-wavelength dipole. This compares favorably with the gain of a three-element Yagi, which is approximately 7 dB. Compared to a dipole antenna, a quad or delta loop antenna has more directivity in both the horizontal and vertical planes. These antennas have a radiation pattern that includes a main lobe, which is the direction of maximum radiated field strength from the antenna. They also exhibit a front-to-back ratio, with more power radiated in the direction of the major lobe compared to the power radiated in exactly the opposite direction.

[Now turn to Chapter 13 and study questions G9B01 through G9B11. Review this section as necessary.]

FEED LINES

A **feed line** (or *transmission line*) is used to transfer power from the transmitter to the antenna. The two basic types of feed lines generally used — **parallel-conductor feed line** and **coaxial cable** — can be constructed in a variety of forms. Both types can be divided into two classes: those in which the majority of the space between the conductors is air, and those in which the conductors are embedded in and separated by a solid plastic or foam insulation (dielectric).

Air-insulated feed lines have the lowest loss per-unit length (usually expressed in dB / 100 ft). Adding a solid dielectric between the conductors increases the losses in the feed line, and decreases the maximum voltage the feed line can withstand before arcing. In general, as frequency increases, feed-line losses become greater. If a 160-meter signal and a 2-meter signal pass through the same coaxial cable, the 2-meter signal will be attenuated a greater amount than the 160-meter signal.

The characteristic impedance of parallel-conductor line depends on the distance between the centers of the conductors and the radius of the conductors. The greater the spacing between the conductors, the higher the characteristic impedance of the feed line. The impedance decreases, however, as the size of the conductors increases. The characteristic impedance of a feed line is not affected by the length of the line.

Solid-Dielectric Feed Lines

Feed lines in which the conductors are separated by a flexible dielectric have several advantages over air-dielectric line: They are less bulky, maintain more uniform spacing between conductors, are generally easier to install and are neater in appearance. Both parallel-conductor and coaxial lines are available with this type of insulation.

One disadvantage of these types of lines is that the power loss per unit length is greater than air-insulated lines because of the dielectric. As the frequency increases, the dielectric losses become greater. The power loss causes heating of the dielectric. If the heating is great enough — as may be the case with high power or a high SWR — the dielectric may actually melt, or arcing may occur inside the line.

One common parallel-conductor feed line has a characteristic impedance of 300 Ω. This type of line is used for TV antenna feed line, and is usually called twin lead. Twin lead consists of two number 20 wires that are molded into the edges of a polyethylene ribbon about a half-inch wide. The presence of the solid dielectric lowers the characteristic impedance of the line as compared to the same conductors in air. **Figure 9-13** shows the construction of some common twin lead feed line.

The characteristic impedance (Z_0) of any feed line can be calculated from a few simple measurements. In the case of a parallel-conductor feed line, you would measure the outside diameter of the individual conductors (d) and the distance between the centers of the two conductors (S). **Figure 9-14** shows the measurements and the equation to calculate Z_0. You don't have to remember this equation, or use it to answer this question, but it shows how you could calculate the characteristic impedance of any parallel-conductor feed line that you might have.

Part of the electric and magnetic fields between the conductors exists outside the solid dielectric. This leads to several operating disadvantages. Dirt or moisture

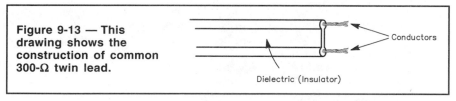

Figure 9-13 — This drawing shows the construction of common 300-Ω twin lead.

Conductors

Dielectric (Insulator)

on the surface of the ribbon tends to change the characteristic impedance and increase the line losses. Weather effects can be minimized, however, by coating the feed line with silicone grease or car wax. In any case, the changes in the impedance will not be very serious if the line is terminated in its characteristic impedance (Z_0). If there is a considerable standing-wave ratio, however, then small changes in Z_0 may cause wide fluctuations of the input impedance.

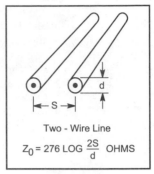

Two - Wire Line

$$Z_0 = 276 \text{ LOG } \frac{2S}{d} \text{ OHMS}$$

Figure 9-14 — This drawing shows the construction of a parallel-conductor feed line. The spacing between the wires and the wire diameter are used to calculate the characteristic impedance of the feed line.

Coaxial Cable

Coaxial cables are available in flexible and semiflexible formats, but the fundamental design is the same for all types. Some coaxial cables have stranded-wire center conductors, while others employ a solid copper conductor. **Figure 9-15** shows the typical construction of coaxial cable. Coaxial cables commonly used by radio amateurs have a characteristic impedance of 50 or 75 Ω. The outer conductor or shield may be a single layer of copper braid, a double layer of braid (more effective shielding), or solid copper or aluminum (most effective shielding). An outer insulating jacket (usually vinyl) provides protection from dirt, moisture and chemicals. Exposure of the dielectric to moisture and chemicals will cause it to deteriorate over time and increase the electrical losses in the line.

The outer conductor or shield serves to contain the electric and magnetic fields inside the cable. This means the impedance of a coaxial line is not affected by weather or other nearby conductors. This is one advantage of coaxial feed lines over parallel conductor feed lines.

The characteristic impedance of a coaxial line depends on the diameter of the center conductor, the inside diameter of the shield braid and the dielectric of the insulating materials between them. The characteristic impedance of coaxial cables increases for larger-diameter shield braids, but it decreases for larger-diameter center conductors.

The larger the diameter, the higher the power-handling capability of the line because of the increased dielectric thickness and conductor size. In general, losses decrease as the cable diameter increases, because there is less power lost in the conductor. As the frequency increases, the conductor and dielectric losses become greater, causing

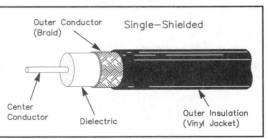

Figure 9-15 — Coaxial cables consist of a conductor surrounded by insulation. The second conductor, called the shield, goes around the insulation. A plastic insulation goes around the entire cable.

Outer Conductor (Braid)   Single—Shielded

Center Conductor   Dielectric   Outer Insulation (Vinyl Jacket)

more attenuation of the signal in the cable. Attenuation is not affected by the characteristic impedance of the line, at least if the spacing between the conductors in the coaxial cable is a small fraction of a wavelength at the operating frequency. If the spacing between conductors is a significant portion of a wavelength, the line will begin to act as a waveguide. At frequencies below about 1.5 GHz this is not a problem.

[Turn to Chapter 13 and study the following questions: G9D01, G9D02, G9D03, G9D07, G9D08 and G9D12. Review this section as needed.]

STANDING-WAVE RATIO (SWR)

In a perfect antenna system, all the power put into the feed line would be radiated by the antenna. A practical antenna system will behave most like this ideal when the feed-line impedance is matched to the antenna feed-point impedance. Some power will always be lost in the feed line, but you should select a low-loss line to minimize this. Power lost in the feed line is converted to heat.

If the antenna feed-point impedance does not exactly match the characteristic impedance of the feed line, some power will be reflected back down the feed line from the antenna. A mismatch is said to exist — the greater the mismatch, the more power will be reflected by the antenna. Some of the reflected power is dissipated as heat. In a lossy line more heat is produced than in a lower-loss line. The reflected power is reflected again at the transmitter, and goes back to the antenna. There, some power will be radiated, and some reflected back to the transmitter. This process continues until, eventually, all the power is either radiated by the antenna or lost as heat in the feed line.

The reflected power creates what is called a standing wave. If we measure either the voltage or current at several points along the feed line, we find that it varies from a maximum value to a minimum value at intervals of one-quarter wavelength, as shown in **Figure 9-16**. The standing wave is created when energy reflected from the antenna meets the forward energy from the transmitter — the two waves alternately cancel and reinforce each other.

The **standing-wave ratio** or **SWR** is defined as the ratio of the maximum voltage to the minimum voltage in the standing wave:

$$SWR = \frac{E_{Max}}{E_{Min}}$$

(Equation 9-7)

An SWR of 1:1 means you have no reflected power. The voltage and current are constant at any point along the feed line, and the line is said to be "flat." If the load is completely resistive, then the SWR can be calculated by dividing the line characteristic impedance by the load resistance, or vice versa — whichever gives a value greater than one:

$$SWR = \frac{Z_0}{R} \quad or \quad SWR = \frac{R}{Z_0}$$

(Equation 9-8)

where:
Z_0 = characteristic impedance of the feed line
R = load resistance (not reactance)

For example, if you feed a 50-Ω antenna with a 50-Ω feed line, the impedances are matched, and you will have an SWR of 50 / 50 or 1:1. Similarly if the antenna

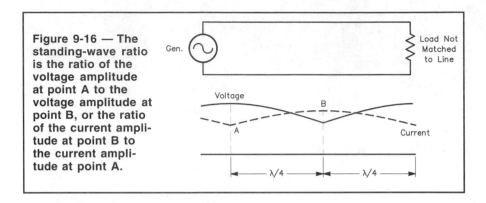

Figure 9-16 — The standing-wave ratio is the ratio of the voltage amplitude at point A to the voltage amplitude at point B, or the ratio of the current amplitude at point B to the current amplitude at point A.

impedance is 10 Ω and you feed it with 50-Ω feed line, the SWR is 50 / 10 or 5:1. If you feed a 200-Ω antenna with 50-Ω feed line, the SWR is 200 / 50 or 4:1.

When a high SWR condition exists, losses in the feed line are increased. This is because of the multiple reflections of the signal between the antenna and transmitter. Each time the transmitted power has to travel up and down the feed line, a little more energy is lost as heat. This effect is not as great as some people believe, however. If the line loss is less than 2 dB (such as for 100 feet of RG-8 or RG-58 cables up to about 30 MHz), the SWR would have to be greater than 3:1 to add an extra decibel of loss because of the SWR.

When a feed line is terminated in a resistance equal to its characteristic impedance, maximum power is delivered to the antenna, and feed line losses are minimized. This would be an ideal condition, although such a perfect match is seldom realized in a practical antenna system.

IMPEDANCE MATCHING

Most currently manufactured Amateur Radio transmitters are designed to be connected to antenna systems with impedances of 50 to 75 Ω. Some transmitters with vacuum-tube power amplifiers that have a tunable output circuit, however, can be adjusted to match impedances over a greater range. Depending on the type of antenna you are using, you may have to improve the match between the feed line and the antenna. A matching network (often called a *Transmatch*) can be used near the transmitter to match the transmitter 50-Ω output impedance to the antenna-system impedance.

When a coaxial cable is used to feed a half-wave dipole, a transformer, called a **balun** (*bal*anced to *un*balanced), can be used to decouple the feed line from the antenna. This prevents the feed line from radiating a signal, which could distort the antenna radiation pattern.

Many amateurs prefer to use parallel-conductor feed line because of its lower loss than coaxial cable. This type of feed line is a **balanced line**, which means that the two conductors have equal but opposite voltages along the line, and neither

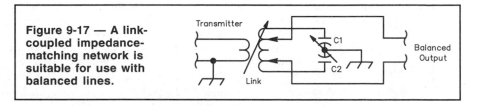

Figure 9-17 — A link-coupled impedance-matching network is suitable for use with balanced lines.

conductor is at ground potential. A balanced fee line will not disturb the balanced nature of a dipole antenna. Modern amateur transceivers use a coaxial-cable connector for the antenna, however. Parallel-conductor feed line requires the use of a balun to change the balanced line condition of the parallel-conductor feed line to the **unbalanced line** condition of the coaxial cable. Coaxial cable is an unbalanced feed line because one conductor — the outer shield — is connected to ground potential.

An inductively coupled impedance-matching network transforms the unbalanced line output (one side grounded to the chassis) at the transceiver to a balanced line output for use with a parallel-conductor feed line. **Figure 9-17** shows such a Transmatch circuit, which *tunes* the balanced feed line to allow the antenna to be used on several bands. The coil taps and capacitor settings are adjusted for lowest SWR. Higher impedance loads are tapped farther out from the center of the coil. The split-stator capacitor (C_1 and C_2) is used because it provides two symmetric sections that are tuned simultaneously with a single control. Normally, a short length of coaxial cable connects the transmitter to an inductively coupled impedance-matching network. Parallel-conductor feed line then connects the matching network to the antenna. An inductively coupled matching network is not normally used when a coaxial-cable feed line goes all the way to a center-fed resonant dipole antenna.

It is important to note that even though a transmitter may "see" a pure 50-Ω load, and the SWR between the transmitter and matching device is 1:1, the SWR measured in the feed line between the matching device and the antenna may not be 1:1. Inserting a matching device between the feed line and the transmitter does not affect the SWR in the feed line between the matching device and the antenna; the matching device only provides a 50-Ω resistive load for the transmitter. The SWR meter should be placed between the transmitter and the Transmatch to adjust the matching network properly.

[Now complete your study of this chapter by turning to Chapter 13 and studying questions G9D04 through G9D06 and questions G9D09 through G9D11. Also study questions G9D13 and G9D14. Review this section as needed. That completes your study of the material in this chapter for your General examination.]

CHAPTER 10
KEYWORDS
KEYWORDS
KEYWORDS

Athermal effects of RF radiation — Health effects of RF radiation related to low-level energy fields that are insufficient to cause ionization or heating effects in body tissue. While research is ongoing, no conclusive evidence has been found to demonstrate that such fields cause serious health effects.

Duty cycle — The ratio between the actual RMS value of an RF signal and the RMS value of a continuous signal having the same PEP value, expressed as a percentage. A duty cycle of 100% corresponds to a continuous carrier, such as a test signal.

Electric field strength — This is the field resulting from the electric charge distribution present on a radiating element. Electric field strength is expressed in volts per meter (V / m).

Far field of an antenna — That region of the electromagnetic field surrounding an antenna where the field strength as a function of angle (the antenna pattern) is essentially independent of the distance from the antenna. In this region (also called the free-space region), the field has a predominantly plane-wave character. That is, locally uniform distributions of electric field strength and magnetic field strength are in a plane perpendicular to the direction of propagation.

Magnetic field strength — This is the field resulting from the currents on a radiating element. Magnetic field strength is expressed in amperes per meter (A / m).

Maximum Permissible Exposure (MPE) limits — The electric field strength, magnetic field strength and plane-wave equivalent power density associated with a radiated electromagnetic wave to which a person may be exposed without harmful effect, and with an acceptable safety factor.

Near field of an antenna — That region of the electromagnetic field immediately surrounding an antenna where the reactive field dominates and where the field strength as a function of angle (the antenna pattern) depends upon the distance from the antenna. It is a region in which the electric and magnetic fields do not have a substantial plane-wave character, but vary considerably from point to point.

Power density — A measure of the power flow through a unit area normal (perpendicular) to the direction of propagation. It is usually expressed in watts per square meter (W / m^2)

Routine RF evaluation — A procedure required of some Amateur Radio stations, to determine that the station meets the maximum permissible exposure (MPE) limits established by FCC Rules.

Specific absorption rate (SAR) — A measure of the rate at which RF energy is absorbed in body tissue. SAR is expressed in units of watts per kilogram (W / kg).

Thermal effects of RF radiation — Body tissues that are subjected to *very high* levels of RF energy may suffer serious heat damage. These effects depend upon the frequency of the energy, the power density of the RF field that strikes the body and factors such as the polarization of the wave. For example, in an extreme case, RF heating of an eye can result in cataract formation or can even cause blindness.

Time averaging — Transmitter power is averaged over times of 6 minutes for controlled RF environments or 30 minutes for uncontrolled RF environments for power density calculations to determine exposure levels.

RF Safety

In 1996, the FCC announced new rules governing exposure to radio frequency (RF) radiation. These new rules set limits on the amount of RF energy to which people may be exposed. The rules also require that some stations be evaluated to determine if they are in compliance with these rules. (Almost all Amateur Radio operation is already in compliance with the rules, but amateurs must be familiar with the requirements.)

The Technician and General question pools put into use on April 15, 2000 include that topic and questions. Your General license exam will include 5 questions, based on these 5 exam topics:

G0A RF safety principles
G0B RF safety rules and guidelines
G0C Routine station evaluation and measurements (FCC Part 97 refers to RF radiation evaluation)
G0D Practical RF-safety applications
G0E RF-safety solutions

The material in this chapter will help you begin to understand the issues involved with RF radiation safety. ARRL's *RF Exposure and You* is a book written to provide you with much more detailed information about the FCC Rules related to RF exposure. That book will also help you understand the process of performing an RF environmental evaluation and includes step-by-step instructions and tables to help you determine that your station meets the maximum permissible exposure requirements. You can order the book directly from the ARRL or purchase a copy from a local ARRL dealer.

RF SAFETY PRINCIPLES

Amateur Radio is basically a safe activity. In recent years, however, there has been considerable discussion and concern about the possible hazards of electromagnetic radiation (EMR), including both RF energy and power-frequency (50 or 60 Hz) electromagnetic fields. FCC regulations set limits on the **maximum permissible exposure (MPE)** allowed from the operation of radio transmitters. These regulations do not take the place of RF-safety practices, however. This section deals with the topic of RF safety.

Extensive research on RF safety is underway in many countries. This section was prepared by members of the ARRL RF Safety Committee and coordinated by Dr Robert E. Gold, WBØKIZ. It summarizes what is now known and offers safety precautions based on the research to date.

All life on Earth has adapted to survive in an environment of weak, natural, low-frequency electromagnetic fields (in addition to the Earth's static geomagnetic field). Natural low-frequency EM fields come from two main sources: the sun, and thunderstorm activity. But in the last 100 years, man-made fields at much higher intensities and with a very different spectral distribution have altered this natural EM background in ways that are not yet fully understood. Much more research is needed to assess the biological effects of EMR.

Both RF and 60-Hz fields are classified as *nonionizing radiation* because the frequency is too low for there to be enough photon energy to ionize atoms. (*Ionizing radiation*, such as X-rays, gamma rays and even some ultraviolet radiation has enough energy to knock electrons loose from their atoms. When this happens, positive and negative ions are formed.) Still, at sufficiently high power densities, EMR poses certain health hazards. It has been known since the early days of radio that RF energy can cause injuries by heating body tissue. (Anyone who has ever touched an improperly grounded radio chassis or energized antenna and received an RF burn will agree that this type of injury can be quite painful. You do not have to actually *touch* the chassis to get an RF burn.) In extreme cases, RF-induced heating in the eye can result in cataract formation and can even cause blindness. Excessive RF heating of the reproductive organs can cause sterility. Other serious health problems can also result from RF heating. These heat-related health hazards are called **thermal effects**. In addition, there is evidence that magnetic fields may produce biologic effects at energy levels too low to cause body heating. The proposition that these **athermal effects** may produce harmful health consequences has produced a great deal of research.

In addition to the ongoing research, much else has been done to address this issue. For example, FCC regulations set limits on exposure from radio transmitters. The Institute of Electrical and Electronics Engineers, the American National Standards Institute and the National Council for Radiation Protection and Measurement, among others, have recommended voluntary guidelines to limit human exposure to RF energy. The ARRL has established an RF Safety Committee, a committee of concerned medical doctors and scientists, serving voluntarily to monitor scientific research in the fields and to recommend safe practices for radio amateurs.

Thermal Effects of RF Energy

Body tissues that are subjected to very high levels of RF energy may suffer serious heat damage. These effects depend upon the frequency of the energy, the power density of the RF field that strikes the body, and even on factors such as the polarization of the wave.

At frequencies near the body's natural resonant frequency, RF energy is absorbed more efficiently, and maximum heating occurs. In adults, this frequency usually is about 35 MHz if the person is grounded, and about 70 MHz if the person's body is insulated from the ground. Also, body parts may be resonant; the adult head, for example is resonant around 400 MHz, while a baby's smaller head resonates near 700 MHz. Human eyes have a free-space resonance in the 1270-MHz range. This resonance may be modified by the presence of other tissues surrounding the eye. Body size and structure thus determine the frequency at which most RF energy is absorbed. As the frequency is increased above resonance, less RF heating generally occurs. However, additional longitudinal resonances occur at about 1 GHz near the body surface. **Specific absorption rate (SAR)** is a term that describes the rate at which RF energy is absorbed into the human body. This quantity, expressed in watts per kilogram (w/kg) best associates the RF energy absorption with the biological effects of RF fields on the human body.

Maximum permissible exposure (MPE) limits are based on whole-body SAR values. This helps explain why these safe exposure limits vary with frequency. The MPE limits define the maximum **electric field strength** and **magnetic field strength** or the plane-wave equivalent **power density** associated with these fields, that a person may be exposed to without harmful effect—and with an acceptable safety factor. The regulations assume that a person exposed to a specified (safe) MPE level will also experience a safe SAR.

Nevertheless, **thermal effects of RF energy** should not be a major concern for most radio amateurs because of the relatively low RF power we normally use and the intermittent nature of most amateur transmissions. Amateurs spend more time listening than transmitting, and many amateur transmissions such as CW and SSB use low-duty-cycle modes. **Duty cycle** takes into account the reduced average transmitted power that results because the transmitter is not operating at full power continuously. This means greater short-term exposure levels can be permitted with low-duty-cycle emissions. (With FM or RTTY, though, the RF is present continuously at its maximum level during each transmission.) In any event, it is rare for radio amateurs to be subjected to RF fields strong enough to produce thermal effects unless they are fairly close to an energized antenna or unshielded power amplifier. Specific suggestions for avoiding excessive exposure are offered later.

The frequency of the radio signal, the power density (and electric and magnetic field strengths) of the energy and the duty cyle of the transmission mode are all important in estimating the effects of RF energy on body tissue. Even the polarization of the radiated wave can have an effect, because the electric field from a vertically polarized wave might have more effect on a person walking around in the field than would a horizontally polarized wave. *Critical angle* is a property of the radiation that is related to how the wave interacts with the ionosphere, but is not a factor in determining the effects of RF energy.

Athermal Effects of RF Radiation

Nonthermal effects of EMR may be of greater concern to most amateurs because they involve lower level energy fields. Research about possible health effects resulting from exposure to the lower level energy fields — the athermal effects — has been of two basic types: epidemiological research and laboratory research.

Scientists conduct laboratory research into biological mechanisms by which EMR may affect animals, including humans. Epidemiologists look at the health patterns of large groups of people using statistical methods. These epidemiological studies have been inconclusive. By their basic design, these studies do not demonstrate cause and effect, nor do they postulate mechanisms of disease. Instead, epidemiologists look for associations between an environmental factor and an observed pattern of illness. For example, in the earliest research on malaria, epidemiologists observed the association between populations with high prevalence of the disease and the proximity of mosquito infested swamplands. It was left to the biological and medical scientists to isolate the organism causing malaria in the blood of those with the disease and identify the same organisms in the mosquito population.

In the case of athermal effects, some studies have identified a weak association between exposure to EMF at home or at work and various malignant conditions including leukemia and brain cancer. A larger number of equally well designed and performed studies, however, have found no association. A risk ratio of between 1.5 and 2.0 has been observed in positive studies (the number of observed cases of malignancy being 1.5 to 2.0 times the "expected" number in the population). Epidemiologists generally regard a risk ratio of 4.0 or greater to be indicative of a strong association between the cause and effect under study. For example, men who smoke one pack of cigarettes per day increase their risk for lung cancer tenfold compared to nonsmokers, and two packs per day increase the risk to more than 25 times the nonsmokers' risk.

Epidemiological research by itself is rarely conclusive, however. Epidemiology only identifies health patterns in groups — it does not ordinarily determine their cause. And there are often confounding factors: Most of us are exposed to many different environmental hazards that may affect our health in various ways. Moreover, not all studies of persons likely to be exposed to high levels of EMR have yielded the same results.

There has also been considerable laboratory research about the biological effects of EMR in recent years. For example, it has been shown that even fairly low levels of EMR can alter the human body's circadian rhythms, affect the manner in which cancer-fighting T lymphocytes function in the immune system, and alter the nature of the electrical and chemical signals communicated through the cell membrane and between cells, among other things.

Much of this research has focused on low-frequency magnetic fields, or on RF fields that are keyed, pulsed or modulated at a low audio frequency (often below 100 Hz). Several studies suggested that humans and animals can adapt to the presence of a steady RF carrier more readily than to an intermittent, keyed or modulated energy source. There is some evidence that while EMR may not directly cause cancer, it may sometimes combine with chemical agents to promote its growth or inhibit the work of the body's immune system.

None of the research to date conclusively proves that low-level EMR causes adverse health effects. Given the fact that there is a great deal of ongoing research to examine the health consequences of exposure to EMF, the American Physical Society (a national group of highly respected scientists) issued a statement in May 1995 based on its review of available data pertaining to the possible connections of cancer to 60-Hz EMF exposure. This report is exhaustive and should be reviewed by anyone with a serious interest in the field. Among its general conclusions were the following:

1. "The scientific literature and the reports of reviews by other panels show no consistent, significant link between cancer and powerline fields."
2. "No plausible biophysical mechanisms for the systematic initiation or promotion of cancer by these extremely weak 60-Hz fields has been identified."
3. "While it is impossible to prove that no deleterious health effects occur from exposure to any environmental factor, it is necessary to demonstrate a consistent, significant, and causal relationship before one can conclude that such effects do occur."

The APS study is limited to exposure to 60-Hz EMF. Amateurs will also be interested in exposure to EMF in the RF range. A 1995 publication entitled *Radio Frequency and ELF Electromagnetic Energies, A Handbook for Health Professionals* includes a chapter called "Biologic Effects of RF Fields." In it, the authors state: "In conclusion, the data do not support the finding that exposure to RF fields is a causal agent for any type of cancer" (page 176). Later in the same chapter they write: "Although the data base has grown substantially over the past decades, much of the information concerning nonthermal effects is generally inconclusive, incomplete, and sometimes contradictory. Studies of human populations have not demonstrated any reliably effected end point." (page 186).

Readers may want to follow this topic as further studies are reported. Amateurs should be aware that exposure to RF and ELF (60 Hz) electromagnetic fields at all power levels and frequencies may not be completely safe. Prudent avoidance of any avoidable EMR is always a good idea. An Amateur Radio operator should not be fearful of using his or her equipment, however. If any risk does exist, it will almost surely fall well down on the list of causes that may be harmful to your health (on the other end of the list from your automobile).

SAFE EXPOSURE LEVELS

How much EM energy is safe? Scientists and regulators have devoted a great deal of effort to deciding upon safe RF-exposure limits. This is a very complex problem, involving difficult public health and economic considerations. The recommended safe levels have been revised downward several times in recent years—and not all scientific bodies agree on this question even today. An Institute of Electrical and Electronics Engineers (IEEE) standard for recommended EM exposure limits went into effect in 1991. It replaced a 1982 IEEE and American National-Standards Institute (ANSI) standard that permitted somewhat higher exposure levels. The newer IEEE standard was adopted by ANSI in 1992.

The IEEE standard recommends frequency-dependent and time-dependent maximum permissible exposure levels. Unlike earlier versions of the standard, the 1991 standard recommends different RF exposure limits in *controlled environments* (that is, where energy levels can be accurately determined and everyone on the premises is aware of the presence of EM fields) and in *uncontrolled environments* (where energy levels are not known or where some persons present may not be aware of the EM fields).

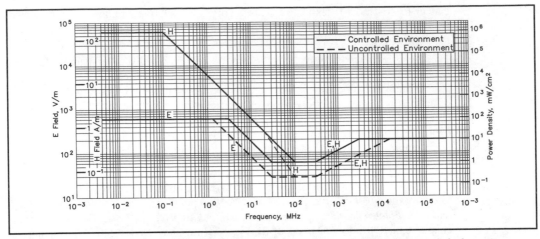

Figure 10-1 — 1991 RF protection guidelines for body exposure of humans. It is known officially as the "IEEE Standard for Safety Levels with Respect to Human Exposure to Radio Frequency Electromagnetic Fields, 3 kHz to 300 GHz." (Note that the exposure levels set by this standard are not the same as the FCC permissible exposure limits.)

The graph in **Figure 10-1** depicts the maximum permissible exposure limits set by the 1991 IEEE standard. It is necessarily a complex graph because the standards differ not only for controlled and uncontrolled environments but also for electric fields (E fields) and magnetic fields (H fields). Basically, the lowest E-field exposure limits occur at frequencies between 30 and 300 MHz. The lowest H-field exposure levels occur at 100 to 300 MHz. The standard sets the maximum E-field limits between 30 and 300 MHz at a power density of 1 mW/cm^2 (61.4 V/m) in controlled environments—but at one-fifth that level (0.2 mW/cm^2 or 27.5 V/m) in uncontrolled environments. The H-field limit drops to 1 mW/cm^2 (0.163 A/m) at 100 to 300 MHz in controlled environments and 0.2 mW/cm^2 (0.0728 A/m) in uncontrolled environments. Higher power densities are permitted at frequencies below 30 MHz (below 100 MHz for H fields) and above 300 MHz, based on the concept that the body will not be resonant at those frequencies and will therefore absorb less energy.

In general, the 1991 IEEE standard requires averaging the power level over time periods ranging from 6 to 30 minutes for power-density calculations, depending on the frequency and other variables. The exposure limits for uncontrolled environments are lower than those for controlled environments, but to compensate for that the standard allows exposure levels in those environments to be averaged over much longer time periods (generally 30 minutes). This long averaging time means that an intermittently operating RF source (such as an Amateur Radio transmitter) will show a much lower power density than a continuous-duty station for a given power level and antenna configuration. **Time averaging** takes into account the total RF exposure averaged over a certain period of time.

Time averaging is based on the concept that the human body can withstand a greater rate of body heating (and thus, a higher level of RF energy) for a short time than for a longer period.

The IEEE standard excludes any transmitter with an output below 7 W because such low-power transmitters would not be able to produce significant whole-body heating. (Recent studies show that hand-held transceivers often produce power densities in excess of the IEEE standard within the head.)

There is disagreement within the scientific community about these RF exposure guidelines. A small but significant number of researchers now believe athermal effects should also be taken into consideration. Several European countries have adopted stricter standards than the recently updated IEEE standard.

[Turn to Chapter 13 now and study questions G0A01 through G0A08 and question G0A12. Also study question G0B03. Review this section as needed.]

THE RF EXPOSURE RULES

The rules and guidelines set *maximum permissible exposure (MPE)* limits for humans who are near radio transmitters. The regulations control *exposure* to RF fields, not the *strength* of RF fields. There is no limit to how strong a field can be as long as no one is being exposed to it, although FCC regulations require that amateurs use the minimum necessary power at all times (§97.313(a)). If the operation of your station would result in exposure over the limits in areas where there are no people *at the time you are operating*, the station is still in compliance.

All radio stations must comply with the requirements for MPEs; even QRP stations running only a few watts or less! The MPEs vary with frequency, as shown in **Table 10-1**.

MPEs are derived from the Specific Absorption Rate (SAR) — the rate at which tissue absorbs RF energy, usually expressed in watts per kilogram (W/kg). The FCC MPEs are not based strictly on IEEE/ANSI C95.1-1992, but rather on a hybrid between that standard and the report written by the NCRP (the National Council on Radiation Protection and Measurements). NCRP is a body commissioned to develop recommendations for federal agencies. In terms of exposure levels, the NCRP report recommends lower MPE levels over some frequency ranges than are found in the IEEE/ANSI standard. The most stringent requirements are from 30 to 300 MHz, because various human-body resonances fall in that frequency range.

MPE limits are specified in maximum electric and magnetic fields for frequencies below 30 MHz, in power density for frequencies above 300 MHz and all three ways for frequencies from 30 to 300 MHz. For compliance purposes, all these limits must be considered separately — if any one is exceeded, the station is not in compliance.

For example, if a 144-MHz amateur station had an H-field exposure level of 0.08 A/m (amperes/meter) and an E-field exposure of 22 V/m (volts/meter), the station is not in compliance for exposure to the general public because the H field limit has been exceeded, even though the E field is below the limits.

MPEs Vary with Frequency

The MPE limits vary with frequency, as shown in Table 10-1. The human body is roughly resonant (for different body sizes and under differing conditions) at frequencies between 30 and 300 MHz. The MPEs are the most stringent over this fre-

Table 10-1

(From §1.1310) Limits for Maximum Permissible Exposure (MPE)

(A) Limits for Occupational/Controlled Exposure

Frequency Range (MHz)	Electric Field Strength (V/m)	Magnetic Field Strength (A/m)	Power Density (mW/cm²)	Averaging Time (minutes)
0.3-3.0	614	1.63	(100)*	6
3.0-30	1842/f	4.89/f	(900/f²)*	6
30-300	61.4	0.163	1.0	6
300-1500	—	—	f/300	6
1500-100,000	—	—	5	6

f = frequency in MHz
* = Plane-wave equivalent power density (see Note 1).

(B) Limits for General Population/Uncontrolled Exposure

Frequency Range (MHz)	Electric Field Strength (V/m)	Magnetic Field Strength (A/m)	Power Density (mW/cm²)	Averaging Time (minutes)
0.3-1.34	614	1.63	(100)*	30
1.34-30	824/f	2.19/f	(180/f²)*	30
30-300	27.5	0.073	0.2	30
300-1500	—	—	f/1500	30
1500-100,000	—	—	1.0	30

f = frequency in MHz
* = Plane-wave equivalent power density (See Note 1.)

Note 1: This means the equivalent far-field strength that would have the E or H-field component calculated or measured. It does not apply well in the near field of an antenna. The equivalent far-field power density can be found in the near or far field regions from the relationships:

$$P_d = |E_{total}|^2 /3770 \text{ mW/cm}^2 \text{ or from } P_d = |H_{total}|^2 \times 37.7 \text{mW/cm}^2$$

quency range. In Table 10-1, there are separate limits for the electric field (E field), the magnetic field (H field) and power density. The electric field is specified in volts per meter (V/m), the magnetic field is specified in amperes per meter (A/m) and the power density is specified in milliwatts per square centimeter. In addition, for some frequencies the term "plane-wave equivalent power density" is used.

Average Exposure

Table 10-1 shows the MPE limits for various frequency ranges. MPEs assume continuous-duty and operation at the average rate. The levels shown assume that the exposure will be continuous over the exposure period. The regulations, however, average the total exposure over 6 minutes for controlled environments and 30 minutes for uncontrolled environments. This average includes both the duty factor of the operating mode and the actual on and off times over the worst-case averaging period. In most cases, the average power of an amateur station is considerably less than its peak power, and the ratio of transmit time to nontransmit time is less than 100%.

Thus, it is permissible to exceed the MPE limits for some time, as long as that is offset by a corresponding reduction in the exposure limits for other times within the averaging period. For example, in an uncontrolled environment, if one is in a field that is 10 times the limit for 3 minutes, this is acceptable as long as one has no exposure for the preceding 27 minutes *and* the following 27 minutes. The 30-minute window is not based on arbitrary half-hour segments by the clock, but is a "sliding" window, such that in *any* 30-minute period, the total exposure must be below the limits. It would not be acceptable to have no exposure for 15 minutes, twice the exposure for 15 minutes, then be exposed at the limit for the next 15 minutes because the total exposure in the worst-case window (the last 30 minutes) exceeds the average MPE levels. There are a number of ways to calculate the time-averaged exposure.

Who Must Comply

All Amateur Radio stations must comply with the MPE limits, regardless of power, operating mode or station configuration. Even a 10-milliwatt station must comply. It is unlikely that low-power stations would exceed the limits, but the rules apply equally to all.

Routine Environmental Evaluations

The FCC RF safety rules are designed to control the maximum permissible human exposure to all RF radiated fields. The core of the requirements under these regulations, then, are the MPE levels. The core of the specific actions that need to be taken by Amateur Radio operators, however, is the requirement for some amateurs to perform a "routine environmental evaluation" for RF exposure. This will establish that the station is being operated in compliance with the FCC RF-Exposure guidelines.

A **routine RF evaluation** is not nearly as onerous as it sounds! This subject is covered in detail in ARRL's *RF Exposure and You*, but it is summarized here, as part of the discussion about the rules. Doing an evaluation will help ensure a safe operating environment for amateurs, their families and neighbors.

The FCC is relying on the demonstrated technical skill of Amateur Radio operators to *evaluate their own stations* (although it is perfectly okay for an amateur to rely on another amateur or skilled professional to perform the evaluation).

Most evaluations will not involve measurements, but will be done with comparisons against typical charts developed by the FCC, relatively straightforward calculations or computer modeling of near-field signal strength. The FCC encourages flexibility in the analysis, and will accept any technically valid approach.

It is not difficult to do the necessary station evaluation. The FCC guidance is contained in *OET Bulletin No. 65: Evaluating Compliance With FCC-Specified Guidelines for Human Exposure to Radio Frequency Radiation* and an Amateur Radio supplement to that bulletin, *Supplement B to OET Bulletin 65: Additional Information for Amateur Radio Stations*. The FCC material contains the basic information hams need to evaluate their stations, including a number of tables showing compliance distances for typical amateur power levels and antennas. The ARRL also publishes a book, *RF Exposure and You*, which includes more information and extensive tables listing common amateur station configurations. Generally, hams will use tables like these to evaluate their stations.

In most cases, hams will be able to use a table that best describes their station's operation to determine the minimum compliance distance for their specific operation. These tables show the compliance distances for uncontrolled environments for a particular type of antenna at a particular height. (The power levels shown in the tables are average power levels, adjusted for the duty cycle of the operating mode being used, and operating on and off time, averaged over 6 minutes for controlled environments or 30 minutes for uncontrolled environments.)

What should you do if you perform the routine RF evaluation on your station and determine that the RF fields exceed the FCC exposure limits in human-accessible areas? Simply take action to prevent human exposure to the excessive RF fields! There are many ways to accomplish that task, and the FCC allows you great latitude to decide what steps are appropriate. For example, you can move your antennas farther away, restrict access to the areas where the exposure would exceed the limits, or reduce power to reduce the field strengths in those areas.

Categorical Exemptions

Some types of amateur stations do not need to be evaluated (but these stations must still comply with the MPE limits!) The FCC has exempted these stations from the evaluation requirement because their output power, operating mode, use, frequency or antenna location are such that they are presumed to be in compliance with the rules. These stations are *not* exempt from the rules, but are presumed to be in compliance without the need for an evaluation.

- Stations using the peak-envelope power levels or less to the antenna as shown in **Table 10-2**.
- Amateur repeaters using 500-W ERP or less.
- Amateur repeaters with antennas not mounted on buildings if the antenna is located more than 10 meters (32.8 feet) high above ground.
- Amateur mobile and portable hand-held stations using push-to-talk or equivalent operation.

Note that Table 10-2 cites power to the antenna. This is not the same as your transmitter output power, although you can conservatively use your transmitter output power to decide if you need to do an evaluation, if you wish. As an example, if you are running 90 W PEP and have a feed line loss of 3 dB, you are losing approximately 50% of your power in the feed line, so you have approximately 45 W PEP to the antenna.

This part of your operation would not have to be evaluated on any band. Note, too, that unlike the MPE, the levels in this table are not average-power levels, but are peak-envelope powers (PEP). If you transmit only one short dit per 30-minute period, and that dit is transmitted at levels above those in the chart, you will still have to do an evaluation. When you did the evaluation, however, you could use average power. Admittedly, this sounds a bit complex, but it is explained in detail in ARRL's *RF Exposure and You*.

Table 10-2 is taken from §97.13(c)(1). That table forms the guideline used to

Table 10-2

Power Thresholds for Routine Evaluation of Amateur Radio Stations

Wavelength Band	Evaluation Required if Power* (watts) Exceeds:
MF	
160 m	500
HF	
80 m	500
75 m	500
40 m	500
30 m	425
20 m	225
17 m	125
15 m	100
12 m	75
10 m	50
VHF (all bands)	50
UHF	
70 cm	70
33 cm	150
23 cm	200
13 cm	250
SHF (all bands)	250
EHF (all bands)	250
Repeater stations (all bands)	*non-building-mounted antennas*: height above ground level to lowest point of antenna < 10 m *and* power > 500 W ERP *building-mounted antennas*: power > 500 W ERP

*Transmitter power = Peak-envelope power input to antenna. For repeater stations **only**, power exclusion based on ERP (effective radiated power).

determine if a routine RF evaluation must be performed. Stations that use more power than the power levels to the antenna shown in Table 10-2 must be evaluated. Most HF transceivers are rated at 100-W PEP output; on 15 meters and below, stations using this power level need not be evaluated. In fact, stations operating on frequencies lower than 21 MHz can operate with even higher power without being required to perform an evaluation. On 17 meters, the threshold is 175 W PEP and on 20 meters it jumps to 225 W PEP. Although amateurs are only permitted to use 200 W PEP on the 30-meter band, the RF evaluation threshold is 425 W PEP on that band! On the 40, 80 and 160-meter bands, the threshold is 500 W PEP.

Operators who wish to use 12 and 10 meters with a 100-watt transceiver could either perform an evaluation for those two bands, or they could reduce power to

75 W PEP on 12 meters and 50 W PEP on 10 meters and forego the evaluation altogether. Most VHF transceivers are rated at 50-W PEP output or less; stations using this power level on VHF (the 6, 2 and 1.25 meter bands) need not be evaluated.

The power levels used to determine if an RF environmental evaluation is required vary with frequency. Compare the MPE levels in Table 10-1 for the various frequencies with the evaluation threshold levels shown in Table 10-2. You will see that the evaluation thresholds vary in much the same way as the MPE limits. Where higher exposure is permitted, the evaluation threshold is higher and where the MPE is lower, so is the evaluation threshold.

Repeater Operation

All amateur repeaters operating at a power of 500 W ERP or less are generally categorically exempt from evaluation. All amateur repeaters whose antennas are not mounted on buildings and that have all parts of their radiating antenna located at least 10 meters (32.8 feet) above ground also are exempt. Amateur repeaters with antennas located on buildings (presumably buildings where people could be located) must be evaluated if they use more than 500 W ERP. ERP is derived by multiplying the power to the antenna by the numerical gain of the antenna over a dipole (6 dBd, for example, represents a numerical equivalent of 3.98). This categorical exemption from evaluation will probably cover many repeater stations.

Multitransmitter Sites

The rules are intended to ensure that operation of transmitters regulated by the FCC doesn't result in exposure in excess of MPE limits. It is fairly easy to make this determination for single transmitters, when there are no other sources of RF to complicate things. Many transmitters operate in proximity to other transmitters, and it is entirely possible for two or more transmitters to all be below their own limit, but the total exposure from them all operating together to be greater than the permitted MPEs. Amateur repeaters are the most likely stations to be located at a site with multiple transmitters operating in several radio services. The multitransmitter rules aren't limited to repeater stations, however.

The FCC regulations and other FCC documents cover the likely situation of multiple transmitters. The bottom line is that, in most cases, all the significant RF transmitters operating at multitransmitter sites generally must be considered when determining if the site's total exposure is in compliance. In addition, all significant emitters are jointly responsible for overall site compliance.

The rules stipulate that in a multi-transmitter environment, a single transmitter operator is jointly responsible with other operators at the site for all areas at the site where the exposure from that transmitter is greater than 5% of what is permitted for that transmitter. (This is 5% of the permitted power density or 5% of the square of the E or H-field value.) Note that this is not the same as 5% of the total exposure, which could sometimes be unknown.

In many cases involving Amateur Radio transmitters, only a relatively small area would be encompassed by that 5% exposure threshold, so joint responsibility might only exist in the immediate vicinity of the amateur antenna. A repeater trustee, for example, might have that 5% level extend only to those areas to 10 feet above and below the antenna up the tower, and thus be responsible for overall site compliance only to that area on the tower. In this case, the responsibility may be only to radio service personnel climbing the tower (generally a controlled exposure environment would apply) or tower maintenance people (who may or may not be trained about RF exposure, so an uncontrolled environment may be more appropriate).

Often, you may not know much about the other transmitters on your site. In that case, you should make the best assumptions you can about the other stations' power, antenna gains and operating duty cycles, and conduct your assessment of site compliance accordingly.

[Turn to Chapter 13 and study questions G0A09, G0A10, G0A11, G0B01, G0B02 and G0B04 through G0B10. Review the material in this section if you have difficulty with any of those questions.]

Paperwork and Proof of Evaluation

Performing a routine RF exposure evaluation is the most important step you can take to ensure that your station is in compliance with the FCC RF safety regulations. Even if your station is exempt from the requirement to perform such an evaluation, you might want to do a simple evaluation. As a minimum, it will be good practice for the time that you make a station change that might require evaluation. The results of your evaluation will certainly demonstrate to yourself, and possibly to your neighbors, that your station operation is well within the guidelines, and is no cause for concern.

Once you determine that your Amateur Radio station complies with the regulations, you may begin operating. There's no need for FCC approval before operating. Other than a short certification on Form 605 station applications when applying for a new license or a license renewal, the regulations do not normally require hams to file proof of evaluation with the FCC. The Commission recommends that each amateur keep a record of the station evaluation procedure and its results, in case questions arise.

When you make changes to your station, you may need to perform a new RF safety evaluation. You should carefully consider the changes and decide if you may have done anything to increase the RF exposure to any area. In many cases you will not have to perform a new evaluation. For example, suppose your station is in compliance and then you decrease your transmitter power from 1000 watts to 500 watts. You should not need to perform a new evaluation when you make this change, unless you made some other changes at the same time that could result in stronger radiated fields. Although your station may not be exempt from the evaluation requirement, since it was in compliance at the higher power, it will also be in compliance at the lower level. If you decrease your transmitter power from 500 watts to 40 watts, your station is even exempt from the requirement to perform an evaluation. Always keep in mind that your station must *comply* with the RF safety rules, however.

If you use a calibrated field-strength meter to measure the RF fields in your station, and those measurements indicate that people might be exposed to more than the MPE limits, there are many steps you can take to reduce the exposure. For example, you should ensure that all of your station equipment is properly grounded and that all equipment covers are tightly fastened. If a piece of equipment is not properly grounded or the cover is off a radio or amplifier, this could be a good source of excessive radiation. You should also be sure to use the minimum transmitter power necessary.

Suppose your station uses an indoor 20-meter dipole. If the calculations in your routine RF evaluation indicate that you or your family might be receiving more radiation exposure than the MPE limits allow, one good step would be to move the antenna to a safe outdoor environment.

In another example, your station evaluation might show that your neighbors could be receiving more than the MPE limits from your Yagi or Quad antenna when it is pointed at their house. One simple solution to this problem is to take steps to ensure that you can't point your antenna at their house while you are transmitting. This could be as simple as a few marks on the rotator control to remind you not to transmit with the antenna pointed in a certain direction. Another step you could take would be to reduce your transmitter power to a level that reduces their exposure to a value that is below the MPE limit.

These examples point out one of the main advantages and also one of the main disadvantages of a high-gain, narrow-beamwidth antenna. From an RF exposure point of view, the main advantage of such an antenna is that it allows you to focus in a direction away from populated areas. That will reduce the possible exposure of people to the RF fields from your signal. The disadvantage of a high-gain, narrow-beamwidth antenna is that individuals in the main beam will receive a greater exposure than when a low-gain antenna is used.

By installing your antenna as high as possible you will maximize the distance from your antenna to any nearby people. That will generally help reduce the RF exposure that your station produces to your neighbors or other people who may be nearby.

No one should be near a transmitting antenna while it is in use. Install ground-mounted transmitting antennas well away from living areas so that people cannot come so close that they receive more than the MPE limits. If there is a possibility of someone walking up to your antenna while you are transmitting, it may be a good idea to install a protective fence around the antenna.

[Turn to Chapter 13 and study questions G0C07, G0C10 and G0C11 Also study questions G0E02, G0E03, G0E05, G0E06, G0E09, G0E10, G0E11 and G0E12. Review this section as needed.]

SOME RF AWARENESS GUIDELINES

The following text was derived from guidelines developed by the ARRL RF Safety Committee. The guidelines represent good common sense anytime you are working with RF circuits. They are especially important safety steps to help you minimize your exposure to the dangers of RF fields.

• Although antennas on towers (well away from people) pose no exposure problem, make certain that the RF radiation is confined to the antenna radiating elements themselves. Provide a single, good station ground (earth), and eliminate radiation from transmission lines. Use good coaxial cable or other feed line properly. Avoid serious imbalance in your antenna system and feed line. For high-powered installations, avoid end fed antennas that come directly into the transmitter area near the operator.

Considering RF exposure, install antennas as high and as far away from populated areas as possible. This is true whether you are installing the antenna outdoors or indoors. In the case of an indoor antenna, you should locate the antenna as far away as possible from any living spaces that will be occupied while you are operating. Minimize feed line radiation into populated areas and if the antenna is a gain antenna, consider keeping it pointed away from populated areas when you are transmitting.

If you must make adjustments to your transmitter while it is operating, it is generally best to use a dummy antenna. In addition to allowing you to make transmitter adjustments without putting a signal on the air, a dummy antenna helps provide a safer test environment because the energy is converted to heat rather than being radiated.

One sure sign that you don't have a good RF ground system is if you feel a tingling sensation or receive minor burns every time you touch your microphone while you are transmitting. Commonly called "RF in the shack," this condition indicates that there is unwanted RF in your station. You and others in the station may be exposed to more than the maximum permissible level of RF radiation. Take steps to provide a better RF ground for your station in that case.

• No person should ever be near any transmitting antenna while it is in use. This is especially true for mobile or ground-mounted vertical antennas. Avoid transmitting with more than 25 W in a VHF mobile installation unless it is possible to first measure the RF fields inside the vehicle. At the 1-kW level, both HF and VHF directional antennas should be at least 35 ft above inhabited areas. Avoid using indoor and attic-mounted antennas if at all possible. If open-wire feeders are used, ensure that it is not possible for people (or animals) to come into accidental contact with the feed line.

All parts of your dipole antenna should be as high and far away from people as possible. Directional, high-gain antennas should be mounted higher than any nearby buildings or other structures that people could have access to while you are transmitting. This will help ensure that you are not directing excessive amounts of RF energy toward people. Ground-mounted antennas must also be installed so no one can be exposed to RF radiation in excess of the MPE limits. You might even consider installing a protective fence around the base of a ground-mounted transmitting antenna. By keeping people a safe distance from the antenna you can ensure that no one will be exposed to excessive amounts of RF energy.

For a variety of reasons, many amateurs use indoor antennas. Sometimes there are deed restrictions or other limitations against outside antennas. Indoor antennas can present a special problem, however, because they may expose people inside the house to strong RF fields. If you must use an attic-mounted antenna, you should perform a routine RF environmental evaluation and be aware of the safe operating distance between the antenna and any location where people might be while you are transmitting. You can reduce power, use lower duty cycle operating modes and reduce your actual transmitting time during any 6-minute period (for the controlled environment of yourself and your family) or 30-minute period (for the uncontrolled environment of your neighbors or the general public).

Typical EME stations will use large high-gain antennas and near maximum power to overcome the huge path loss for signals traveling to the Moon, reflecting off the poor reflecting surface and traveling all the way back to Earth. With typical EME antenna gains of 20 dB or more, the minimum safe distance between the antenna and any person can easily exceed several hundred feet. Needless to say, most amateurs will have to be very careful about pointing such an antenna array at the horizon, in the direction of any neighbors! The high effective radiated power (ERP) may produce hazardous RF fields in controlled and uncontrolled areas. Of course it is also possible that such an EME station might cause TVI/RFI problems for any neighbors at which the antennas would be pointed. If your antenna system is not in a clear location, any reflections from nearby conductive objects could even affect the tuning of the antenna array.

The best place to mount the antenna for a mobile installation is in the center of a metal roof. This is most practical for a VHF or UHF station, but may not be possible for an HF station because of the overall height of the vehicle and antenna. The roof will provide an excellent shield to prevent the driver and passengers from being exposed to excessive amounts of RF energy.

- Don't operate high-power amplifiers with the covers removed, especially at VHF/UHF.

When the metal covers are properly fastened in place, any RF that might radiate from the circuit will be contained inside the cabinet. The RF only leaves the cabinet through the coaxial cable connection to a properly connected feed line.

If you build RF equipment, such as a home made transmatch or even a transmitter or amplifier, be sure to construct a metal cabinet with secure covers. Aluminum or other metal conductive materials make good enclosures.

- In the UHF/SHF region, never look into the open end of an activated length of waveguide or microwave feed-horn antenna or point it toward anyone. (If you do, you may be exposing your eyes to more than the maximum permissible exposure level of RF radiation.) Never point a high-gain, narrow-beamwidth antenna (a paraboloid, for instance) toward people. Use caution in aiming an EME (moonbounce) array toward the horizon; EME arrays may deliver an effective radiated power of 250,000 W or more.

You should be especially careful with RF in the range of 1270 MHz because this is close to a resonant frequency for your eyes and other structures in your head and body. Take steps to ensure that the antenna is far from your eyes!

- With hand-held transceivers, keep the antenna away from your head and use the lowest power possible to maintain communications. Use a separate microphone and hold the rig as far away from you as possible. This will reduce your exposure to the RF energy.
- Don't work on antennas that have RF power applied.

You should also take precautions to ensure that no one can activate the transmitter while you are making repairs to your antenna or adjusting the feed line. One way to do this is to turn off the transmitter power supply and disconnect the antenna feed line. This also applies to work on a microwave feed horn or waveguide.

- Don't stand or sit close to a power supply or linear amplifier when the ac power is turned on. Stay at least 24 inches away from power transformers, electrical fans and other sources of high-level 60-Hz magnetic fields.

[Turn to Chapter 13 now and study questions G0D01 through G0D12 and questions G0E01, G0E04, G0E07 and G0E08. Review this section if any of those questions give you difficulty.]

FIELD STRENGTH MEASUREMENTS AND CALCULATIONS

The FCC maximum permissible exposure limits are given in terms of electric and magnetic field strengths and power density or plane-wave-equivalent power density. So how do you determine if the transmitted signal from your station is within these RF exposure limits? You must analyze, measure or otherwise determine your transmitted field strengths and power density. This means there are a number of ways you can perform the required routine RF radiation evaluation.

One way to do this is by making direct measurements of the electric and magnetic field strengths around your antenna while transmitting a signal. If you happen to have a calibrated field-strength meter with a calibrated field-strength sensor (antenna), you can make accurate measurements. Unfortunately, such calibrated meters are expensive and not normally found in an amateur's tool box. The relative field- strength meters many amateurs use are not accurate enough to make this type of measurement.

Even if you do have access to a laboratory-grade calibrated field-strength meter, you must be aware of factors that can upset your readings. Reflections from ground and nearby conductors (power lines, other antennas, house wiring, etc) can easily confuse field strength readings. For example, the measuring probe and the person making the measurement can interact with the antenna radiation if they are in the near-field zone. In addition, you must know the frequency response of the test equipment and probes, and use them only within the appropriate range. Even the orientation of the test probe with respect to the test antenna polarization is important.

Why should we be concerned with the separation between the source antenna and the field-strength meter, which has its own receiving antenna? One important reason is that if you place a receiving antenna very close to an antenna when you measure the field strength, *mutual coupling* between the two antennas may actually alter the radiation from the antenna you are trying to measure.

This sort of mutual coupling can occur in the region very close to the antenna under test. This region is called the *reactive* **near-field region of an antenna**. The term "reactive" refers to the fact that the mutual impedance between the transmitting and receiving antennas can be either capacitive or inductive in nature. The reactive near field is sometimes called the "induction field," meaning that the magnetic field usually is predominant over the electric field in this region. The antenna acts as though it were a rather large, lumped-constant inductor or capacitor, storing energy in the reactive near field rather than propagating it into space.

For simple wire antennas, the reactive near field is considered to be within about a half wavelength from an antenna's radiating center. For making field-strength measurements, we do not want to be too close to the antenna being measured.

The strength of the reactive near field decreases in a complicated fashion as you increase the distance from the antenna. Beyond the reactive near field, the antenna's radiated field is divided into two other regions: the *radiating near field* and the *radiating far field*. Nearly any metal object or other conductor that is located within the radiating near field can alter the radiation pattern of the antenna. Conductors such as telephone wiring or aluminum siding on a building that is located in the radiating near field will interact with the theoretical electric and magnetic fields to add or subtract intensity. This results in areas of varying field strength. Although you have measured the fields in the general area around your antenna and found that your station meets the MPE limits, there may still be "hot spots" or areas of higher field strengths within that region. In the near field of an antenna, the field strength varies in a way that depends on the type of antenna and other nearby objects as you move farther away from the antenna.

Because the boundary between the fields is rather "fuzzy," experts debate where one field begins and another leaves off, but the boundary between the radiating near and far fields for a full-sized dipole or Yagi antenna is generally accepted as:

$$D \approx \frac{2L^2}{\lambda}$$ (Equation 10-1)

where:
   D = distance between the antenna and the boundary
   L = largest dimension of the physical antenna, expressed in the same units of
       measurement as the wavelength
   λ = wavelength of the RF energy

Remember, many specialized antennas do not follow the rule of thumb in Equation 10-1 exactly. In general, however, the wavelength of the energy and the physical size of the antenna are the factors that help determine the boundary between the near and far fields of an antenna. **Figure 10-2** shows the three fields in front of a simple wire antenna.

The radiating **far field of the antenna** forms the traveling electromagnetic waves. Far-field radiation is distinguished by the fact that the power density is proportional

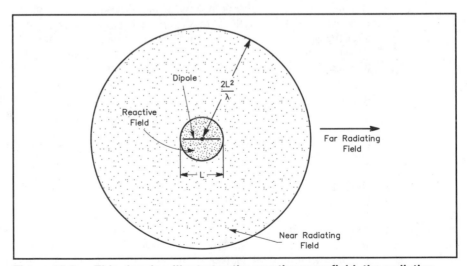

Figure 10-2 — This drawing illustrates the reactive near field, the radiating near field and the far field around a half wavelength dipole antenna.

to the inverse square of the distance. (That means if you double the distance from the antenna, the power density will be one fourth as strong.) The electric and magnetic fields are perpendicular to each other in the wave front, and are in time phase. The total energy is equally divided between the electric and magnetic fields. Beyond several wavelengths from the antenna, these are the only fields we need to consider. For accurate measurement of radiation patterns, we must place our measuring instruments at least several wavelengths away from the antenna under test.

In the far field there is a simple relationship between the electric (E) field and the magnetic (H) field of an electromagnetic wave. The strength of the E field divided by the strength of the H field is a constant 377 ohms, sometimes called the *impedance of free space*.

$$\frac{E}{H} = \text{Impedance of Free Space} = 377 \, \Omega \qquad \text{(Equation 10-2)}$$

where:
    E = electric field strength in volts per meter
    H = magnetic field strength in amperes per meter

This calculation is only valid in the far field of the antenna. In the near field the relationship is not this simple. This calculation may prove useful to you as you analyze your station for compliance with the FCC MPE limits. If you know the E or H field strength at some point in the far field then you can calculate the other value at that same point.

If you don't have the necessary measuring equipment, you can perform some calculations to determine the electric and magnetic field strengths from your station. This is normally done using computer programs, which take into account the gain and directivity of an antenna. There are a variety of programs available. The most reliable ones are based on the Numerical Electromagnetic Code (NEC) program that uses the Method of Moments analysis. (MININEC is a variation of the original program, shortened to run on personal computers.) While these are powerful analysis tools, you must be careful to accurately model the antenna and all conductors that might be within a few wavelengths of the antenna for reliable results.

If you measure or calculate the electric or magnetic field strength or power density at some point in the far field of your antenna, then you can calculate the field strengths or power density at other points in the far field. Rather than learning the extensive math involved with a complete calculation, though, let's just explore some simple relationships that you can use to make some quick estimates.

We will start by considering a sphere surrounding our antenna at some point in the far field of the antenna. All of the power radiated from our antenna must flow through that sphere. Next, imagine a second sphere with a radius of twice the first one. The same total power also flows through the second sphere. The second sphere has four times as much surface area as the first one, because its radius is double the first radius. See **Figure 10-3**. (The area of a sphere is calculated by multiplying the square of the radius times 4 pi: $A = 4 \pi R^2$.) We can use this relationship to write a simple equation:

$$\text{Power Density}_1 \times (\text{Distance}_1)^2 = \text{Power Density}_2 \times (\text{Distance}_2)^2 \quad \text{(Equation 10-3)}$$

If we know the power density at some distance and want to find the power

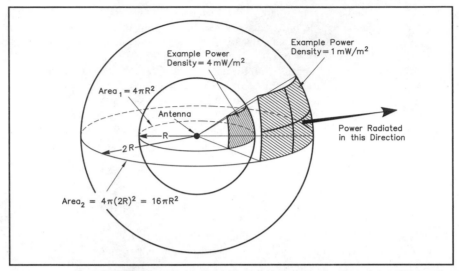

Figure 10-3 — This drawing illustrates that the power density of the RF energy radiated from an antenna decreases as the square of the distance. This is because the area of a sphere increases as the square of the radius (A = 4πR²).

density at a second distance we can simply solve this equation for Power Density$_2$.

$$\text{Power Density}_2 = \text{Power Density}_1 \times \frac{\left(\text{Distance}_1\right)^2}{\left(\text{Distance}_2\right)^2} \qquad \text{(Equation 10-4)}$$

For example, suppose you calculate the free-space far-field power density of the signal from your 18-MHz Yagi antenna to be 10 mW/m² at a distance of 3 wavelengths. What will be the power density at a distance of 6 wavelengths?

$$\text{Power Density}_2 = \text{Power Density}_1 \times \frac{\left(\text{Distance}_1\right)^2}{\left(\text{Distance}_2\right)^2} = 10\frac{\text{mW}}{\text{m}^2} \times \frac{\left(3\lambda\right)^2}{\left(6\lambda\right)^2}$$

$$\text{Power Density}_2 = 10\frac{\text{mW}}{\text{m}^2} \times \frac{9\lambda^2}{36\lambda^2} = 10\frac{\text{mW}}{\text{m}^2} \times \frac{1}{4} = 2.5\frac{\text{mW}}{\text{m}^2}$$

Equation 10-4 provides a simple mathematical formula that will help you make this type of calculation for any power density and distances in the far field of an antenna. You might also recognize that you can multiply by the ratio of the square of the distances. Use the method that makes more sense to you.

As another example, suppose you measure the free-space far-field power density of an antenna to be 9 mW/m² at a distance of 5 wavelengths. What will be the power density at a distance of 15 wavelengths? We can use Equation 10-4 again.

$$\text{Power Density}_2 = \text{Power Density}_1 \times \frac{\left(\text{Distance}_1\right)^2}{\left(\text{Distance}_2\right)^2} = 9\frac{mW}{m^2} \times \frac{\left(5\lambda\right)^2}{\left(15\lambda\right)^2}$$

$$\text{Power Density}_2 = 9\frac{mW}{m^2} \times \frac{25\lambda^2}{225\lambda^2} = 9\frac{mW}{m^2} \times \frac{1}{9} = 1\frac{mW}{m^2}$$

You might also reach this answer by recognizing that the ratio of the distances squared is 9. Then you can divide by that value to get the new power density.

Power density is related to the square of the electric and magnetic field strengths. This means that the electric and magnetic field strengths will decrease linearly with distance. We can write another equation for field strength, then:

$$\text{Field Strength}_1 \times \text{Distance}_1 = \text{Field Strength}_2 \times \text{Distance}_2 \qquad \text{(Equation 10-5)}$$

We can also solve this equation for Field Strength$_2$ and write a new equation.

$$\text{Field Strength}_2 = \text{Field Strength}_1 \times \frac{\text{Distance}_1}{\text{Distance}_2} \qquad \text{(Equation 10-6)}$$

In this case you can simply multiply by the ratio of the distances.

For example, suppose you measure the electric field strength 5 wavelengths from a 10-MHz dipole antenna, and you get a reading of 1.0 millivolts per meter. What will the field strength be at a distance of 10 wavelengths?

$$\text{Field Strength}_2 = \text{Field Strength}_1 \times \frac{\text{Distance}_1}{\text{Distance}_2} = 1.0\frac{mV}{m} \times \frac{5\lambda}{10\lambda} = 1.0\frac{mV}{m} \times \frac{1}{2}$$

$$\text{Field Strength}_2 = 0.50\frac{mV}{m}$$

As another example, suppose you measure the free-space far-field strength of a 28-MHz Yagi antenna to be 4.0 mV/m at a distance of 5 wavelengths from the antenna. What will the field strength be 20 wavelengths from the antenna?

$$\text{Field Strength}_2 = \text{Field Strength}_1 \times \frac{\text{Distance}_1}{\text{Distance}_2} = 4.0\frac{mV}{m} \times \frac{5\lambda}{20\lambda} = 4.0\frac{mV}{m} \times \frac{1}{4}$$

$$\text{Field Strength}_2 = 1.0\frac{mV}{m}$$

As a final example of this type of calculation, let's suppose you calculate the free-space far-field electric field strength of a 1.8-MHz dipole antenna at a distance of 4 wavelengths from the antenna. If the field strength is 9 microvolts per meter, what will it be at a distance of 12 wavelengths from the antenna?

$$\text{Field Strength}_2 = \text{Field Strength}_1 \times \frac{\text{Distance}_1}{\text{Distance}_2} = 9.0\frac{\mu V}{m} \times \frac{4\lambda}{12\lambda} = 9.0\frac{\mu V}{m} \times \frac{1}{3}$$

$$\text{Field Strength}_2 = 3.0\frac{\mu V}{m}$$

If you know a magnetic field strength you can also use Equation 10-6 to calculate a new field strength at some other distance.

[Congratulations! You have completed your study of the RF safety material for your General license exam. This also completes your study for all of the material on the exam, if you have followed the book from beginning to end. Before you are done, however, you have a few more questions to review. Turn to Chapter 13 now and study questions G0C01 through G0C06 and questions G0C08 and G0C09. If you have any difficulty with those questions, review this section.]

Setting Up Your HF Station

For many new General licensees, this will be your first venture onto the HF amateur bands. Now you'll have the opportunity to experience first-hand the excitement of contacting another amateur continents away. Or perhaps it is the lure of exchanging slow-scan television pictures or operating radioteletype that has enticed you to earn the General license. In any case, you probably have a million questions about selecting equipment, installing antennas and operating practices.

The ARRL publishes a library of technical and operating books to assist you in becoming an active Amateur Radio operator. We can't duplicate all that information here, but we collected some material for this chapter that we hope will start you on a successful path. What other books might you want in your library? The ultimate reference books are the "ARRL Big Three:" *The ARRL Handbook*, *The ARRL Antenna Book*, and *The ARRL Operating Manual.* You may not want or need the detailed technical information these books contain right now, however. ARRL books like *On the Air with Ham Radio*, *ARRL's HF Digital Handbook*, *APRS—Moving Hams on Radio and the Internet*, and *Your Mobile Companion* contain operating guidelines to help you get started on specific modes or operating styles. *Simple and Fun Antennas for Hams* contains a wealth of proven antenna-design information for simple antennas you can build with basic materials. *The ARRL Antenna Compendium* series contains many new and interesting tips and antenna projects you can build. *Understanding Basic Electronics* is written in an easy-to-understand style for electronics beginners and those who want to review their knowledge of basic electronics. You can purchase these and the many other ARRL books from your local Amateur Radio equipment dealer or directly from the ARRL, 225 Main Street, Newington, CT 06111-1494, Phone 888-277-5289 (toll-free). You can also order directly from our World Wide Web site at: **www.arrl.org/**

HF EQUIPMENT

Our discussion of equipment here is limited. In this chapter we will try to give you some basic information to help you set up your first HF station.

Beginners often call the ARRL Technical Information Service to ask which radio or antenna is best. In truth, the choice of a radio is very subjective; the best radio for you is determined by the space available for your station, your budget, your personal operating goals and inclinations.

Start by considering your needs. Make a list: How much room do you have for your station? What modes and bands would you like to operate? Do you want to invest the time and effort to learn about complicated features? From the list, you can derive requirements for your equipment. For example, if CW operation interests you, narrow filters are important. Blocking and intermodulation-distortion dynamic ranges are important for contesting or weak-signal work. DXers (those who pursue contacts with foreign countries) need convenient split-frequency operation (and the more frequency memories, the better).

No one can point to one best radio, or make the buying decision for you. With that warning out of the way, here are descriptions of some typical HF stations suitable for beginners. By all means, collect many such opinions from your ham friends before forming your own opinion and making a purchase.

For general HF operation, a basic modern transceiver offers the following qualities:

Modes — SSB and CW (10-m FM is a common option)

Frequency control
• stable VFO, PTO or frequency synthesizer,
• a digital frequency display
• easy split-frequency operation by means of dual VFOs, frequency memories or an external VFO
• RIT for small receiver-frequency offsets

Bandwidth
• 2 to 3-kHz crystal IF filter for SSB
• 250 to 600-Hz crystal IF filter for CW

Power Output — Up to 100 W from broadband solid-state (transistor) finals

Recent Evolution of Amateur Radio Equipment

The power amplifier (PA) stages of modern transceivers all use transistors. The broadband designs eliminate the frequent adjustment and tuning required of older HF transceivers that used tube-type PAs. Digital circuits are prominent, as frequency synthesizers, multiple VFOs, switched-capacitor filters and microprocessor controllers appear in many radios. Some of the newer radios even include digital signal processing (DSP) circuitry! All these radios are suitable for packet radio, AMTOR and RTTY, which require fast transmit-receive switching and high transmitter duty cycles. Many radios can even be controlled by a personal computer.

Selecting Older Radios

Many older radios hold some sentimental value, yet they may be nightmares under modern operating conditions. The frequency accuracy and stability of amateur equipment has increased steadily over the years. With this increased stability have come narrower receiver bandwidths that allow more stations to operate in less spectrum. While older radios may well perform their intended function today, they may not fit into modern Amateur Radio practices.

As an example, a very inexpensive HF CW station might include a Heath DX-40 transmitter with a VF-1 VFO and an HR-10B receiver. This station may seem attractive, since it may have a price tag under $200. This gear is typical of inexpensive stations of the 1950s and '60s. Worldwide CW communication is possible with such a simple station, but operation would be limited: It does not cover the 30, 17 and 12-meter amateur bands. The transmitter does not operate SSB (voice). The VFO offers insufficient band spread and excessive drift by today's standards. Receiver bandwidth is about 12 kHz. In a crowded band you would hear five or six signals at once. Many enjoyable, if noncompetitive, contacts could be made with the station, but $500 or less would likely buy a transceiver that covers all HF bands with CW and SSB (possibly FM, too) capabilities and a receiver bandwidth of 500 to 2000 Hz.

In general, Amateur Radio equipment has improved so much in the last two decades that only enthusiasts with a special interest should consider buying equipment made before the early 1980s. Hence, most operators will be happier with an HF radio made since 1985 or so.

There are some notable exceptions to the time line, and chief among those are radios made by Collins Radio and R. L. Drake Company. The Collins Radio permeability-tuned oscillators and filters are legendary. Collins radios such as the "S-line" are not as easy to operate as the average modern transceiver, but they were top performers in their day, and are still quite useful today.

A Drake T4XC or an R4C may appear at a bargain price. Even though the transmitter uses sweep tubes, it merits consideration. Drake produced stable VFOs with a permeability tuned oscillator and quality radios.

EQUIPMENT FEATURES

The array of features available on amateur equipment can be bewildering. Here are brief explanations of some common features and their usefulness.

Frequency Tuning and Display

Of all the controls on a rig, you use the frequency control most often. Choose a dial mechanism you are comfortable reading and tuning. Make sure the knob operates smoothly. Avoid radios that feel sloppy and slip or skip.

Two different types of tuning-dial mechanisms are available on ham transceivers and separate transmitter/receiver combinations: *digital* and *analog*. Most manufacturers now use digital displays. These are generally quite accurate and easier to read than analog dials.

Analog circular dials are usually marked at 1-kHz intervals. Some circular dials tune "backwards": clockwise motion decreases frequency rather than increasing it. Also, some circular dials have different frequency indicators for different

Ten-Tec's Argonaut 516 is a good basic radio with simple-to-use controls, but with plenty of other features.

modes; that can be confusing.

You may hear of a "slide-rule" analog dial: a rectangular display with a moving pointer that indicates frequency on one or more scales. Slide-rule dials are common on broadcast receivers, but they are not precise enough for most amateur operation.

Other Receive Features

Sensitivity is the ability of a receiver to detect weak signals. *Selectivity* is the ability of a receiver to separate two closely spaced signals. This determines how well you can receive one signal that is very close to another. Selectivity is generally more important than sensitivity. The ability to isolate the signal you are receiving from all the others nearby directly affects how much you will enjoy your time on the air.

Some receivers have a *crystal calibrator*, sometimes called a marker generator. It produces a signal every 25, 50 or 100 kHz. You can use the calibrator to make sure the dial or display is accurate.

Noise blankers are most useful for filtering ignition noise produced by cars and trucks. Some noise blankers can help reduce power-line noise as well. Some newer radios offer adjustable or multiple noise blankers to combat various types of electrical impulse noise.

Nearly all modern transceivers allow you to select between several IF filters, such as a narrow or wide SSB filter or a narrow or wide CW filter. This can be a helpful interference-fighting feature, and can also prove useful when you want a narrow filter for RTTY operation with the radio set in the upper or lower sideband position.

Generally, the better radios pack in more interference-fighting features, such as the ability to adjust the tuning of the IF passband. Such features add cost, so you may have to make some compromises. Extra money spent on receive features is often worthwhile, however.

Transmit Features

Speech processing increases the average power of a single-sideband signal. This can be done in the audio stages or the RF stages. RF processing is better. Used properly, a speech processor can greatly improve the readability of a signal. Almost all SSB transmitters and transceivers built since 1980 include a speech processor. On some rigs, there is no adjustment for the speech processor: just an on/off control. Other gear has one or more variable controls to set the speech processing level.

Some radios also include a *speech monitor*. This feature allows you to listen to a sample of the transmitted audio. You will probably have to use headphones so the monitored audio doesn't get back into the microphone. It can be helpful to monitor your transmitted signal as you adjust the speech processor, for example. A speech monitor is helpful so you can listen to the transmitted signal when you are operating radioteletype or slow-scan TV, too.

A *voice-operated transmit (VOX)* circuit puts a transceiver into the transmit mode automatically when you speak into the microphone, then back to receive when you stop talking. There are usually three VOX controls: *gain*, *delay* and *anti-VOX*. Gain sets the circuit sensitivity and determines how loud a sound must be before it turns on the circuit. Delay determines how long the transmitter stays on after you stop talking. The Anti-VOX setting prevents received audio from tripping the switch circuit. These controls are only adjusted occasionally, so they may be located on the back, top or bottom of a transceiver.

The ICOM IC-706MKIIG features a removable control head, operation on all bands from 1.8 through 450 MHz (12 in all!), DSP noise reduction and an SWR Graph mode—all in a compact 2.3 × 6.6 × 7.9 inch package.

The Kenwood TS-570S is a marvel of modern technology. The rig includes a general-coverage receiver, a 100-W transmitter with built-in electronic keyer and even digital signal processing (DSP) circuitry. All this fits in a package the size of a large shoebox.

Synthesized transceivers offer almost unlimited frequency flexibility. Many transceivers with phase-locked loop (PLL) frequency synthesizers use digital memory to allow you to select one of two variable-frequency oscillators (VFOs). For added convenience, synthesized transceivers often have *memories* to store your favorite frequencies and offer automatic *scanning* up and down a band, or among memory channels.

Many radios now have a built-in electronic *keyer* for CW operation. Even if your radio includes this feature, you may want an external keyer, such as a memory keyer that allows you to store several messages that can be transmitted with a simple button press. Keyers that use relays with floating (ungrounded) contacts in the output circuit are more flexible than those with transistor output stages because transmitter key-jack voltage polarity is unimportant with a relay. But relay contacts can stick, and the trend is to use transistorized output stages in modern keyers. If you use an electronic keyer with a transistor switch in the output circuit, transmitter keying polarity and voltage must be considered.

Modern transistorized rigs may use either *positive* or *negative* polarity at the key jack. Some transistorized rigs require that the keying voltage go to 0 V; others need only a TTL false level (0.8 V maximum) to key properly.

Before you connect a keyer to any rig, check the key-jack voltage and polarity. Check the equipment manuals, and make sure the keyer will work with the rig. There are ways to change a keyer output circuit if needed. (You can use a transistor or optoisolator, for instance). See *The ARRL Handbook* for more information about keyers and transmitter keying.

A Perspective on Performance Measurements

Receivers

Comparing the performance of one radio to another is difficult at best. The features of one may outweigh those of a second, even though its performance is lower under some conditions. Although the final decision on which radio to buy is usually based on personal preference, there are ways to compare receiver performance characteristics. The most important parameters are noise floor, intermodulation distortion and blocking (gain compression).

The *noise floor* determines the minimum discernible signal (MDS) that the operator can hear. A signal at the MDS level increases the receiver audio output by 3 dB over the background noise. (The specific tests performed on equipment reviewed at the ARRL Lab are described under "Measuring Receiver Performance" in *The ARRL Handbook*.)

Intermodulation distortion (IMD) is the production of spurious responses that results from the mixing of desired and undesired signals in a receiver. IMD occurs in any receiver when signals of sufficient magnitude are present. ARRL Product Reviews in our membership journal, *QST*, present IMD performance as the IMD dynamic range (IMD DR). IMD DR is the difference, in dB, between the noise floor and the strength of two equal incoming signals that produce a third-order product 3 dB above the noise floor.

Blocking dynamic range (Blocking DR) is the difference, in dB, between the noise floor and a signal that causes 1 dB of gain compression in the receiver. It indicates the signal level, above the noise floor, that begins to cause desensitization.

What do these measurements mean? When the IMD DR is exceeded, false signals begin to appear along with the desired signal. When the Blocking DR is exceeded, the receiver begins losing its ability to amplify weak signals. Typically, the IMD DR is 20 dB or more below the Blocking DR, so false signals appear well before sensitivity is significantly decreased.

Phase noise has been a hot topic in recent years. It is low-level noise resulting from phase differences between the local oscillator (LO) output and a perfect sine wave of the same frequency. It effectively widens the LO signal and produces false

The ICOM IC-746 Pro HF/VHF transceiver uses digital signal processing (DSP) filters to select an appropriate received-signal bandwidth for any operating mode. This radio operates on all ham bands from 160 meters through 2 meters (except the newest 60-meter band).

signals similar to IMD effects. A discussion of phase noise is beyond the scope of this book. The technical side of the problem is fully treated in J. Grebenkemper's "Phase Noise and its Effects on Amateur Communications" in March and April 1988 *QST*.

A receiver should provide the signal we are trying to copy at a particular moment, and no others. If the station is near sources of strong signals, both IMD DR and Blocking DR are important. If it is far from other transmitters, MDS may be the most significant factor. When circumstances lie between those two extremes, other features may drive the decision.

Transmitters

ARRL Product Review transmitter test procedures check for spurious radiation in the forms of harmonics and IMD. The limits for both problems are set by the FCC, and graphics of the spectral test results appear with each review.

BUYING USED EQUIPMENT

There are many different sources of used ham gear. You can buy gear from equipment retailers and local hams. You may see what you want listed in the pages of ham magazines (see the *QST* Ham Ads) or used-equipment flyers. Other sources are auctions, flea markets, hamfests, the Internet and even garage sales.

ARRL Product Reviews in *QST* are an excellent source of information about equipment you may wish to purchase. Extensive laboratory measurements and on-the-air tests are made and results reported. The ARRL Technical Information Service can tell you when the review of a particular radio was published. (Call 860-594-0200, write to ARRL Headquarters, 225 Main Street, Newington, CT 06111 or by e-mail to tis@arrl.org) You can purchase back issues or photocopies of the articles for radios in which you are interested. ARRL members can view past Product Review columns on the *ARRL Web* members only page at **www.arrl.org/members-only/**.

Local retailers are usually safe sources for used equipment. Many dealers route used equipment through their service shop before selling it. This ensures that the gear is working properly. Some dealers even allow you to operate the equipment before you buy it. This is the best way to determine a rig's capabilities and condition. Dealers usually offer a warranty (30 days is typical). Ask about the warranty *before* you buy.

Individual sellers usually offer no guarantees. Ask to see the unit in operation. You might even wish to take along a more knowledgeable ham who can help you make the decision. A local ham can be a good source of information on used equipment.

Mail-order purchases from individuals require simple precautions: Try to be sure that you'll have the right to return the equipment if you don't like what you receive. Shipment by truck freight with the right of inspection permits you to examine the package contents before you accept delivery. If you don't like what you receive, simply refuse delivery. You may be asked to pay by bank check or money order, rather than personal check.

Auctions, hamfests and flea markets are common sources of used equipment.

A careful buyer can get excellent equipment there, but you should study a little before you shop. What are the capabilities of the equipment, its current market value, cost of repairs and availability of repair parts? If you don't know these things, ask at your local club. Maybe someone more knowledgeable would be willing to go along with you.

Keep an eye on the classified sections of radio magazines. Know the going prices for radios that interest you, and never pay a high price without a demonstration. Try to buy equipment that was of high quality when it was new; it is more likely to serve you well when it is old.

When you are not able to test equipment before you buy it, a thorough visual inspection by an experienced eye will generally suffice if the price is right. You can usually tell a lot from the external appearance of equipment. If it looks physically abused, chances are it has been treated badly on the air as well.

When you find a good prospect, open the cabinet. Look for evidence of missing parts, fire or damage. Purchase modified equipment if you trust the seller, or if the modifications are well documented and can be easily removed. (Modifications are only as good as the skill of the person who installed them.) There is one other important point to keep in mind when you buy used equipment from any source. Be sure you get an owner's manual with the radio! Some hams may even have the service or shop manual. A shop manual can be a valuable addition, because it generally has more complete service procedures and troubleshooting guidelines.

If you are not getting at least an owner's manual with the radio, be cautious about making the purchase. Manuals for some pieces of equipment are available from the manufacturer. Several companies sell manuals for used or surplus equipment. See the advertising section of *QST*.

STATION LOCATION

Give some thought to station location. Hams put their equipment in many places. Some use the basement or attic, while others choose the den, kitchen, closet or a spare bedroom. Some hams with limited space build their station into a small closet. A fold-out shelf and folding chair form the operating position!

Where you put your station depends on the room you have available and on your personal tastes. There are, however, several things to keep in mind while searching for the best place. The photos show several ways amateur stations can be arranged.

One often-overlooked requirement for a good station location is adequate electrical service. Eventually you will have several pieces of station equipment and accessories, many of which will require power to operate. Be sure at least one, and preferably several, electrical outlets are located near your operating position. Be sure the outlets provide the proper voltage and current for your rig. The power supplies for most modern radios require only a few amperes at 120 V. You may run into problems, however, if your shack is on the same circuit as the air conditioner or washing machine. The total current drawn at any one time must not exceed rated limits. Someday you'll probably upgrade and may want to purchase a linear amplifier. If so, you should have a 240-V line in the shack, or the capability to get one.

Another must for your station is a good ground connection. A good ground

Many modern radios are very small and don't need much deck space. The Yaesu FT-897 is only $3^3/_{16}$ inches high by $7^7/_8$ inches wide and $10^5/_{16}$ inches deep.

Many operators prefer to stack some equipment on a shelf unit to save table space.

An armoire makes a good hiding place for your equipment. When the doors are closed it is just another piece of furniture in the room. The drawers provide convenient storage places for log books, headphones and other accessories.

not only reduces the possibility of electrical shock but also improves the performance of your station. By connecting all of your equipment to ground, you will help to avoid stray RF current in the shack. Stray RF can cause equipment to malfunction. A good ground can also help reduce the possibility of interference. The wire connecting your station to an earth ground should be as short as possible.

Basement and first-floor locations generally make it easier to provide a good ground connection for your station. You can also find a way to put your ham shack on the second or third floor or even on the top floor of a high-rise apartment, for that matter.

Your radio station will require a feed line of some sort to connect the antenna to the radio. So you will need a convenient means of getting the feed line into the shack. There are many ways of doing this, but one of the easiest and most effective requires only a window. Many hams simply replace the glass pane in a nearby window with a clear acrylic panel. The panel can be drilled to accept as many feed lines as needed. Special threaded "feedthrough" connectors make the job easy. If you decide to relocate your station, the glass pane can be replaced. This will restore the window to its original condition.

A simpler, more temporary approach is to pass the feed lines through the open window and close it gently. Do this carefully, because crushed coaxial cable will give you nothing but trouble. Be sure to secure the window with a block of wood or some other "locking" device. Your new equipment, seen through an open window, may tempt an unwanted visitor!

Another important requirement for your station is comfort. The space should be large enough so you can spread out as needed. Operating from a telephone booth isn't much fun! Because you will probably be spending some time in your shack (an understatement!), be sure it will be warm in the winter and cool in the summer. It should also be as dry as possible. High humidity can cause such equipment problems as high-voltage arcing and switch-contact failure.

If your station is in an area often used by other family members, be sure they know what they shouldn't touch. You should have some means of ensuring that no "unauthorized" person can use your equipment. One way to do this is to install a key-operated on-off switch in the equipment power line. When the switch is turned off and you have the key in your pocket, you will be sure that no one can misuse your station.

You will eventually want to operate late at night to snag the "rare ones" on 80 and 40 meters. Although a Morse code or voice contact is music to your ears, it may not endear you to a sleeping family. Putting your station in a bedroom shared with others may not be the best idea. You can keep the "music" to yourself, however, by using a good pair of headphones when operating your station.

ARRANGING YOUR EQUIPMENT

Before you set everything up and hook up the cables, think about where you want each piece of equipment. While there is no one best layout for a ham station, some general rules do apply. Of course, the location you've chosen for your station may limit your choices a bit. For example, if you're going to put the station in the basement, you may have a lot of space. If you'll be using a corner of the bedroom or den, however, you may have to keep the equipment in a small area.

Generally, the piece of gear that requires the most adjustment is the transceiver (or receiver if you have one). Make sure you can conveniently reach its controls, and keep in mind which hand you're going to use to make the adjustments. It doesn't make

If you don't have room to dedicate to your Amateur Radio station, you might consider using simple, lightweight equipment. This Elecraft K2 station can be set up on a table for an operating session, and stored away when not in use.

much sense to put the transceiver on the left side of the desk or table if you're going to adjust it with your right hand. Once you've found the best location for the transceiver, you can position the rest of your equipment around it.

If you'll be using Morse code, placement of the telegraph key is also very important. It must be easy to reach with the hand you send with, and placed so your arm will be supported when you're using it. Try to position it away from anything dangerous, such as sharp corners and edges, rough surfaces and electrical wires.

If you have room for a desk or table long enough to hold all your gear, you might prefer to keep everything on one surface. If you're pressed for space, however, a shelf built above the desk top is a good solution. Make certain the shelf is high enough to permit free air circulation above your transceiver. Leave at least 3 inches of space between the top of the tallest unit on the desktop and the bottom of your shelf. In building the shelf, keep in mind the weight of the equipment you're going to place on it. Try to restrict the shelf space to units you don't have to adjust very often such as the station clock, SWR meter and antenna rotator control box.

After you have a good idea of where you want everything, you can start connecting the cables. Often a dummy load or dummy antenna is also connected. Many hams use an antenna switch to select the dummy load or one of several antennas.

PRACTICAL ANTENNAS

Hams use many different kinds of antennas. There is no one best kind, so you will have to select an antenna type based on the available space, supports, cost and other factors. In this section we will give you some pointers on building and installing half-wavelength dipole antennas and quarter-wavelength vertical antennas for the HF bands.

The Half-Wave Dipole Antenna

Probably the most common amateur antenna is a wire cut to $1/2$ λ at the operating frequency. The feed line attaches across an insulator at the center of the wire. This is the half-wave dipole. We often refer to an antenna like this as a *dipole antenna*. (*Di* means two, so a dipole has two equal parts. A dipole could be a length other than $1/2$ λ.) The feed line connects to the center, so each side of a dipole is $1/4$-λ long.

You probably remember the equation to calculate the length of a half-wavelength dipole antenna.

$$\text{Length (in feet)} = \frac{468}{\text{f (in MHz)}}$$

(Equation 11-1)

The factor 468 in this equation takes into account antenna end effects, the slower velocity of the electromagnetic wave in wire than in free space and a few other factors. The result is a length that gives an approximate length for the antenna. You will probably have to adjust the final length, depending on the height of the antenna, nearby objects and other factors associated with your location. Notice

Table 11-1

Half-Wavelength Dipole Lengths for the Amateur HF Bands

Wavelength (m)	Frequency (MHz)	Length (ft)*
160	1.8	260
	2.0	234
80	3.5	133.7
	4.0	117
40	7.0	66.9
	7.3	64.1
30	10.1	46.3
	10.15	46.1
20	14.0	33.4
	14.35	32.6
17	18.068	25.9
	18.168	25.8
15	21.0	22.3
	21.45	21.8
12	24.89	18.8
	24.99	18.7
10	28.0	16.7
	29.7	15.8

* It's a good idea to add some extra wire to each leg to allow for adjustment.

that the frequency is given in megahertz and the antenna length is in feet for this equation. **Table 11-1** lists calculated lengths for $\frac{1}{2}$-λ dipoles at each end of the HF bands. This table shows how the length varies over the band. You should calculate the length for the section of the band you plan to operate on most.

Figure 11-1A shows the construction of a basic $\frac{1}{2}$-λ dipole antenna. Parts B through D show enlarged views of how to attach the insulators. You can use just about any kind of copper or copper-clad steel wire for your dipole. Most hardware or electrical supply stores carry suitable wire. Ordinary house wire, or two-conductor small-appliance wire (called "zip cord") can work well. Avoid using zip cord on 40 and 80 meters, however.

House wire and stranded wire will stretch with time, so a heavy gauge copper-clad steel wire is best. This wire consists of a copper jacket over a steel core. Such construction provides the strength of steel combined with the excellent conducting properties of copper. You can sometimes find copper-clad steel wire at a radio store. This wire is used for electric fences to keep farm animals in their place, so another place to try is a farm supply store.

Remember, you want a good conductor for the antenna, but the wire must also be strong. The wire must support itself *and* the weight of the feed line connected at the center. Although you can make a dipole antenna from almost any size wire, 12 or 14 gauge is usually best.

Cut your dipole according to the dimension found by Equation 11-1, but leave a little extra length to wrap the ends around the insulators. You'll need a feed line to connect it to your transmitter. If you plan to use coaxial cable to feed your dipole, look for some with a heavy braided shield. If possible, get good quality cable that has at least 95% shielding. If you stick with name brand cable, you'll get a good quality feed line. **Figure 11-2** shows the steps required to prepare the cable end for attachment to the antenna wires at the center insulator.

The final items you'll need for your dipole are three insulators. You can purchase them from your local radio or hardware store. **Figure 11-3** shows some common insulator types. You can also make your own insulators from plastic or Teflon blocks. See **Figure 11-4**. One insulator goes on each end and another holds the two wires together in the center. **Figure 11-5** shows some examples of how the feed line can attach to the antenna wires at a center insulator.

Antenna Location

Once you have assembled your dipole, find a good place to put it. *Never* put your antenna under, or over the top of electrical power lines. If they ever come into

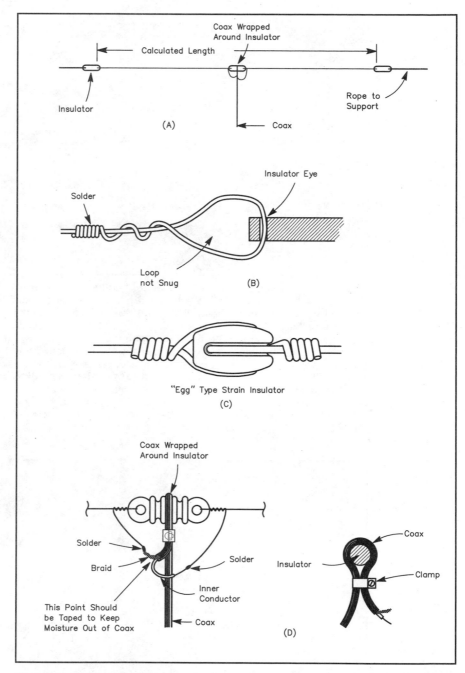

Figure 11-1—Simple half-wave dipole antenna construction. B and C show how to connect the wire ends to various insulator types. D shows the feed-line connection at the center.

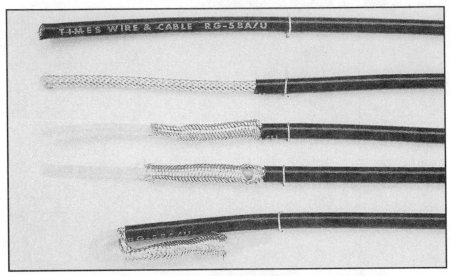

Figure 11-2—Preparing coaxial cable for connection to antenna wire.
A—Remove the outer insulation with a sharp knife or wire stripper. If you nick the braid, start over. B—Push the braid in accordion fashion against the outer jacket. C—Spread the shield strands at the point where the outer insulation ends. D— Fish the center conductor through the opening in the braid. Now strip the center conductor insulation back far enough to make the connection and tin (flow solder onto) both center conductor and shield. Be careful not to use too much solder, which will make the conductors inflexible. Also be careful not to apply too much heat, or you will melt the insulation. A pair of pliers used as a heat sink will help. The outer jacket removed in step A can be slipped over the braid as an insulator, if necessary. Be sure to slide it onto the braid before soldering the leads to the antenna wires.

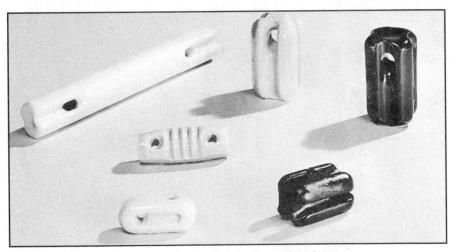

Figure 11-3—Various commercially made antenna insulators.

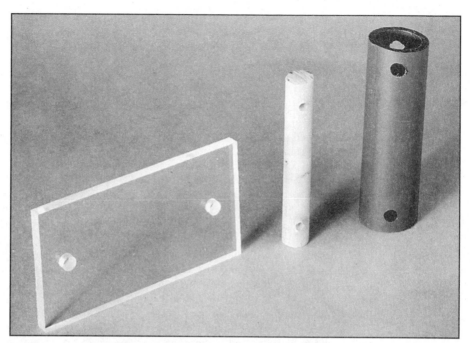

Figure 11-4—Some ideas for homemade antenna insulators.

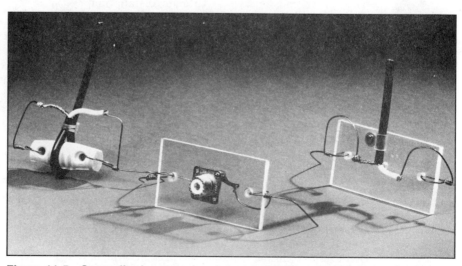

Figure 11-5—Some dipole center insulators have connectors for easy feeder removal. Others have a direct solder connection to the feed line.

contact with your antenna, you could be electrocuted. Avoid running your antenna parallel to power lines that come close to your station. Otherwise you may receive unwanted electrical noise. Sometimes power-line noise can cover up all but the strongest signals your receiver hears. You'll also want to avoid running your antenna too close to metal objects. These could be rain gutters, metal beams, metal siding, or even electrical wiring in the attic of your house. Metal objects tend to shield your antenna, reducing its capability.

The key to good dipole operation is height. How high? One wavelength above ground is good, and this ranges from about 35 feet on 10 meters to about 240 feet on 80 meters. Install the antenna as high as possible, to get it clear of buildings and trees. Of course very few people can get their antennas 240 feet in the air, so 40 to 60 feet is a good average height for an 80-meter dipole. Don't despair if you can get your antenna up only 20 feet or so, though. Low antennas can work well. Generally, the higher above ground and surrounding objects you can get your antenna, the greater the success you'll have. You'll find this to be true even if you can get only part of your antenna up high.

Normally you will support the dipole at both ends. The supports can be trees, buildings, poles or anything else high enough. Sometimes, however, there is just no way you can put your dipole high in the air at both ends. If you're faced with this problem, you have two reasonably good alternatives. You can support your dipole in the middle or at one end.

You can support the antenna in the middle, with both ends drooping toward the ground or with one end higher than the other in a sloping configuration. If you don't have the room to install a dipole in the standard form, don't be afraid to experiment a little. You can get away with bending the ends to fit your property, or even making a horizontal V-shaped antenna. Many hams have enjoyed countless hours of successful operating with antennas bent in a variety of shapes and angles.

Antenna Installation

After you've built your antenna and chosen its location, how do you get it up? There are many schools of thought on putting up antennas. Can you support at least one end of your antenna on a mast, tower, building or in an easily climbed tree? If so, you have solved some of your problems. Unfortunately, this is not always the case. Hams use several methods to get antenna support ropes into trees. Most methods involve a weight attached to a rope or line. You might be able to tie a rope around a rock and throw it over the intended support. This method works for low antennas. Even a major league pitcher, however, would have trouble getting an antenna much higher than 40 feet with this method.

A better method is to use a bow and arrow, a fishing rod or even a slingshot to launch the weight and rope. See **Figures 11-6** and **11-7**. You'll find that strong, lightweight fishing line is the best line to attach to the weight. (Lead fishing weights are a good choice.) Regular rope is too heavy to shoot any great distance. When you have successfully cleared the supporting tree, remove the weight. Then tie the support rope to the fishing line and reel it in.

If your first attempt doesn't go over the limb you were hoping for, try again. Don't just reel in the line, however. Let the weight down to the ground first and take it off the line. Then you can reel the line in without getting the weight tangled in the branches.

Figure 11-6—There are many ways to get an antenna support rope into a tree. These hams use a bow and arrow to shoot a lightweight fishing line over the desired branch. Then they attach the support rope to the fishing line and pull it up into the tree.

You can put antenna supports in trees 120 feet and higher with this method. As with any type of marksmanship, make sure all is clear downrange before shooting. Your neighbors will not appreciate stray arrows, sinkers or rocks falling in their yards!

When your support ropes are in place, attach them to the ends of the dipole and haul it up. Pull the dipole reasonably tight, but not so tight that it is under a lot of strain. Tie the ends off so they are out of reach of passersby. Be sure to leave enough rope so you can let the dipole down temporarily if necessary.

Just one more step and your antenna installation is complete. After routing the coaxial cable to your station, cut it to length and install the proper connector for your rig. Usually this connector will be a PL-259, sometimes called a UHF connector. **Figure 11-8** shows how to attach one of these fittings to RG-8 or RG-11 cable. Follow the step-by-step instructions exactly as illustrated and you should have no trouble. Be sure to place the coupling ring on the cable *before* you install the connector body! If you are using RG-58 or RG-59 cable, use an adapter to fit the cable to the connector. **Figure 11-9** illustrates the steps for installing the connector with adapter. The PL-259 is standard on most rigs. If you require another kind of connector, consult your radio instruction manual or *The ARRL Handbook* for installation information.

Tuning the Antenna

When you build an antenna, you cut it to the length given by an equation. This length is just a first approximation. Nearby trees, buildings or large metal objects and height above ground all affect the antenna resonant frequency. An SWR meter

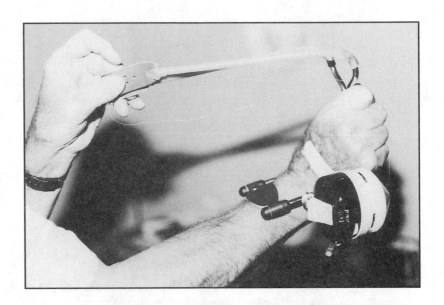

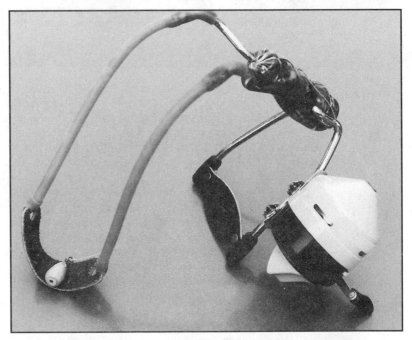

Figure 11-7—Another method for getting an antenna support into a tree. Small hose clamps attach a casting reel to the wrist bracket of a slingshot. Monofilament fishing line attached to a 1-ounce sinker is easily shot over almost any tree. Remove the sinker and rewind the line for repeated shots. When you find a suitable path through the tree, use the fishing line to pull a heavier line over the tree.

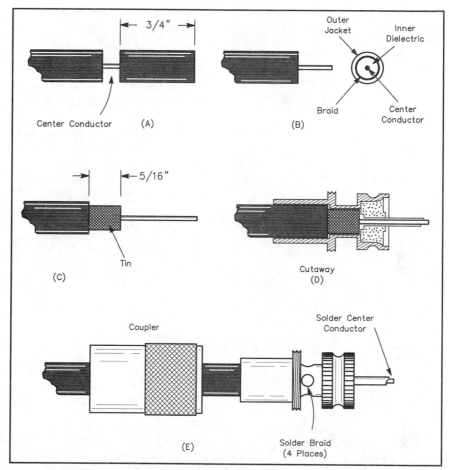

Figure 11-8—The PL-259 or UHF connector is almost universal for amateur HF work. It is also popular for equipment operating in the VHF range. Steps A through E illustrate how to install the connector properly. Despite its name, the UHF connector is rarely used on frequencies above 225 MHz.

can help you determine if you should shorten or lengthen the antenna. The correct length provides the best impedance match for your transmitter.

The first step is to measure the SWR at the bottom, middle and top of the band. On 80 meters, for example, you would check the SWR at 3.626, 3.700 and 3.749 MHz. For operation in the phone portion of the band (usually called 75 meters) you should check at 3.851 MHz and 3.999 MHz. Notice that we selected test frequencies that stay inside the appropriate General frequency limits. (A friend with a higher license class can help you check the SWR over a wider frequency range.) Graph the readings, as shown in **Figure 11-10**. You could be lucky—no further antenna adjustments may be necessary, depending on your transmitter.

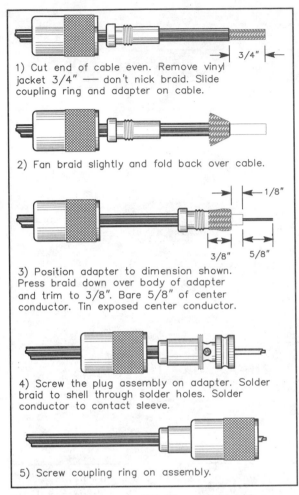

1) Cut end of cable even. Remove vinyl jacket 3/4" — don't nick braid. Slide coupling ring and adapter on cable.

2) Fan braid slightly and fold back over cable.

3) Position adapter to dimension shown. Press braid down over body of adapter and trim to 3/8". Bare 5/8" of center conductor. Tin exposed center conductor.

4) Screw the plug assembly on adapter. Solder braid to shell through solder holes. Solder conductor to contact sleeve.

5) Screw coupling ring on assembly.

Figure 11-9—If you use RG-58 or RG-59 with a PL-259 connector, you should use an adapter, as shown here. Thanks to Amphenol Electronic Components, RF Division, Bunker Ramo Corp, for this information.

Many tube-type transmitters include an output tuning network. They will usually operate fine with an SWR of 3:1 or less. Most solid-state transmitters (using all transistors and integrated circuits) do not include such an output tuning network. These no-tune radios begin to shut down— the power output drops off — with an SWR much higher than 1.5:1. In any event, most hams like to prune their antennas for the lowest SWR they can get at the center of the band. With a full-size dipole 30 or 40 feet high, your SWR should be less than 2:1. If you can get the SWR down to 1.5:1, great! It's not worth the time and effort to do any better than that.

If the SWR is lower at the low-frequency end of the band, your antenna is probably too long. Making the antenna *shorter* will *increase* the resonant frequency. Disconnect the transmitter and try shortening your antenna at each end. The amount to trim off depends on two things. First is which band the antenna is operating on, and, second, how much you want to change the resonant frequency. Let's say the antenna is cut for the 80-meter band. You'll probably need to cut 8 or 10 inches off each end to move the resonant frequency 50 kHz. You may have to trim only an inch or less for small frequency changes on the 10-meter band. Measure the SWR again (remember to recheck the calibration). If the SWR went down, keep shortening the antenna until the SWR at the center of the band is less than 2:1.

If the SWR is lower at the high-frequency end of the band, your antenna was probably too short to begin with. If so, you must add more wire until the SWR is acceptable. Making the antenna *longer* will *decrease* the resonant frequency. Before you solder more wire on the antenna ends, try attaching a 12-inch wire on each end. Use alligator clips to attach a length of wire. You don't need to move the insulators yet. Clip a wire on each end and again measure the SWR. Chances are the antenna will now be too long. You will need to shorten it a little at a time until the SWR is below 2:1.

Once you know how much wire you need to add, cut two pieces and solder them to the ends of the antenna. When you add wire, be sure to make a sound

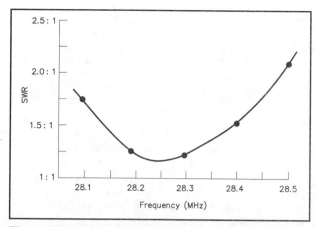

Figure 11-10—This graph shows how the SWR might vary across an amateur frequency band. The point of lowest SWR here is near the center of the band, so no further antenna-length adjustments are necessary.

mechanical connection before soldering. Remember that these joints must bear the weight of the antenna and the feed line. After you solder the wire, reinstall the insulators at the antenna ends, past the solder connections.

If the SWR is very high, you may have a problem that can't be cured by simple tuning. A very high SWR may mean that your feed line is open or shorted. Perhaps a connection isn't making good electrical contact. It could also be that your antenna is touching metal. A metal mast, the rain gutter on your house or some other conductor would add considerable length to the antenna. If the SWR is very high, check all connections and feed lines, and be sure the antenna clears surrounding objects.

Now you have enough information to construct, install and adjust your dipole antenna. You'll need a separate dipole antenna for each band you expect to operate on. Sometimes a 40-meter dipole will also work on 15 meters, though. Check the SWR on 15 meters—you may have a two-for-one dipole!

The Quarter-Wave Vertical Antenna

The quarter-wavelength vertical antenna is simple and popular. It requires only one support and can be very effective. On the HF bands (80 to 10 meters) it is often used for DX work. Vertical antennas send radio energy equally well in all compass directions. This is why we sometimes call them *nondirectional antennas* or *omnidirectional antennas*. They also tend to concentrate the signals toward the horizon. Vertical antennas do not generally radiate strong signals straight up, like horizontal dipoles do. Because they concentrate the signals toward the horizon, we say they have *gain* as compared to a dipole. Gain always refers to a comparison with another antenna. A dipole is one common comparison antenna for stating antenna gain.

Use Equation 11-2 to find $\frac{1}{4}$ λ for the radiator. The frequency is given in megahertz and the length is in feet in this equation.

$$\text{Length (in feet)} = \frac{234}{\text{f (in MHz)}}$$

(Equation 11-2)

Table 11-2 gives approximate lengths for the radiator and each ground radial of a $\frac{1}{4}$-λ vertical at each end of the HF bands.

As with $\frac{1}{2}$-λ dipoles, the resonant frequency of a $\frac{1}{4}$-λ vertical decreases as the length increases. Shorter antennas have higher resonant frequencies.

The $1/4$-λ vertical also has radials. For operation on 80 through 10 meters, the vertical may be at ground level, and the radials placed on the ground. The key to successful operation with a ground-mounted vertical antenna is a good radial system. The best radial system uses *many* ground radials. Lay them out like the spokes of a wheel, with the vertical at the center. Some hams have buried ground radial systems containing over 100 individual wires.

Ideally, these wires would be $1/4$ λ long or more at the lowest operating frequency. With such a system, earth or ground losses will be negligible. When the antenna is mounted at ground level, radial length is not very critical, however. Studies show that with fewer radials you can use shorter lengths, but with a corresponding loss in antenna efficiency.[1, 2] Some of your transmitter power does no more than warm the earth beneath your antenna. With 24 radials, there is no point in making them longer than about $1/8$ λ. With 16 radials, a length greater than 0.1 λ is unwarranted. Four radials should be considered an absolute minimum. Don't put the radials more than about an inch below the ground surface.

Compared with 120 radials of 0.4 λ, antenna efficiency with 24 radials is roughly 63%. For 16 radials, the efficiency is roughly 50%. So it pays to put in as many radials as you can.

If you place the vertical above ground, you reduce earth losses drastically. Here, the wires should be cut to $1/4$ λ for the band you plan to use. Above ground, you need only a few radials—two to five. If you install a multiband vertical antenna above ground, use separate ground radials for each band you plan to use. These lengths are more critical than for a ground-mounted vertical. For elevated verticals, you should have a minimum of two radials for each band. You can mount a vertical on a pipe driven into the ground, on the chimney or on a tower.

The radials at the bottom of a vertical antenna mounted above ground form a surface that acts like the ground under the antenna. These anten-

[1]J. O. Stanley, "Optimum Ground Systems for Vertical Antennas," *QST*, December 1976, pp 14-15.
[2]B. Edwards, "Radial Systems for Ground-Mounted Vertical Antennas," *QST*, June 1985, pp 28-30. This article is reproduced in *Vertical Antenna Classics*, available from the ARRL.

Table 11-2
Quarter-Wavelength Vertical Antenna Lengths for the Amateur HF Bands

Wavelength (m)	Frequency (MHz)	Length (ft)
160	1.8	130
	2.0	117
80	3.5	66.9
	4.0	58.5
40	7.0	33.4
	7.3	32.1
30	10.1	23.2
	10.15	23.1
20	14.0	16.7
	14.35	16.3
17	18.068	13.0
	18.168	12.9
15	21.0	11.1
	21.45	10.9
12	24.89	9.4
	24.99	9.4
10	28.0	8.4
	29.7	7.9

nas are sometimes called *ground-plane antennas*.

Once a vertical antenna is several feet above ground, there is little advantage in more height. (This assumes your antenna is above nearby obstructions.) For sky-wave signals, a height of 15 feet for the base is almost as good as 50. This isn't the case for a horizontal antenna, where height is important for working DX. Only if you can get the vertical up 2 or 3 wavelengths does the low-angle radiation begin to improve. Even then the improvement is only slight.

CONCLUSION

The information presented in this chapter is intended to help you begin assembling your Amateur Radio station. It won't answer every question you may have, however. A local ham or radio club can prove very helpful. In addition, the many books published by the ARRL provide a wealth of technical and operating assistance.

Perhaps the most important piece of advice we can leave you with is to experiment, and have fun.

The Ham's International Language

To qualify for a General license, you must pass the Element 2 and Element 3 written exams about FCC Rules, electronics theory and amateur operating practices. In addition, you will have to show your ability to communicate using the international Morse code. The code exam (Element 1) is given at 5 words per minute (wpm).

To earn the Technician license, you do not have to pass a Morse code exam. That license does not include privileges on the amateur high-frequency bands, however. Those bands provide direct worldwide communications. As a Technician class licensee, you can earn those privileges and join the excitement of communicating with other Amateur Radio operators around the world. You simply have to pass the 5-wpm Morse code exam. That can be your first step towards a General class license.

This book won't teach you the Morse code. In fact, you won't even find a copy of the Morse code printed in this book. That's because Morse code is best learned by sound, not by sight. We don't want to confuse you with printed dots and dashes. That will only slow you down later. (If you really want to see a printed copy of the Morse code dots and dashes, you can find it in *The ARRL Handbook* and *Morse Code: The Essential Language*, both published by the ARRL. Most dictionaries also include a copy of the Morse code, as do many encyclopedias.)

As we said, the code is best learned by listening to sounds, and for that you'll need a method that teaches you the Morse code characters by producing the sounds for you. ARRL's *Your Introduction to Morse Code* is available as a set of two cassette tapes or two audio CDs. *Your Introduction to Morse Code* teaches you all of the Morse code characters required by the FCC. Practice text at 5 wpm ensures

How can one ham who speaks only English communicate with another ham who speaks no English? Using Morse code, of course! This chapter gives you some helpful hints and study suggestions for learning the skill of communicating by Morse code.

that you are prepared for the 5-wpm code exam.

There are several good computer programs that will teach you the code and give you unlimited practice without repeating the same text. For the IBM PC and compatible computers, the ARRL sells the MICA *Ham University* program, which includes a Windows-based Morse-code training program. You can tell the computer what code speed you want it to use for the characters and words it sends. *Ham University* drills you on individual characters and combinations of characters. You can practice with a mixture of random characters or with words.

The audio CDs, cassette tapes and computer program are available directly from ARRL Headquarters or from the many dealers who sell ARRL publications. Write to ARRL, 225 Main Street, Newington, CT 06111 or call toll free at 888-277-5289 to order. You can even order directly over the Internet on *ARRLWeb:* **www.arrl.org/**

CODE IS FUN!

Using the code is an exciting way to communicate. Many long-time hams beam with pride when they proclaim that they don't even own a microphone! You are fluent in another complete language when you know the code. You can chat with hams from all around the world using this common language. With the practice you will gain by making on-the-air contacts, your speed will increase quickly.

Morse code goes back to the very beginning of radio, and is still one of the most effective radio-communication methods. We send Morse code by interrupting the continuous-wave signal generated by a transmitter, and so we call it CW for short.

It takes far less power to establish reliable communications with CW than it does with voice (phone). On phone, we sometimes need high power and elaborate antennas to communicate with distant stations, or *DX* in the ham's lingo. On CW, less power and more modest stations will provide the same contacts. There is great satisfaction in being able to communicate using Morse code. This is similar to the satisfaction you might feel from using any acquired skill.

Conserve Time and Spectrum Space

Another advantage of CW over phone is its very narrow bandwidth. Morse code makes efficient use of spectrum space. The group of frequencies where hams operate — the *ham bands* — are narrow portions of the whole spectrum. Many stations use the bands, and because they are so crowded, interference is sometimes a problem. A CW signal occupies only about one-tenth the bandwidth of a phone signal. This means as many as 10 CW signals can fit into the space taken up by one phone signal.

Over the years, radiotelegraph operators have developed a vocabulary of three-letter Q signals, which other radio-telegraphers throughout the world understand. For example, the Q signal *QRM* means "you are being interfered with." Just imagine how hard it would be to communicate that thought to someone who didn't understand one word of English. You learned some common Q signals for your Technician license. *The ARRL Handbook* and *The ARRL Operating Manual* also have lists of Q signals.

Another advantage to using Q signals is speed. It's much faster to send three letters than to spell out each word. That's why you'll use these Q signals even when you're chatting by code with another English-speaking ham. Speed of transmission is also the reason radiotelegraphers use a code "shorthand." For example, to acknowledge that you heard what was transmitted to you, send the letter R. This means "I have received your transmission okay."

Standard Q signals and shorthand abbreviations reduce the total time necessary to send a message. When radio conditions are poor and signals are marginal, a short message is much more likely to get through.

Many hams prefer to use CW in traffic nets. (A net is a regular on-the-air meeting of a group of hams.) Using these nets, hams send messages across the country for just about anyone. When they send messages using CW, there is no confusion about the spelling of names such as Lee, Lea or Leigh.

CW: Sometimes the Only Choice

When WA6INJ's jeep went over a cliff in a February snowstorm, he was able to call for help using his mobile rig. This worked well at first, but as the search for him continued, he became unable to speak. The nearly frozen man managed to tap out Morse code signals with his microphone push-to-talk button. That was all his rescuers had to work with to locate him. Morse code saved this ham's life.

Some hams like to bounce VHF and UHF signals off the surface of the moon to another ham station on the Earth. Because of its efficiency, hams use CW in most of this *moonbounce* work. They could use voice signals, but this increases the power and antenna-gain requirements quite a bit.

On some frequencies, amateurs communicate by bouncing their signals off an auroral curtain in the northern sky. (Stations in the Southern Hemisphere would use an auroral curtain in the southern sky.) Phone signals become so distorted in the process of reflecting off an auroral curtain that they are difficult or impossible to understand. CW is the most effective way to communicate using signals bounced off an aurora.

You will feel a special thrill and a warm satisfaction when you use the Morse code to communicate with someone. This feeling comes partly from sending messages to another part of the world without regard to language barriers. Sending and receiving Morse code is a skill that helps set amateurs apart from other people. It provides a common bond between amateurs worldwide.

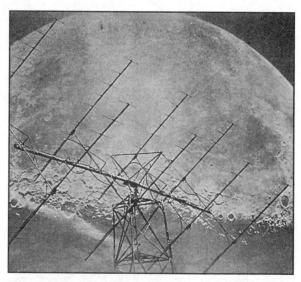

Figure 12-1—Even the moon isn't immune from ever-ambitious hams, and code is the most efficient type of signal to bounce off its surface and be reflected back to Earth.

GETTING ACQUAINTED WITH THE CODE

The basic element of a Morse code character is a dot. The length of the dot determines how long a dash should be. The dot length also determines the length of the spaces between elements, characters and words. **Figure 12-2** shows the proper timing of each piece. Notice that a dash is three times as long as a dot. The time between dots and dashes in a character is equal to the length of a dot. The time between letters in a word is equal to three dot lengths and the space between words is seven dot lengths.

The lengths of Morse code characters are not all the same, of course. Samuel Finley Breese Morse (1791-1872) developed the system of dots and dashes in 1838. He assigned the shortest combinations to the most-used letters in plain-language text. The letter E has the shortest sound, because it is the most-used letter. T and I are the next most-used characters, and they are also short. Character lengths get longer for letters used less often.

An analysis of English plain-language text shows that the average word (including the space after the word) is 50 units long. By a unit we mean the time of a single dot or space between the parts of a character. The word PARIS is 50 units long, so we use it as a standard word to check code speed accurately. For example, to transmit at 5 words per minute (wpm), adjust your code-speed timing to send PARIS five times in one minute. To transmit at 10 wpm, adjust your timing to send PARIS 10 times in one minute.

As you can see, the correct dot length (and the length of dashes and spaces) changes for each code speed. As a result, the characters sound different when the speed changes. This leads to problems for a person learning the code. Also, at slower speeds, the characters seem long and drawn out. The slow pace encourages students to count dots and dashes, and to learn the code through this counting method.

Learning the code by counting dots and dashes introduces an extra translation in your brain. (Learning the code by memorizing Morse-code-character dot/dash patterns from a printed copy introduces a similar extra translation.) That extra translation may seem okay at first. As you try to increase your speed, however, you will

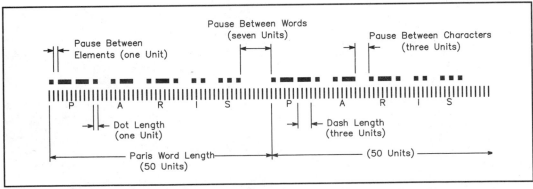

Figure 12-2—Whether you are a beginner or an expert, good sending depends on maintaining proper time ratios or "weight" between the dots, dashes and spaces, as shown.

soon find out what a problem it is. You won't be able to count the dots and dashes and then make the translation to a character fast enough!

Learning Morse Code

Many studies have been done, and various techniques tried, to teach Morse code. The method that has met with the most success is called the *Farnsworth method*. That is the method used in the ARRL package, *Your Introduction to Morse Code*. With this technique, we send each character at a faster speed (we use a 15-wpm character speed at ARRL). At speeds in this range, the characters — and even some short words — begin to take on a distinctive rhythmic pattern.

With this faster character speed, we use longer spaces between characters and words to slow the overall code speed. *Your Introduction to Morse Code* starts with code sent at an overall speed of 5 wpm. (You can still measure this timing by using the word PARIS, as described earlier.) Once you learn the character sounds, and can copy at 5 wpm, it will be easy to increase your speed. Just decrease the spaces between letters and words, and your code speed increases without changing the rhythmic pattern of the characters. If you use a computer program to learn the code, or another method, we recommend this same technique, with the code characters sent at 15 wpm or faster.

Learn to recognize that rhythmic pattern, and associate it directly with the character. You'll learn the code in the shortest possible time, and it will be much easier to increase your code speed. Decreasing the space between characters and words provides a natural progression to increase your code speed.

Morse code is a communication method that depends on sounds. To understand the communication you must hear the sounds and interpret their meaning. This is why most code-teaching methods repeat the sounds for you to listen to and associate the characters with the sounds. It is also why you will not find a copy of the Morse code dots and dashes printed in this book. Audio CDs, cassette tapes, computer programs and even classroom or individual practice with a code-practice oscillator and a code key all rely on sounds to teach you the Morse code characters. (A code key is sometimes called a hand key, straight key or a telegraph key.)

Several other techniques have also been successful for teaching Morse code. Some of these methods involve memorizing a printed copy of the code. There are even a few commercial packages that picture the dots and dashes of the code characters in various "creative" patterns to help you remember them. You can learn Morse code by following any of these methods. Most people will learn faster, and will be able to increase their code speed easier, however, by using a "listening" method. For those who seem unable to learn the Morse code using the audio CDs or tapes in *Your Introduction to Morse Code* or a computer program, one of these "printed" methods may prove helpful. Save that as a last resort, however. Practice faithfully with *Your Introduction to Morse Code* or a computer program every day for at least three to four weeks. Then, if you have not learned many of the characters, you may want to try one of the visual, or printed, methods.

The 26 letters of the alphabet and the numbers 0 through 9 each have a different sound. There are also different sounds for the period, comma, question mark, double dash (=), fraction bar (/) and some procedural signals that hams use. The + sign, which hams call $\overline{AR}$, means "end of message." $\overline{SK}$ means "end of work," or

Morse Code From the Heart

The power of modern medicine kept one ham alive; the power of Morse code kept him in touch with loved ones.

Reprinted from July 1990 QST.

Have you ever felt Morse code rather than heard it? Although I've been a licensed amateur for 34 years and have had many wonderful experiences in this hobby, none can compare to the one I'll never forget. It happened on Monday, August 26, 1985.

My husband Ralph, W8LCU, went to the hospital the Friday before for an ECG and heart catheterization. He was told that an immediate quadruple bypass was needed, and he could not leave the hospital. Arrangements were made for the following Monday. Ralph had always been healthy and never showed an inkling of heart problems, so this came as quite a shock to us. The doctors told us that after his surgery he would not be speaking or doing much else, until perhaps the next day or until he was off the respirator, heart pump and everything else that goes with this type of surgery.

Over the weekend we had many things to discuss, and Amateur Radio never entered my mind.

Monday evening, in the intensive care unit (ICU) after surgery, I held Ralph's hand. He began tapping on my palm. I didn't think much of it, but his eyes opened and seemed to be telling me something. He moved his fingers to my wrist and suddenly, as clear as if I were hearing CW on the radio, came the letters P A (tears coming down my face) I N (pause) I S   H E L L. "Pain is hell."

Before I could say or do anything, four nurses were in the room. I told them what Ralph had just said to me in Morse code. I think they were ready to have me taken to the mental ward, but they gave him a shot. Then they stood by in disbelief as he tapped out a short message to our son and daughter, and a final I LUV U before drifting off to sleep.

Meanwhile, in another hospital across the state, a friend of ours, Vanessa, KA8THR, was back in her room after major surgery. Her first words to her husband, Chuck, N8EOJ, were "How is Ralph doing?" Sure enough, through the wonders of CW, hand-held transceivers, repeaters and many hams along the way, relays of Ralph's progress were sent to N8EOJ all day long. When Ralph tapped CW to me, within moments hams all across the state knew all was well.

The ICU and other units were buzzing for days as doctors came into Ralph's room to find out about this "new" form of communication. They told Ralph they thought more people should know Morse code, themselves included. Of course, Ralph told them that with a little study and determination, anyone can do it.

Needless to say, CW will always be important to me, either in an emergency or just for plain fun.

Ralph recovered from his surgery, and I, while I was an Advanced at the time, have gone on to earn my Amateur Extra license. — *Donna Burch, W8QOY* "end of contact." Hams sometimes refer to the double dash as $\overline{BT}$ and the fraction bar as $\overline{DN}$. These two-letter combinations are written with a line over the letters to indicate that two letters are sent as one character to form these symbols. That's a total of 43 character sounds that you will have to learn for your 5-wpm Morse code exam. You'll learn the sounds of all these characters as you practice with your code cassettes or computer program.

Morse Code: The Essential Language, by L. Peter Carron, Jr, W3DKV (published by the ARRL) contains other suggestions for learning Morse code. That book also describes the history of Morse code. It includes several stories about lives saved because of emergency messages transmitted over the radio.

Sounding the Code

Some people find it helpful to say the sounds of Morse code characters, especially when they are first learning the code. Instead of saying the names of the Morse code elements, dot and dash, we use the sounds "dit" and "dah." If the dot is at the beginning or in the middle of the character, we sound it out as "di" instead of "dit."

Listen to the difference between the sounds you make saying the word "dit" and saying "dah." If you can tell the difference between those sounds, you have all the ability you need to learn the code. Being able to receive Morse code is really nothing more than being able to recognize a sound. Try it yourself. Say "didah." Now say "didahdit." Can you hear the difference? Congratulations! You now know the sounds for the letters A and R, and are on your way to learning the Morse code.

Using this method, you hear the sound "didah" and associate that sound with the letter A. With practice, you'll learn all the sounds and associate them with the correct letters, numbers and punctuation.

Learning to Write

To learn the code, you train your hand to write a certain letter, number or punctuation mark whenever you hear a specific sound. You are forming a habit through your practice. After all, forming a habit is nothing more than doing something the same way time after time. Eventually, whenever you want to do that thing, you automatically do it the same way; it has become a habit.

You will need lots of practice writing the specific characters each time you hear a sound. Eventually you'll copy the code without thinking about it. In other words, you'll respond automatically to the sound by writing the corresponding character.

As you copy code sent at faster speeds, you may find that your ability to write the letters limits you. Practice writing the characters as quickly as possible. If you normally print, look for ways to avoid retracing lines. Don't allow yourself to be sloppy, though, because you may not be able to read your writing later.

Many people find that script, or cursive writing is faster than printing. Experiment with different writing methods, and find one that works for you. Then practice writing with that method so you don't have to think about forming the characters when you hear the sounds.

Some Study Suggestions

The secret to easy and painless mastery of the Morse code is regular practice. Set aside two 15 to 30-minute periods every day to practice the code. If you try longer sessions, you may become over tired, and you will not learn as quickly.

Likewise, if you only practice every other day or even less often, you will tend to forget more between practice sessions. It's a good idea to work your practice sessions into your daily routine. For instance, practice first thing in the morning and before dinner. Daily practice gives quick results.

Learn the sound of each letter. Morse code character elements sound like dits and dahs, so that's what we call them. Each character has its own pattern of dits and dahs, so learn to associate the sound of that pattern with the character. Don't try to remember how many dots and dashes make up each character. Practice until you automatically recognize each Morse code character.

Feel free to review. You're learning a new way of communicating, by using the Morse code. If you are having trouble with a particular character, spend some extra time with it. If you are using ARRL's *Your Introduction to Morse Code*, replay the problem CD track or rewind the tape and play that section again. If you are using one of the computer programs, spend some extra time drilling on the problem character. After you've listened to the practice on one character two or three times, however, go on to the next one. You will get more practice with the problem character later, because you will be constantly reviewing all of the characters learned so far. You can even come back to the problem character again in a later practice session. Be sure to listen to the practice for at least 3 characters during each practice session, however.

If you hear a Morse code character you don't immediately know, just draw a short line on your copy paper. Then get ready for the next letter. If you ignore your mistakes now, you'll make fewer of them. If you sit there worrying about the letter you missed, you'll miss a lot more! You shouldn't expect to copy perfectly while you are learning the code. You will get better with more practice.

When you think you've recognized a word after copying a few letters, concentrate all the more on the actual code sent. If you try to anticipate what comes next, your guess may turn out to be wrong. When that happens, you'll probably get confused and miss the next few letters as well. Write each letter just after it is sent. With more practice, you'll learn to "copy behind," hearing and writing whole words at one time. For now, when you are just learning the code, concentrate on writing each character as it is sent. This helps reinforce the association of a sound and a character in your mind.

One trick that some people use while learning the code is to whistle or hum the code while walking or driving. You can also say the sounds di, dit and dah to sound out the characters. Send the words on street signs, billboards and store windows. This extra practice may be just the help you need to master the code!

Morse code is a language. Eventually you'll begin to recognize common syllables and words. With practice you will know many complete words, and won't even listen to the individual letters. When you become this familiar with the code, it really starts to be fun!

Don't be discouraged if you don't seem to be breaking any speed records. Some people have an ear for code and can learn the entire code in a week or less. Others require a month or more to learn it. Be patient, continue to practice and you will reach your goal.

To pass the Element 1 code examination, you must be able to understand a plain-language message sent at 5 wpm. As you know, in the English language some words are just one or two letters long, others 10 or more letters long. To standardize the code test, the FCC defines a word as a group of five letters. Numbers and punctuation marks normally count as two characters for this purpose. You will receive 25 characters in one minute for 5-wpm code.

COMFORTABLE SENDING

There's more to the code than just learning to receive it; you'll also have to learn to send it. To accomplish this, you'll need a telegraph key and a code-practice oscillator. You can get these items at most electronics-parts stores, or you can build your own simple oscillator. The "Assembling A Code-Practice Oscillator" sidebar gives you step-by-step instructions for building one simple oscillator that even includes a key for you to practice with. The key with this project is not the best key you could use, but it will allow you to do some practice sending. You may want to consider adding a real telegraph key to the project — the circuit board has two holes to connect wires to a straight key.

Many experienced amateurs prefer to use an electronic keyer to send Morse code. An electronic keyer produces properly timed dots and dashes, because it uses one circuit to produce dots and another circuit to produce dashes. In general, it is probably better to learn to send Morse code with a hand key at first. Some students may have good success with a keyer. Commercial keyers range from simple, basic units to full-featured Morse code machines. While they are comparatively expensive, some of the full-featured machines offer features that are quite helpful to a beginner. The keyers shown in **Figure 12-3** can send random-character code practice at any speed you desire. While the AEA keyer is no longer

Figure 12-3 — Electronic keyers shown here can store messages in memory to be sent at the touch of a button and also offer various training features to help you learn Morse code and increase your speed. The MFJ keyer sends 5-character random-letter groups, random 1 to 8-character groups and plain text in the format of an on-the-air Amateur Radio contact (QSO). In addition to sending random characters, the AEA MM-3 can connect to a computer to send text from a keyboard or text file. The MM-3 also contains a program that allows you to "contact" other stations in the keyer and exchange QSO information.

Figure 12-4 — Older than radio itself, code still reigns as the most efficient and effective communications mode; many hams use it almost exclusively. A modern straight key, the device most beginners use, is in the foreground.

available new, you may find one on the use-equipment market. MFJ also has several keyer models available. There are numerous electronic keyers and kits available as new or used equipment.

Figure 12-4 shows two different standard straight key models (one in the foreground and one to the left). There is a semiautomatic "bug" in the background and a popular Bencher paddle used with modern electronic keyers on the right. You'll want to obtain some type of code key to practice with before you're ready for your 5-wpm code test. You'll need the key for some of your on-the-air operating after the license arrives from the FCC as well! Most new hams start with an inexpensive straight key.

Just as with receiving, it is important that you be comfortable when sending. It

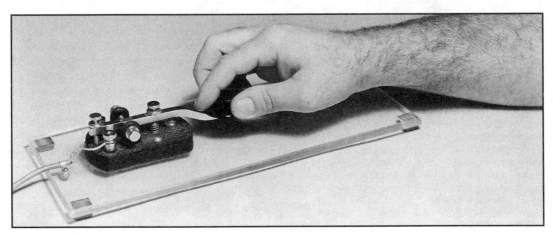

Figure 12-5 — Proper forearm support, with the wrist off the table, a gentle grip and a smooth up-and-down motion make for clean, effortless sending.

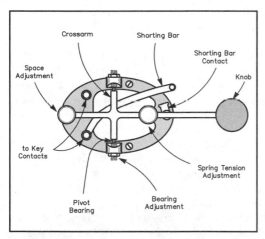

Figure 12-6 — A few simple adjustments to suit your style allow you to send for hours without fatigue. The contact spacing and spring tension should be set to provide the most comfortable feel.

helps to rest your arm on the table, letting your wrist and hand do all the work. Grasp the key lightly with your fingertips. Don't grip it tightly. If you do, you'll soon discover a few muscles you didn't know you had, and each one will ache. With a light grasp, you'll be able to send for long periods without fatigue. See **Figure 12-5**.

Another important part of sending code is the proper adjustment of your telegraph key. There are only two adjustments that you will normally have to make on a straight key, but you'll find they are very important. **Figure 12-6** illustrates these adjustments.

The first adjustment is the spacing between the contacts, which determines the distance the key knob must move to send a letter. Adjust the contacts so the knob moves about the thickness of a dime ($^1/_{16}$ inch). Try it. If you're not satisfied with this setting, try a wider space. If that doesn't do it, try reducing the space. Eventually, you'll find a spacing that works best for you. Don't be surprised if your feelings change from time to time, however, especially when your sending speed increases.

The second adjustment you must make is the spring tension that keeps the contacts apart. Just as there is no "correct" contact spacing, there is no correct tension adjustment for everyone. You will find, however, that adjusting the spacing may also require a tension adjustment. Adjust the tension to provide what you think is the best "feel" when sending.

Some straight keys have ball bearing pivot points on either side of the crossarms. These normally need no adjustment, but you should be sure the crossarms move freely in these pivots. If the bearings are too tight the key will bind or stick. If they are too loose there will be excessive play in the bearings. The side screws also adjust the crossarms from side to side so the contact points line up.

All of these adjustment screws have lock nuts, so be sure you loosen them before you make any adjustments. Tighten the lock nuts securely after making the adjustment.

Some better-quality keys have a shorting bar like the one shown in Figure 12-6. This shorting bar can be used to close the key contacts for transmitter tuning or adjustments.

You'll probably want to fasten the key to a piece of wood or other heavy weight to prevent it from sliding around as you send. You might even want to fasten the key directly to the table. It's best to experiment with different positions before permanently attaching the key to any surface, however.

Some operators use a board that extends under their forearm and allows their arm to hold the key in position. This way the key can be moved aside for storage, or to clean off the table for other activities. A piece of $^1/_4$-inch Plexiglas about 12 inches long works well for this type of key base. This technique also allows you to reposition the key to find the most comfortable sending position.

Learning to send good code, like learning to receive, requires practice. A good way to start is to send along with your code practice. Try to duplicate the sounds as much as possible.

Always remember that you're trying to send a complete sound, not a series of dots and dashes. With that in mind, take a moment to think about the sound you're trying to send. It consists of dits, dahs and pauses (or spaces). The key to good sending, then, is timing. If you're not convinced, listen to some code on the bands. What transmissions would you prefer to receive? Why? That's right — because they have good timing.

Some people send code that is very easy and enjoyable to copy. Others are not so good, and require a lot of concentration to understand. Since your fist is your on-the-air signature, try to make it as easy to read as possible. Learn to send code that is easy to copy. Then you'll have a fist you can be proud of. It can make a big difference in your success as an amateur operator.

One of the best ways to learn to send and receive code is to work with another person. That person may be another member of your own family, a friend who is also studying the code or a licensed ham. If you're attending an organized class, you may be able to get together with another student several times a week.

If you must work alone, make good use of your tape recorder. Try recording your sending. After waiting a day or two, try receiving what you sent. The wait between sending and receiving will help prevent you from writing down the message from memory. Not only will this procedure provide practice in receiving, but it will also let you hear exactly how your sending sounds. If your timing is off, you'll hear it. If you're having trouble with a specific letter or number or punctuation mark, you'll soon know about it. If you can't understand your own sending, neither will anyone else!

Regardless of whether you're sending or receiving, the key to your success with Morse code is regular practice. After you've learned all the characters you will want to continue your regular practice to gain confidence and increase your speed.

If you can, listen to actual contacts between hams. Make use of the code-practice material transmitted daily by W1AW, the ARRL station heard nationwide from Newington, Connecticut. There is a W1AW schedule in the Introduction chapter of this book.

EXAM DAY

If you have prepared properly, the code exam should be easy — a pleasant experience, not an agonizing one. The key to a successful exam is to relax. One of the biggest reasons people fail the code examination is because they become nervous. You can be confident of passing, however, because you have studied hard and are properly prepared. The Introduction chapter has more information about what to expect on exam day.

The exam isn't *that* important. You can always take the test again soon. It won't kill you even if you don't pass. But you are going to pass! You can copy 5 wpm at home with only a couple of errors every few minutes. The code test will be even easier.

Think positively: "I will pass!" Relax. Before you know it, you'll be grinning from ear to ear as you hear the words, "You've passed!"

Once you have passed the Morse code exam, you will want to practice your new skills on the air. You will soon find yourself wishing you could copy faster. Your speed will increase as you spend more time on the air operating CW. W1AW code practice and computer programs can give you additional practice at faster code speeds. If you followed the plan, and learned Morse code using 15 wpm characters, then the space between characters decreases gradually to increase your overall code speed. You'll be copying faster code before you know it! W1AW code practice is also available as MP3 files that you can download from the Internet. Copy the Morse code practice using the Windows Media Player or Real Audio Player on your computer or a portable MP3 player. To download the code practice files go to **www.arrl.org/w1aw/morse.html**.

Good luck. We hope to hear you on the air!

Assembling a Code-Practice Oscillator

It isn't difficult to construct a code-practice oscillator. A complete oscillator that mounts on a small piece of wood is shown in Figure A. Figure B shows all the parts for this project laid out ready for assembly. The circuit board for this project can be ordered from FAR Circuits, 18 N. 640 Field Court, Dundee, IL 60118-9269.

Contact FAR Circuits for the latest pricing or visit their web site at **www.farcircuits.net/**.

Please read all instructions carefully before mounting any parts. Check the parts-placement diagram for the location of each part.

Figure A

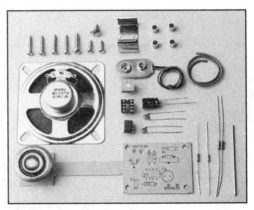

Figure B

☐ Check each box as that part is installed and soldered.

Quantity	Description	RadioShack Part Number	Component Number	Used in Step Number
Capacitors				
☐ 1	0.01-µF	272-131	C1	3
☐ 1	0.01-µF	272-131	C2	6
☐ 1	220-µF, 35-V electrolytic	272-1029	C3	5
Resistors				
☐ 1	10-kilohm, ¼ W (brown-black-orange stripes)	271-1335	R2	4
☐ 1	47-kilohm, ¼ W (yellow-violet-orange stripes)	271-1342	R3	8
☐ 1	10-kilohm, ¼ W (brown-black-orange stripes)	271-1335	R1	9
☐ 1	47-ohm, ¼ W (yellow-violet-black stripes)		R5	11

Miscellaneous

☐ 1	100-kilohm potentiometer	271-284	R4	7
☐ 1	8-pin IC socket	276-1995		2
☐ 1	7555 CMOS Timer IC (or 555 Timer IC)	276-1718	U1	13
☐ 1	Loudspeaker — 2-inch, 8-ohm		LS1	11
☐ 1	Six to 10 inches of insulated wire, about 18 or 22 gauge			11
☐ 1	9-V battery connector	270-325		10
☐ 1	9-V battery		BT1	10
☐ 1	U-shaped battery holder	270-326		12
☐ 1	Brass rod, 2 inches long, approximately 18 gauge (about the diameter of a coat hanger). Available at hobby shops.			
☐ 4	¼-inch spacers			12
☐ 1	2 × 4 × ½-inch piece of wood for base			12
☐ 5	No. 6 wood screws, ¾-inch long			12
☐ 2	No. 6 wood screws, ⅜-inch long			12
☐ 1	Five-lug tie point, used to mount speaker (optional)	274-688		11

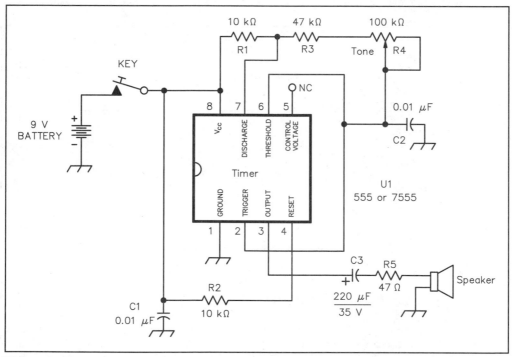

Figure C—Schematic diagram of a code-practice oscillator.

Assembly instructions

☐ **Check each box as that step is completed.**

☐ **Step 1: Attach the brass rod. Check the parts-placement diagram (Figure E) for location.**

Clean the brass rod with sandpaper or steel wool. Bend one end of the rod slightly less than 90 degrees. Lay the circuit board on the table with foil side up. Place the hooked end of the brass rod over the large hole near the handle (see Figure D). Make sure the rod extends out over the handle area. Solder the rod to the board on the foil side. The end of the brass rod should not extend past the marked oval on the handle. This is your contact point. If it does extend beyond this point, cut the rod off just before the end of the oval.

☐ **Step 2: Solder the IC socket to the board.**

The socket for the IC is placed on the component (non-foil) side of the board first. Do not plug the IC into the socket now. After all the other parts are soldered to the board you will be instructed to plug the IC into the socket (Step 13). Identify the notched end of the socket. Insert the socket into the circuit board. Turn the board over and gently spread the pins on the socket so they make contact with the foil side of the board. Solder the socket in place.

☐ **Step 3: Place C1 (0.01-µF capacitor) on the component side of the board.**

Thread the wire leads on C1 through the holes on the board. (See Figure F.) Solder the wires onto the foil side of the board. Cut the extra wire off above the solder joint.

☐ **Step 4: Place R2 (10-kilohm resistor) on the component side of the board.**

Prepare resistors for mounting by bending each lead (wire) of the resistor to approximately a 90° angle. (See Figure G.) Insert the leads into the board holes and bend them over to hold the resistor in place. Solder the leads to the foil and trim them close to the foil.

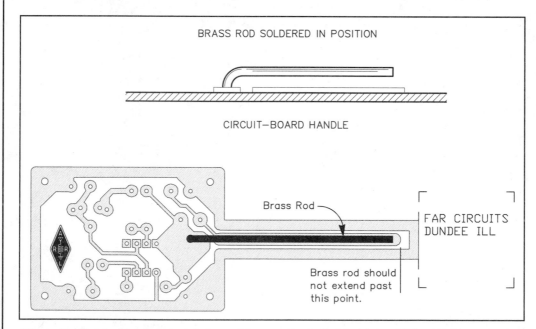

Figure D—Solder the brass rod in position on the foil side of the PC board. You can also use this figure as a circuit-board etching pattern if you want to make your own circuit board, since the pattern is printed full size.

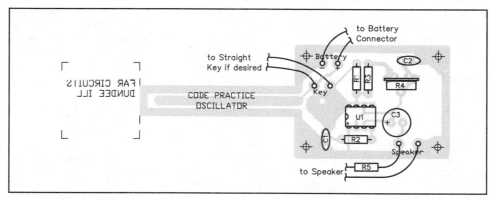

Figure E—Parts-placement diagram.

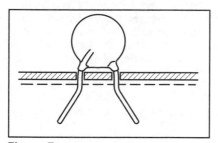

Figure F

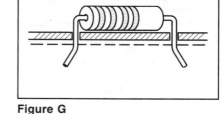

Figure G

☐ **Step 5: Place C3 (220-μF, 35-volt electrolytic capacitor) on the component side of the board.**

This capacitor has a plus (+) side and a negative (−) side. The (−) side is placed on the board facing away from the handle. (Notice the + sign printed on the circuit board at this location.) Insert the capacitor leads into the circuit board holes, solder them in place and trim off the extra wire.

☐ **Step 6: Place C2 (0.01-μF capacitor) on the component side of the board.**

Thread the wire leads from C2 through the holes on the circuit board. (See Step 3 and Figure F.) Solder the wires onto the board. Cut the extra wire off above the solder joint.

☐ **Step 7: Place R4 (100-kilohm potentiometer) on the component side of the circuit board.**

This component has three pins. All three pins must be plugged into the holes on the board. (It fits only one way.) Solder them in place and cut off any excess lead length.

☐ **Step 8: Place R3 (47-kilohm resistor) on the component side of the circuit board.**

Bend the wires on the resistor to plug it into the board. (See Step 4 and Figure G.) Plug the resistor into the board, spread the wires and solder it in place. Trim off the extra wire lengths.

☐ **Step 9: Place R1 (10-kilohm resistor) on the component side of the board.**

Bend the wires on the resistor to plug it into the board. (See Step 4 and Figure G.) Plug the resistor into the board, spread the wires and solder it in place. Trim the excess wire lengths.

☐ **Step 10: Hook up the battery connector leads.**

The battery connector consists of two wires, one red and one black, attached to a snap-on cap. Remove $1/4$ inch of plastic insulation from the end of both wires. The black wire is negative and the red wire is positive. The positive and negative battery connections are marked on the component side of the board. Be sure the red wire goes in the hole marked "+" and the black wire goes in the hole marked "–". Solder the wires in place and trim any excess length close to the solder joint.

☐ **Step 11: Hook up the speaker.**

If you are using the tie lug to hold the speaker in place, solder the speaker lugs to tie-point lugs on each side of the center post. (If you are not using the tie lug then you will solder the wires directly to the speaker lugs.) Cut the speaker wire into two equal lengths. Remove $1/4$ inch of plastic insulation from each end of both wires. Solder one end of each wire to one of the tie-point solder lugs below the speaker terminals. Solder one end of R5 (47-ohm resistor) to the circuit board as shown in Figure E. Solder one wire to the other end of this resistor and the second speaker wire to the circuit board.

☐ **Step 12: Attach the circuit board to the wood base.**

Place the completed circuit board on the wood. Trace through the four corner holes with a pencil. Take the circuit board off the wood and lay it aside. Place the spacers on the wood, standing upright. Carefully put the circuit board on top of the spacers. Put the $3/4$-inch screws through the holes in the circuit board and through the spacers and screw them into the board until snug. Be sure not to overtighten the screws, or you may crack the circuit board. Attach the speaker to the end of the board opposite the handle with a $3/8$-inch screw through the tie-point mounting hole, or mount the speaker to the board using two screws and holes in the outside edge of the speaker's metal frame. Attach the U-shaped metal battery holder to the wooden base with a $3/8$-inch screw.

☐ **Step 13: Plug the integrated circuit (IC) into the socket, being careful to position it so the notch or dot on one end of the IC is toward the handle.**

CAUTION — The static electricity from your body could destroy the IC. Before touching the IC, be sure you have discharged any static that may be built up on your body. While sitting at your table or workbench, touch a metal pipe or other large metal object for a few seconds. Carefully remove the IC from its foam padding. Hold it by the black body and avoid touching the wires. Plug it into the socket, being sure that the notched end of the IC is facing toward the handle. The notch on the IC should line up with the notch on the socket.

Attach the battery to the snap-on battery connector and place it in the U-shaped battery holder. This unit uses electricity only when the telegraph key handle is pushed down. No ON/OFF switch is necessary, and you may leave the battery connected at all times.

You're done! The oscillator should produce a tone when you press the key. If your oscillator does not work, check all your connections carefully. Make sure the IC is positioned correctly in the socket, and that you have a fresh battery. If it still doesn't work check all your solder connections.

Once you have the oscillator working, you're ready to use it to practice Morse code. If you are studying with a friend, you can use the oscillator to send code to each other. If you are studying alone, tape record your sending and play it back later. Can you copy what you sent? How would it sound on the air? Good luck!

General Class (Element 3) Question Pool— with Answers

DON'T START HERE!

This chapter contains the complete question pool for the Element 3 General class written exam. To earn a General class license, you must also pass (or have credit for passing) the Element 2 (Technician) written exam, and the Element 1 Morse code exam.

Credit for Current or Previous Licenses

You have credit for passing the Element 2 exam if you have a Technician license. You also have credit for the Element 3 exam if your Technician license was issued before March 21, 1987. In that case you also have credit for the Element 1, 5 word-per-minute Morse code exam. To upgrade to a General class license you will only have to appear before an Volunteer Examiner Team at an Exam session and show them all the paperwork to prove your license credit. By filing an NCVEC Quick Form 605 with the Examiner Team, and paying the test fee, they will issue you a Certificate of Successful Completion of Examination for your upgrade to General.

Table 13-1
Exam Elements Needed to Qualify for a General Class License

Current License	Exam Requirements	Study Materials
None	Morse code (Element 1)	*Your Introduction to Morse Code*
	Technician (Element 2)	*Now You're Talking!*, 5th Ed or *ARRL Technician Class Video Course*, 4th Ed or *ARRL's Tech Q & A*, 2nd Ed
	General (Element 3)	*The ARRL General Class License Manual*, 5th Ed or *ARRL's General Q & A*, 2nd Ed
Novice	Technician (Element 2)	*Now You're Talking!*, 5th Ed or *ARRL Technician Class Video Course*, 4th Ed or *ARRL's Tech Q & A*, 2nd Ed
	General (Element 3)	*The ARRL General Class License Manual*, 5th Ed or *ARRL's General Q & A*, 2nd Ed
Technician issued on or after Feb 14, 1991	Morse code (Element 1)	*Your Introduction to Morse Code*
	General (Element 3)	*The ARRL General Class License Manual*, 5th Ed or *ARRL's General Q & A*, 2nd Ed
Technician issued before Feb 14, 1991	General (Element 3)	*The ARRL General Class License Manual*, 5th Ed *ARRL's General Q & A*, 2nd Ed
Technician Plus or Technician with Morse code credit	General (Element 3)	*The ARRL General Class License Manual*, 5th Ed or *ARRL's General Q & A*, 2nd Ed
Technician issued before Mar 21, 1987	None*	

*Individuals who qualified for the Technician license before March 21, 1987 will be able to upgrade to General class by providing documentary proof to a Volunteer Examiner Coordinator, paying an application fee and completing NCVEC Quick Form 605.

If you hold an expired or unexpired Novice license or an expired Technician license that was issued before February 14, 1991, you will be given credit for the Morse code exam. You will also receive credit for the Morse code exam if you possess an FCC-issued commercial radiotelegraph operator license or permit that is valid or has been expired less than five years.

If you do not have any of those licenses, you will have to pass the Element 1 Morse code exam and the Element 2 and Element 3 written exams. *Study ARRL's* Now You're Talking! *for the Technician exam. ARRL's* Your Introduction to Morse Code *or other Morse code training products will help you prepare for the Element 1 exam.* The requirements for the General class license are summarized in **Table 13-1**.

Before you read the questions and answers printed in this chapter, be sure to read the text in the previous chapters. Use these questions as review exercises, when the text tells you to study them. (Paper clips make excellent place markers to help you find your place in the text and the question pool as you study.) Don't try to memorize all the questions and answers.

The material presented in this book has been carefully written and presented to guide you step-by-step through the learning process. By understanding the electronics principles and Amateur Radio concepts as they are presented, your insight into our hobby and your appreciation for the privileges granted by an Amateur Radio license will be greatly enhanced.

This question pool, released by the Volunteer Examiner Coordinators' Question Pool Committee in December 2003, will be used on exams beginning July 1, 2004. The pool is scheduled to be used until June 30, 2008. Changes to FCC Rules and other factors may result in some questions being "locked out" or removed from the pool. It is also possible that the question pool will be revised earlier because of significant Rules changes or schedule revisions. Such changes will be announced in *QST* and other Amateur Radio publications.

How Many Questions?

The FCC specifies that an Element 3 exam must include 35 questions. This question pool is divided into ten sections, called subelements. (A subelement is a portion of the exam element, in this case Element 3.) Although it's not an FCC requirement, the VEC Question Pool Committee specifies the number of questions from each subelement that should appear on your test. For example, there must be six questions from the Commission's Rules section, Subelement G1 and three questions from the Radio-Wave Propagation section, Subelement G3. **Table 13-2** summarizes the number of questions to be selected from each subelement on the General class (Element 3) written exam. The number of questions to be used from each subelement also appears at the beginning of that subelement in the question pool.

The Volunteer Examiner Coordinators' Question Pool Committee has broken the subelements into smaller groups. Each subelement has the same number of groups as there are questions from that subelement on the exam. This means there are six groups for the Commission's Rules subelement and two groups for the Electrical Principles subelement. The Question Pool Committee intends one question from each smaller group be used on the exam. This is not an FCC requirement, however.

There is a list of topics printed in **bold** type at the beginning of each small group. This list of topics for each subelement forms the *syllabus*, or study guide topics for that section. The entire General class syllabus is printed at the end of the Introduction chapter, before Chapter 1 of this book.

Table 13-2
General Class Exam Content

Subelement	Topic	Number of Questions
G1	Commission's Rules	6
G2	Operating Procedures	6
G3	Radio-Wave Propagation	3
G4	Amateur Radio Practices	5
G5	Electrical Principles	2
G6	Circuit Components	1
G7	Practical Circuits	1
G8	Signals and Emissions	2
G9	Antennas and Feed Lines	4
G0	RF Safety Practices	5
	Total	35

The small groups are labeled alphabetically within each subelement. The six groups in the Commission's Rules subelement are labeled G1A through G1F. One exam question comes from the Circuit Components subelement, so that subelement has a section labeled G6A.

The question numbers used in the question pool relate to the syllabus or study guide printed at the end of the Introduction, before Chapter 1 of this book. The syllabus is an outline of topics covered by the exam. Each question number begins with a G to indicate the question is from the General class question pool. Next is a number to indicate which of the ten subelements the question is from. These numbers will range from 1 to 0. Following this number is a letter to indicate which group the question is from in that subelement. Each question number ends with a two-digit number to specify its position in the set. So question G2A01 is the first question in the A group of the second subelement. Question G9C08 is the eighth question in the C group of the ninth subelement.

Who Picks the Questions?

The FCC allows Volunteer Examiner Teams to select the questions that will be used on amateur exams. If your test is coordinated by the ARRL/VEC, your test will be prepared by the VEC, or by using a computer program supplied by the VEC. All VECs and Examiners must use the questions, answers and distracters (incorrect answers) printed here. The positions of the answers may be scrambled on the exam, though. This means that you can't be sure the correct answer to question G1A01 will always be C. The correct answer may appear in location A, B, C or D.

This question pool contains more than ten times the number of questions necessary to make up an exam. This ensures that the examiners have sufficient questions to choose from when they make up an exam.

Who Gives the Test?

All Amateur Radio licesne exams are given by teams of three or more Volunteer Examiners (VEs). Each of the examiners is accredited by a Volunteer Examiner Coordinator (VEC) to give exams under their program. A VEC is an organization that has entered into an agreement with the FCC to coordinate the efforts of VEs. The VEC reviews the paperwork for each exam session, and then forwards the information to the FCC. The FCC then issues new or upgraded Amateur Radio licnses to those who qualify for them.

Question Pool Format

The rest of this chapter contains the entire Element 3 question pool. We have printed the answer key to these questions along the edge of the page. There is a line to indicate where you should fold the page to hide the answer key while you study. (Fold the edge of the page *under* rather than *over* the page, so you don't cover part of the questions on the page.) After making your best effort to answer the questions, you can look at the answers to check your understanding.

We have also included page references along with the answers. These page numbers indicate where in this book you will find the text discussion related to each question. If you have any problems with a question, refer to the page listed for that question. You may have to study beyond the listed page number to review all the related material. With the "Commission's Rules" questions of Subelement 1, we have also included references to the sections of Part 97 of the FCC Rules. Chapter 1 of this book includes the text of the rules for these questions, along with an explanation to help you answer the questions. Chapter 1 does *not* include a complete copy of Part 97, however. For the complete text of the FCC Rules governing Amateur Radio, we recommend a copy of *The ARRL FCC Rule Book*. That book includes the complete text for Part 97, along with text to explain all the rules. *The ARRL FCC Rule Book* is updated regularly to reflect changes to the Rules.

Good luck with your studies. With a bit of time devoted to reviewing this book, you'll soon be ready for your General class exam!

Answer Key |
Page num-
bers tell you
where to look
in this book
for more
information.

Subelement
G1
Numbers in
[square
brackets]
indicate
sections in
Part 97, the
Amateur
Radio Rules.

G1A01
(C)
[97.301d]
Page 1-3

G1A02
(A)
[97.301d]
Page 1-3

G1A03
(D)
[97.301d]
Page 1-3

G1A04
(A)
[97.301d]
Page 1-3

ELEMENT 3 (GENERAL CLASS) QUESTION POOL

As released by the National Conference of Volunteer Examiner Coordinators' Question Pool Committee, December 2003. This question pool is to be used for all Element 3 exams starting July 1, 2004.

Subelement G1 — Commission's Rules
[6 Exam Questions — 6 Groups]

G1A General control operator frequency privileges

G1A01
What are the frequency privileges for a General Class control operator in the 160-meter band (ITU Region 2)?
A. 1800 - 1900-kHz
B. 1900 - 2000-kHz
C. 1800 - 2000-kHz
D. 1825 - 2000-kHz

G1A02
What are the frequency privileges for a General Class control operator in the 75/80-meter band (ITU Region 2)?
A. 3525 - 3750-kHz and 3850 - 4000-kHz
B. 3525 - 3775-kHz and 3875 - 4000-kHz
C. 3525 - 3750-kHz and 3875 - 4000-kHz
D. 3525 - 3775-kHz and 3850 - 4000-kHz

G1A03
What are the frequency privileges for a General Class control operator in the 40-meter band (ITU Region 2)?
A. 7025 - 7175-kHz and 7200 - 7300-kHz
B. 7025 - 7175-kHz and 7225 - 7300-kHz
C. 7025 - 7150-kHz and 7200 - 7300-kHz
D. 7025 - 7150-kHz and 7225 - 7300-kHz

G1A04
What are the frequency privileges for a General Class control operator in the 30-meter band?
A. 10100 - 10150-kHz
B. 10100 - 10175-kHz
C. 10125 - 10150-kHz
D. 10125 - 10175-kHz

G1A05
What are the frequency privileges for a General Class control operator in the 20-meter band?
A. 14025 - 14100-kHz and 14175 - 14350-kHz
B. 14025 - 14150-kHz and 14225 - 14350-kHz
C. 14025 - 14125-kHz and 14200 - 14350-kHz
D. 14025 - 14175-kHz and 14250 - 14350-kHz

G1A06
What are the frequency privileges for a General Class control operator in the 15-meter band?
A. 21025 - 21200-kHz and 21275 - 21450-kHz
B. 21025 - 21150-kHz and 21300 - 21450-kHz
C. 21025 - 21150-kHz and 21275 - 21450-kHz
D. 21025 - 21200-kHz and 21300 - 21450-kHz

G1A07
What are the frequency privileges for a General Class control operator in the 12-meter band?
A. 24890 - 24975-kHz
B. 24890 - 24990-kHz
C. 24900 - 24990-kHz
D. 24900 - 24975-kHz

G1A08
What are the frequency privileges for a General Class control operator in the 10-meter band?
A. 28000 - 29700-kHz
B. 28025 - 29700-kHz
C. 28100 - 29600-kHz
D. 28125 - 29600-kHz

G1A09
What are the frequency privileges for a General Class control operator in the 17-meter band?
A. 18068 - 18300-kHz
B. 18025 - 18200-kHz
C. 18100 - 18200-kHz
D. 18068 - 18168-kHz

G1A10
What are the frequency segments for General class licensees within the 75/80-meter band in ITU Region 2 for CW emissions?
A. 3500 - 3750 kHz and 3800 - 4000 kHz
B. 3700 - 3750 kHz and 3850 - 4000 kHz
C. 3525 - 3750 kHz and 3850 - 4000 kHz
D. 3525 - 4000 kHz

G1A05
(B)
[97.301d]
Page 1-3

G1A06
(D)
[97.301d]
Page 1-3

G1A07
(B)
[97.301d]
Page 1-3

G1A08
(A)
[97.301d]
Page 1-3

G1A09
(D)
[97.301d]
Page 1-3

G1A10
(C)
[97.305a]
Page 1-5

G1A11
(C)
[97.305c]
Page 1-5

G1A11
What are the frequency segments within the 10-meter band for phone emissions?
A. 28000 - 28300 kHz
B. 29000 - 29700 kHz
C. 28300 - 29700 kHz
D. 28000 - 29000 kHz

G1B   Antenna structure limitations; good engineering and good amateur practice; beacon operation; restricted operation; retransmitting radio signals

G1B01
(C)
[97.15a]
Page 1-13

G1B01
Provided it is not at or near a public-use airport, what is the maximum height above ground an antenna structure may rise without requiring its owner to notify the FAA and register with the FCC?
A. 50 feet
B. 100 feet
C. 200 feet
D. 300 feet

G1B02
(B)
[97.101a]
Page 1-11

G1B02
If the FCC Rules DO NOT specifically cover a situation, how must you operate your amateur station?
A. In accordance with standard licensee operator principles
B. In accordance with good engineering and good amateur practice
C. In accordance with station operating practices adopted by the VECs
D. In accordance with procedures set forth by the International Amateur Radio Union

G1B03
(B)
[97.203g]
Page 1-14

G1B03
Which of the following types of stations may transmit only one-way communications?
A. Repeater station
B. Beacon station
C. HF station
D. VHF station

G1B04
(A)
[97.113b]
Page 1-14

G1B04
Which of the following does NOT need to be true when an amateur station is being used to gather news information for broadcast purposes?
A. The information is more quickly transmitted by Amateur Radio
B. The information must involve the immediate safety of life of individuals or the immediate protection of property
C. The information must be directly related to the event
D. The information cannot be transmitted by other means

G1B05

Under what limited circumstances may music be transmitted by an amateur station?

A. When it produces no dissonances or spurious emissions
B. When it is used to jam an illegal transmission
C. When it is transmitted on frequencies above 1215 MHz
D. When it is an incidental part of a space shuttle retransmission

G1B06

When may an amateur station in two-way communication transmit a message in a secret code in order to obscure the meaning of the communication?

A. When transmitting above 450 MHz
B. During contests
C. Never
D. During a declared communications emergency

G1B07

What are the restrictions on the use of abbreviations or procedural signals in the amateur service?

A. Only "Q" codes are permitted
B. They may be used if they do not obscure the meaning of a message
C. They are not permitted because they obscure the meaning of a message to FCC monitoring stations
D. Only "10-codes", are permitted

G1B08

Which of the following amateur station transmissions is NOT prohibited by the FCC Rules?

A. The playing of music
B. The use of obscene or indecent words
C. False or deceptive messages or signals
D. Retransmission of space shuttle communications

G1B09

What should you do to prevent your station from retransmitting music or signals from a non-amateur station?

A. Turn up the volume of your transceiver
B. Speak closer to the microphone to increase your signal strength
C. Turn down the volume of background audio
D. Adjust your transceiver noise blanker

G1B10

Which of the following is NOT an FCC requirement regarding beacon stations?

A. All transmissions must use audio frequency shift keying (AFSK)
B. Only one signal per band is permitted from a given location
C. The transmitter power of the beacon station must not exceed 100 watts
D. The control operator of the beacon station must hold a Technician, Technician Plus, General, Advanced or Extra Class operator license

---

G1B05
(D)
[97.113e]
Page 1-13

G1B06
(C)
[97.113a4]
Page 1-13

G1B07
(B)
[97.113a4]
Page 1-13

G1B08
(D)
[97.113a4,
97.113e]
Page 1-13

G1B09
(C)
[97.113a4,
97.113e]
Page 1-13

G1B10
(A)
[97.203]
Page 1-14

G1C01
(A)
[97.313c1]
Page 1-6

G1C01
What is the maximum transmitting power an amateur station may use on 3690 kHz?
A. 200 watts PEP output
B. 1000 watts PEP output
C. 1500 watts PEP output
D. The minimum power necessary to carry out the desired communications, with a maximum of 2000 watts PEP output

G1C02
(C)
[97.313a,b]
Page 1-6

G1C02
What is the maximum transmitting power an amateur station may use on 7080 kHz?
A. 200 watts PEP output
B. 1000 watts PEP output
C. 1500 watts PEP output
D. 2000 watts PEP output

G1C03
(A)
[97.313c1]
Page 1-6

G1C03
What is the maximum transmitting power an amateur station may use on 10.140 MHz?
A. 200 watts PEP output
B. 1000 watts PEP output
C. 1500 watts PEP output
D. 2000 watts PEP output

G1C04
(B)
[97.313c1]
Page 1-6

G1C04
What is the maximum transmitting power an amateur station may use on 21.305 MHz?
A. The minimum power necessary to carry out the desired communications, with a maximum of 200 watts PEP output
B. The minimum power necessary to carry out the desired communications, with a maximum of 1500 watts PEP output
C. 1000 watts PEP output
D. 2000 watts PEP output

G1C05
(C)
[97.313a,b]
Page 1-6

G1C05
What is the maximum transmitting power an amateur station may use on 24.950 MHz?
A. 200 watts PEP output
B. 1000 watts PEP output
C. 1500 watts PEP output
D. 2000 watts PEP output

G1C06
What is the maximum transmitting power an amateur station may
use on 7255 kHz?
A. The minimum power necessary to carry out the desired
   communications, with a maximum of 200 watts PEP output
B. The minimum power necessary to carry out the desired
   communications, with a maximum of 1500 watts PEP output
C. 1000 watts PEP output
D. 2000 watts PEP output

G1C07
What is the maximum transmitting power an amateur station may
use on 14.300 MHz?
A. The minimum power necessary to carry out the desired
   communications, with a maximum of 1500 watts PEP output
B. 200 watts PEP output
C. 1000 watts PEP output
D. 2000 watts PEP output

G1C08
What is the maximum transmitting power a station with a General
Class control operator may use on 28.400 MHz?
A. The minimum power necessary to carry out the desired
   communications, with a maximum of 200 watts PEP output
B. The minimum power necessary to carry out the desired
   communications, with a maximum of 1000 watts PEP output
C. The minimum power necessary to carry out the desired
   communications, with a maximum of 1500 watts PEP output
D. 2000 watts PEP output

G1C09
What is the maximum transmitting power a station with a General
Class control operator may use on 28.150 MHz?
A. The minimum power necessary to carry out the desired
   communications, with a maximum of 200 watts PEP output
B. The minimum power necessary to carry out the desired
   communications, with a maximum of 1000 watts PEP output
C. The minimum power necessary to carry out the desired
   communications, with a maximum of 1500 watts PEP output
D. The minimum power necessary to carry out the desired
   communications, with a maximum of 2000 watts PEP output

G1C10
What is the maximum transmitting power an amateur station may
use on 1825 kHz?
A. 200 watts PEP output
B. The minimum power necessary to carry out the desired
   communications, with a maximum of 1000 watts PEP output
C. 2000 watts PEP output
D. The minimum power necessary to carry out the desired
   communications, with a maximum of 1500 watts PEP output

G1C06
(B)
[97.313]
Page 1-6

G1C07
(A)
[97.313]
Page 1-6

G1C08
(C)
[97.313]
Page 1-6

G1C09
(C)
[97.313]
Page 1-6

G1C10
(D)
[97.313]
Page 1-6

G1C11
(D)
[97.303 (s)]
Page 1-6

G1C11
Which of the following is NOT a requirement when a station is transmitting on the 60-meter band?
A. All transmissions may only use Upper Sideband (USB)
B. The 3-dB bandwidth of a signal shall not exceed 2.8 kHz, when centered on any of the five FCC-authorized transmitting frequencies
C. Transmissions shall not exceed an effective radiated power (ERP) of 50 W PEP
D. Antenna height shall not exceed 50 feet above mean sea level (AMSL)

G1D Examination element preparation; examination administration; temporary station identification

G1D01
(B)
[97.507a2]
Page 1-16

G1D01
What examination elements may you prepare when you hold a General class operator license?
A. None
B. Elements 1 and 2 only
C. Element 1 only
D. Elements 1, 2 and 3

G1D02
(C)
[97.509b3i]
Page 1-16

G1D02
What license examinations may you administer when you are an accredited VE holding a General Class operator license?
A. None
B. General only
C. Technician and Morse code
D. Technician, General and Amateur Extra

G1D03
(A)
[97.501e]
Page 1-18

G1D03
What minimum examination elements must an applicant pass for a Technician Class operator license?
A. Element 2 only
B. Elements 1 and 2
C. Elements 2 and 3
D. Elements 1, 2 and 3

G1D04
(B)
[97.501d]
Page 1-18

G1D04
What minimum examination elements must an applicant pass for a Technician Class operator license with Morse code credit to operate on the HF bands?
A. Element 2 only
B. Elements 1 and 2
C. Elements 2 and 3
D. Elements 1, 2 and 3

G1D05
What are the requirements for administering a Technician Class operator examination?
A. Three VEC-accredited General Class or higher VEs must be present
B. Two VEC-accredited General Class or higher VEs must be present
C. Two General Class or higher VEs must be present, but only one need be VEC accredited
D. Any two General Class or higher VEs must be present

G1D06
When may you participate as an administering VE for a Technician Class operator license examination?
A. Once you have notified the FCC that you want to give an examination
B. Once you have a Certificate of Successful Completion of Examination (CSCE) for General class
C. Once you have prepared telegraphy and written examinations for the Technician license, or obtained them from a qualified supplier
D. Once you have been granted your FCC General class or higher license and received your VEC accreditation

G1D07
If you are a Technician Class operator with a CSCE for General Class operator privileges, how do you identify your station when transmitting on 14.035 MHz?
A. You must give your call sign and the location of the VE examination where you obtained the CSCE
B. You must give your call sign, followed by the slant mark "/", followed by the identifier "AG"
C. You may not operate on 14.035 MHz until your new license arrives
D. No special form of identification is needed

G1D08
If you are a Technician Class operator with a CSCE for General Class operator privileges, how do you identify your station when transmitting phone emissions on 14.325 MHz?
A. No special form of identification is needed
B. You may not operate on 14.325 MHz until your new license arrives
C. You must give your call sign, followed by any suitable word that denotes the slant mark and the identifier "AG"
D. You must give your call sign and the location of the VE examination where you obtained the CSCE

G1D05
(A)
[97.509a,b]
Page 1-19

G1D06
(D)
[97.509b3i]
Page 1-16

G1D07
(B)
[97.119f2]
Page 1-19

G1D08
(C)
[97.119f2]
Page 1-19

G1D09
(A)
[97.119f2]
Page 1-19

G1D09
If you are a Technician Class operator with a CSCE for General Class operator privileges, when must you add the special identifier "AG" after your call sign?
A. Whenever you operate using your new frequency privileges
B. Whenever you operate
C. Whenever you operate using Technician frequency privileges
D. A special identifier is not required as long as your General class license application has been filed with the FCC

G1D10
(D)
[97.119f2]
Page 1-19

G1D10
If you are a Technician Class operator with a CSCE for General Class operator privileges, on which of the following band segments must you include the special identifier "AG" after your call sign?
A. Whenever you operate from 18068 - 18168-kHz
B. Whenever you operate from 14025 - 14150-kHz and 14225 - 14350-kHz
C. Whenever you operate from 10100 - 10150-kHz
D. All of these choices are correct

G1D11
(D)
[97.509b3i]
Page 1-16

G1D11
When may you participate as a VE in administering a Morse code examination?
A. Once you have notified the FCC that you want to give an examination
B. Once you have a Certificate of Successful Completion of Examination (CSCE) for General class
C. Once you have prepared telegraphy and written examinations for the Technician license, or obtained them from a qualified supplier
D. Once you have been granted your FCC General Class or higher operator license and received your VEC accreditation

G1D12
(C)
[97.119f2]
Page 1-19

G1D12
If you are a Technician licensee with Morse code credit and hold a CSCE for Element 3, what is one way you could identify your station when transmitting phone emissions on a General class amateur frequency?
A. Give your call sign followed by the words "general class"
B. No special identification is needed, since your license upgrade would already be shown in the FCC's database
C. Give your call sign followed by the words "temporary AG"
D. You must wait until your new license arrives by mail from the FCC before using general class frequencies

G1E Local control; repeater and harmful interference definitions; third party communications

G1E01
As a General Class control operator at the station of a Technician Class operator, how must you identify the station while transmitting on 7250 kHz?
A. With your call sign, followed by the word "controlling" and the Technician call sign
B. With the Technician Class operator's station call sign, followed by the slant bar "/" (or any suitable word) and your own call sign
C. With your call sign, followed by the slant bar "/" (or any suitable word) and the Technician call sign
D. A Technician station should not be operated on 7250-kHz, even with a General control operator

G1E01
(B)
[97.119e]
Page 1-20

G1E02
Under what circumstances may a 10-meter repeater retransmit the 2-meter signal from a station having a Technician Class control operator?
A. Under no circumstances
B. Only if the station on 10 meters is operating under a Special Temporary Authorization allowing such retransmission
C. Only during an FCC-declared general state of communications emergency
D. Only if the 10-meter control operator holds at least a General class license

G1E02
(D)
[97.205a]
Page 1-8

G1E03
What kind of amateur station simultaneously retransmits the signals of other stations on a different channel?
A. Repeater station
B. Space station
C. Telecommand station
D. Relay station

G1E03
(A)
[97.3a37]
Page 1-8

G1E04
What name is given to a form of interference that seriously degrades, obstructs or repeatedly interrupts a radiocommunication service?
A. Intentional interference
B. Harmful interference
C. Adjacent interference
D. Disruptive interference

G1E04
(B)
[97.3a22]
Page 1-8

G1E05
What types of messages for a third party may be transmitted by an amateur station to a foreign country?
A. Messages for which the amateur operator is paid
B. Messages facilitating the business affairs of any party
C. Messages of a technical nature or remarks of a personal character
D. No messages may be transmitted to foreign countries for third parties

G1E05
(C)
[97.115, 97.117]
Page 1-11

G1E06
(A)
[97.3a23]
Page 1-8

G1E06
Should a repeater cause harmful interference to another repeater when a frequency coordinator has recommended the operation of one station only, who is responsible for resolving the interference?
A. The licensee of the uncoordinated repeater
B. Both repeater licensees
C. The licensee of the recommended repeater
D. The frequency coordinator

G1E07
(C)
[97.303]
Page 1-4

G1E07
What does it mean where the FCC rules say that the amateur service is a secondary user and another service is a primary user?
A. Nothing special; all users of the frequency band have equal rights to operate
B. Amateur stations are only allowed to use the frequency band during emergencies
C. Amateur stations are allowed to use the frequency band only if they do not cause harmful interference to primary users
D. Amateur stations must increase transmitter power to overcome any interference caused by primary users

G1E08
(D)
[97.303]
Page 1-4

G1E08
What action must you take while using the 30-meter band when a station assigned to the band's primary service causes interference?
A. Notify the FCC's regional Engineer in Charge of the interference
B. Increase your transmitter's power to overcome the interference
C. Attempt to contact the station and request that it stop the interference
D. Change frequencies; you may be causing harmful interference to the other station, in violation of FCC rules

G1E09
(C)
[97.119b2]
Page 1-20

G1E09
While you are using a language other than English in making a contact, what language must you use when identifying your station?
A. The language being used for the contact
B. The language being used for the contact, provided the US has a third-party communications agreement with that country
C. English
D. Any language of a country that is a member of the International Telecommunication Union

G1E10
(A)
[97.303s]
Page 1-7

G1E10
What protection from harmful interference caused by primary service users do amateur radio stations have while operating in the 60-meter band?
A. None
B. Stations in the mobile and fixed service must not interfere with amateur stations
C. Stations in the mobile and fixed service must not interfere if an amateur station is already on the frequency
D. Stations in the mobile and fixed service must not interfere with amateur stations if they are located in ITU Region 2

G1E11
What operating restrictions must amateur radio stations observe while operating in the 60-meter band?
A. They must not cause harmful interference to stations operating in other radio services
B. They must transmit no more than 30 minutes during each hour to minimize harmful interference
C. They must use lower sideband, suppressed-carrier, only
D. They must not exceed 2.0 kHz of bandwidth

G1E11
(A)
[97.303s]
Page 1-7

G1E12
What must be done at an amateur radio station while it is transmitting third party messages?
A. Keep a station log of when the message was handled
B. Use local or remote station control
C. Identify both stations that handle the message
D. Use local, remote or automatic station control

G1E12
(B)
[97.109e]
Page 1-9

G1F Certification of external RF-power-amplifiers; standards for certification of external RF-power amplifiers; HF data emission standards

G1F01
External RF power amplifiers designed to operate below what frequency may require FCC certification?
A. 28 MHz
B. 35 MHz
C. 50 MHz
D. 144 MHz

G1F01
(D)
[97.315a]
Page 1-15

G1F02
Without a grant of FCC certification, how many external RF amplifiers of a given design capable of operation below 144 MHz may you build or modify in one calendar year?
A. None
B. 1
C. 5
D. 10

G1F02
(B)
[97.315a]
Page 1-15

G1F03
Which of the following standards must be met where FCC certification of an external RF amplifier is required?
A. The amplifier must not be able to amplify a 28-MHz signal to more than ten times the input power
B. The amplifier must not be capable of reaching its designed output power when driven with less than 50 watts
C. The amplifier must not be able to be operated for more than ten minutes without a time delay circuit
D. The amplifier must not be able to be modified by an amateur operator

G1F03
(B)
[97.317a3]
Page 1-15

G1F04
(D)
[97.317b,c]
Page 1-15

G1F04
Which of the following would NOT disqualify an external RF power amplifier from a FCC certification grant?
A. The capability of being modified by the operator for use outside the amateur service bands
B. The capability of achieving full output power when driven with less than 50 watts
C. The capability of achieving full output power on amateur service frequencies between 24 and 35 MHz
D. The capability of being switched by the operator to all amateur service frequencies below 24 MHz

G1F05
(D)
[97.305c,
97.307f3]
Page 2-15

G1F05
What is the maximum symbol rate permitted for RTTY emissions transmitted on frequency bands below 10 meters?
A. 56 kilobauds
B. 19.6 kilobauds
C. 1200 bauds
D. 300 bauds

G1F06
(C)
[97.307f5]
Page 2-15

G1F06
What is the maximum symbol rate permitted for packet emissions on the 2-meter band?
A. 300 bauds
B. 1200 bauds
C. 19.6 kilobauds
D. 56 kilobauds

G1F07
(C)
[97.307f4]
Page 2-15

G1F07
What is the maximum symbol rate permitted for RTTY or data emissions on the 10-meter band?
A. 56 kilobauds
B. 19.6 kilobauds
C. 1200 bauds
D. 300 bauds

G1F08
(B)
[97.307f5]
Page 2-15

G1F08
What is the maximum symbol rate permitted for RTTY or data emissions on the 6- and 2-meter bands?
A. 56 kilobauds
B. 19.6 kilobauds
C. 1200 bauds
D. 300 bauds

G1F09
(A)
[97.307f5]
Page 2-16

G1F09
What is the maximum authorized bandwidth for RTTY, data or multi-plexed emissions using an unspecified digital code transmitted on the 6- and 2-meter bands?
A. 20 kHz
B. 50 kHz
C. The total bandwidth shall not exceed that of a single-sideband phone emission
D. The total bandwidth shall not exceed 10 times that of a CW emission

G1F10
What must an external RF amplifier exhibit in order to receive a FCC grant of certification?
A. It must not be capable of operation on any frequency between 24 MHz and 35 MHz
B. Its wiring must be accessible to permit modification of the amplifier
C. It must have an internal RF sensing switch to place the amplifier in the transmit mode
D. Its manual must provide instructions for modification of the amplifier

G1F10
(A)
[97.317b]
Page 1-15

G1F11
What is the maximum power gain that a 10-meter RF amplifier can have to receive FCC certification?
A. 6 dB
B. 3 dB
C. 4 dB
D. 10 dB

G1F11
(A)
[97.317b1]
Page 1-15

SUBELEMENT G2
OPERATING PROCEDURES
[6 EXAM QUESTIONS — 6 GROUPS]

Subelement G2

G2A Phone operating procedures

G2A01
Which sideband is commonly used for 20-meter phone operation?
A. Upper
B. Lower
C. Amplitude compandored
D. Double

G2A01
(A)
Page 2-2

G2A02
Which sideband is commonly used on 3925-kHz for phone operation?
A. Upper
B. Lower
C. Amplitude compandored
D. Double

G2A02
(B)
Page 2-2

G2A03
Which sideband is commonly used for 40-meter phone operation?
A. Upper
B. Lower
C. Amplitude compandored
D. Double

G2A03
(B)
Page 2-2

G2A04
(D)
Page 2-2

G2A04
Which sideband is commonly used for 10-meter phone operation?
A. Double
B. Lower
C. Amplitude compandored
D. Upper

G2A05
(A)
Page 2-2

G2A05
Which sideband is commonly used for 15-meter phone operation?
A. Upper
B. Lower
C. Amplitude compandored
D. Double

G2A06
(C)
Page 2-2

G2A06
Which sideband is commonly used for 17-meter phone operation?
A. Amplitude compandored
B. Lower
C. Upper
D. Double

G2A07
(C)
Page 2-1

G2A07
Which of the following modes of voice communication is most commonly used on the High Frequency Amateur bands?
A. Frequency modulation (FM)
B. Amplitude modulation (AM)
C. Single sideband (SSB)
D. Phase modulation (PM)

G2A08
(D)
Page 2-1

G2A08
Why is the single sideband mode of voice transmission used more frequently than Amplitude Modulation (AM) on the HF amateur bands?
A. Single sideband transmissions use less spectrum space
B. Single sideband transmissions are more power efficient
C. No carrier is transmitted with a single sideband transmission
D. All of these choices are correct

G2A09
(B)
Page 2-2

G2A09
Which of the following statements is true of a lower sideband transmission?
A. It is called lower sideband because the lower sideband is greatly attenuated
B. It is called lower sideband because the lower sideband is the only sideband transmitted, since the upper sideband is suppressed
C. The lower sideband is wider than the upper sideband
D. The lower sideband is the only sideband that is authorized on the 160-, 75- and 40-meter amateur bands

G2A10
Which of the following statements is true of an upper sideband transmission?
A. Only the upper sideband is transmitted, since the lower sideband is suppressed
B. The upper sideband is greatly attenuated as compared with the carrier
C. The upper sideband is greatly attenuated as compared with the lower sideband
D. Only the upper sideband may be used for phone transmissions on the amateur bands with frequencies above 14 MHz

G2A10
(A)
Page 2-2

G2A11
Why do most amateur stations use lower sideband on the 160-, 75- and 40-meter bands?
A. The lower sideband is more efficient at these frequency bands
B. The lower sideband is the only sideband legal on these frequency bands
C. Because it is fully compatible with an AM detector
D. Current amateur practice is to use lower sideband on these frequency bands

G2A11
(D)
Page 2-2

G2B Operating courtesy

G2B01
If you are the net control station of a daily HF net, what should you do if the frequency on which you normally meet is in use just before the net begins?
A. Reduce your output power and start the net as usual
B. Increase your power output so that net participants will be able to hear you over the existing activity
C. Conduct the net on a clear frequency 3 to 5-kHz away from the regular net frequency
D. Cancel the net for that day

G2B01
(C)
Page 2-3

G2B02
If a net is about to begin on a frequency which you and another station are using, what should you do?
A. As a courtesy to the net, move to a different frequency
B. Increase your power output to ensure that all net participants can hear you
C. Transmit as long as possible on the frequency so that no other stations may use it
D. Turn off your radio

G2B02
(A)
Page 2-3

G2B03
(C)
Page 2-3

G2B03
If propagation changes during your contact and you notice increasing interference from other activity on the same frequency, what should you do?
A. Tell the interfering stations to change frequency, since you were there first
B. Report the interference to your local Amateur Auxiliary Coordinator
C. Move your contact to another frequency
D. Turn on your amplifier to overcome the interference

G2B04
(B)
Page 2-2

G2B04
When selecting a CW transmitting frequency, what minimum frequency separation from a contact in progress should you allow to minimize interference?
A. 5 to 50 Hz
B. 150 to 500 Hz
C. 1 to 3 kHz
D. 3 to 6 kHz

G2B05
(B)
Page 2-2

G2B05
When selecting a single-sideband phone transmitting frequency, what minimum frequency separation from a contact in progress should you allow (between suppressed carriers) to minimize interference?
A. 150 to 500 Hz
B. Approximately 3 kHz
C. Approximately 6 kHz
D. Approximately 10 kHz

G2B06
(B)
Page 2-2

G2B06
When selecting a RTTY transmitting frequency, what minimum frequency separation from a contact in progress should you allow (center to center) to minimize interference?
A. 60 Hz
B. 250 to 500 Hz
C. Approximately 3 kHz
D. Approximately 6 kHz

G2B07
(A)
Page 2-4

G2B07
What is a band plan?
A. A voluntary guideline beyond the divisions established by the FCC for using different operating modes within an amateur band
B. A guideline from the FCC for making amateur frequency band allocations
C. A plan of operating schedules within an amateur band published by the FCC
D. A plan devised by a club to best use a frequency band during a contest

G2B08
What is another name for a voluntary guideline that guides amateur activities and extends beyond the divisions established by the FCC for using different operating modes within an amateur band?
A. A "Band Plan"
B. A "Frequency and Solar Cycle Guide"
C. The "Knowledgeable Operator's Guide"
D. The "Frequency Use Guidebook"

G2B09
When choosing a frequency for Slow-Scan TV (SSTV) operation, what should you do to comply with good amateur practice?
A. Review FCC Part 97 Rules regarding permitted frequencies and emissions
B. Follow generally accepted gentlemen's agreement band plans
C. Before transmitting, listen to the frequency to be used to avoid interfering with an ongoing communication
D. All of these choices

G2B10
When choosing a frequency for radioteletype (RTTY) operation, what should you do to comply with good amateur practice?
A. Review FCC Part 97 Rules regarding permitted frequencies and emissions
B. Follow generally accepted gentlemen's agreement band plans
C. Before transmitting, first listen to the frequency to be used to avoid interfering with an ongoing communication
D. All of these choices

G2B11
When choosing a frequency for HF Packet operation, what should you do to comply with good amateur practice?
A. Review FCC Part 97 Rules regarding permitted frequencies and emissions
B. Follow generally accepted gentlemen's agreement band plans
C. Before transmitting, first listen on the frequency to be used to avoid interfering with an ongoing communication
D. All of these choices

G2B12
What is a considerate way to avoid harmful interference when using phone?
A. Ask if the frequency is in use, and say your call sign
B. Call MAYDAY to make sure that the frequency is clear
C. Call CQ for two minutes and see if anyone responds
D. Turn on your amplifier, then go ahead and transmit

G2B08
(A)
Page 2-4

G2B09
(D)
Page 2-4

G2B10
(D)
Page 2-4

G2B11
(D)
Page 2-4

G2B12
(A)
Page 2-4

G2B13
What is a considerate way to avoid harmful interference when using Morse code or CW?
A. Send the letter "V" 12 times and then listen for a response
B. Call CQ for two minutes and see if anyone responds
C. Send "QRL? de" followed by your call sign and listen for a response
D. Turn on your amplifier, then go ahead and transmit

G2C   Emergencies, including drills and emergency communications

G2C01
What means may an amateur station in distress use to attract attention, make known its condition and location, and obtain assistance?
A. Only Morse code signals sent on internationally recognized emergency channels
B. Any means of radiocommunication, but only on internationally recognized emergency channels
C. Any means of radiocommunication
D. Only those means of radiocommunication for which the station is licensed

G2C02
During a disaster in the US, when may an amateur station make transmissions necessary to meet essential communication needs and assist relief operations?
A. When normal communication systems are overloaded, damaged or disrupted
B. Only when the local RACES net is activated
C. Never; only official emergency stations may transmit in a disaster
D. When normal communication systems are working but are not convenient

G2C03
If a disaster disrupts normal communications in your area, what may the FCC do?
A. Declare a temporary state of communication emergency
B. Temporarily seize your equipment for use in disaster communications
C. Order all stations across the country to stop transmitting at once
D. Nothing until the President declares the area a disaster area

G2C04
If a disaster disrupts normal communications in an area what would the FCC include in any notice of a temporary state of communication emergency?
A. Any additional test questions needed for the licensing of amateur emergency communications workers
B. A list of organizations authorized to temporarily seize your equipment for disaster communications
C. Any special conditions requiring the use of non-commercial power systems
D. Any special conditions and special rules to be observed by stations during the emergency

G2C05
During an emergency, what power output limitations must be observed by a station in distress?
A. 200 watts PEP
B. 1500 watts PEP
C. 1000 watts PEP during daylight hours, reduced to 200 watts PEP during the night
D. There are no limitations during an emergency

G2C06
During a disaster in the US, what frequencies may be used to obtain assistance?
A. Only frequencies in the 80-meter band
B. Only frequencies in the 40-meter band
C. Any frequency
D. Any United Nations approved frequency

G2C07
If you are communicating with another amateur station and hear a station in distress break in, what is the first thing you should do?
A. Continue your communication because you were on frequency first
B. Acknowledge the station in distress and determine its location and what assistance may be needed
C. Change to a different frequency so the station in distress may have a clear channel to call for assistance
D. Immediately cease all transmissions because stations in distress have emergency rights to the frequency

G2C08
Why do stations in the Radio Amateur Civil Emergency Service (RACES) participate in training tests and drills?
A. To provide orderly and efficient operations for the civil defense organization they serve in the event of an emergency
B. To ensure that members attend monthly on-the-air meetings of the civil defense organization they serve
C. To ensure that RACES members are able to conduct tests and drills
D. To acquaint members of RACES with other members they may meet in an emergency

G2C04
(D)
Page 2-5

G2C05
(D)
Page 2-5

G2C06
(C)
Page 2-5

G2C07
(B)
Page 2-4

G2C08
(A)
Page 2-8

G2C09
(C)
Page 2-4

G2C09
When are you prohibited from helping a station in distress?
A. When that station is not transmitting on amateur frequencies
B. When the station in distress offers no call sign
C. You are not ever prohibited from helping any station in distress
D. When the station is not another amateur station

G2C10
(B)
Page 2-5

G2C10
When FCC declares a temporary state of communication emergency, what must you do?
A. Stay off the air until 30 days after FCC lifts the emergency notice
B. Abide by the limitations or conditions set forth in the FCC notice
C. Only communicate with stations within 2 miles of your location
D. Nothing; wait until the President declares a formal emergency before taking further action

G2C11
(C)
Page 2-5

G2C11
During a disaster in the US, which of the following emission modes must be used to obtain assistance?
A. Only SSB
B. Only SSB and CW
C. Any mode
D. Only CW

G2C12
(B)
Page 2-4

G2C12
What information should anyone who sends a distress transmission give to stations who answer?
A. The ITU region and grid square locator of the emergency
B. The location and nature of the distress
C. The time that the emergency occurred and the names of the persons involved
D. The agencies to notify and the name of the emergency coordinator

G2C13
(A)
Page 2-5

G2C13
What frequency should be used to send a distress call?
A. Whatever frequency has the best chance of communicating the distress message
B. 3873 kHz at night or 7285 kHz during the day
C. Only frequencies that are within your operating privileges
D. Only frequencies used by police, fire or emergency medical services

G2D Amateur auxiliary to the FCC's Compliance and Information Bureau; antenna orientation to minimize interference; HF operations, including logging practices

G2D01
What is the Amateur Auxiliary to the FCC's Compliance and Information Bureau?
A. Amateur volunteers who are formally enlisted to monitor the airwaves for rules violations
B. Amateur volunteers who conduct amateur licensing examinations
C. Amateur volunteers who conduct frequency coordination for amateur VHF repeaters
D. Amateur volunteers who use their station equipment to help civil defense organizations in times of emergency

G2D01
(A)
Page 2-8

G2D02
What are the objectives of the Amateur Auxiliary to the FCC's Compliance and Information Bureau?
A. To conduct efficient and orderly amateur licensing examinations
B. To encourage amateur self-regulation and compliance with the rules
C. To coordinate repeaters for efficient and orderly spectrum usage
D. To provide emergency and public safety communications

G2D02
(B)
Page 2-8

G2D03
Why are direction-finding "Fox Hunts" important to the Amateur Auxiliary?
A. Fox Hunts compel amateurs to upgrade their licenses
B. Fox Hunts provide an opportunity to practice direction-finding skills
C. Someone always receives an FCC Notice of Apparent Liability (NAL) when a Fox Hunt is concluded
D. Fox Hunts allow amateurs to work together with Environmental Protection Agencies

G2D03
(B)
Page 2-8

G2D04
What is an azimuthal projection map?
A. A map projection centered on the North Pole
B. A map projection centered on a particular location, used to determine the shortest path between points on the surface of the earth
C. A map that shows the angle at which an amateur satellite crosses the equator
D. A map that shows the number of degrees longitude that an amateur satellite appears to move westward at the equator with each orbit

G2D04
(B)
Page 2-9

G2D05
(A)
Page 2-9

G2D05
What is the most useful type of map to use when orienting a directional HF antenna toward a distant station?
A. Azimuthal projection
B. Mercator
C. Polar projection
D. Topographical

G2D06
(C)
Page 2-9

G2D06
A directional antenna pointed in the long-path direction to another station is generally oriented how many degrees from its short-path heading?
A. 45 degrees
B. 90 degrees
C. 180 degrees
D. 270 degrees

G2D07
(B)
[97.103b]
Page 2-10

G2D07
If a visiting amateur transmits from your station on 14.325 MHz, which of these is NOT true?
A. You must first give permission for the visiting amateur to use your station
B. You must keep in your station log the call sign of the visiting amateur together with the time and date of transmissions
C. The FCC may think that you were the station's control operator, unless your station records show otherwise
D. You both are equally responsible for the proper operation of the station

G2D08
(D)
Page 2-10

G2D08
Why should I keep a log if the FCC doesn't require it?
A. To help with your reply, if FCC requests information on who was control operator of your station for a given date and time
B. Logs provide information (callsigns, dates & times of contacts) used for many operating contests and awards
C. Logs are necessary to accurately verify contacts made weeks, months or years earlier, especially when completing QSL cards
D. All of these choices

G2D09
(D)
Page 2-10

G2D09
What information is normally contained in a station log?
A. Date and time of contact
B. Band and/or frequency of the contact
C. Call sign of station contacted and the RST signal report given
D. All of these choices

G2D10
(C)
Page 2-11

G2D10
Which of the following is a good reason to keep a log of your station's activities?
A. It is required by the FCC's rules
B. It is a tradition from the earliest days of amateur radio
C. It can aid you in resolving interference complaints
D. It can be a source of great enjoyment when reviewed in later years

G2D11
Which HF antenna would best be used to focus your signal to minimize interference?
A. A bidirectional antenna
B. A horizontal antenna positioned broadside to the desired direction
C. A unidirectional antenna
D. An omnidirectional antenna at low power

G2D12
Which of the following is required by the FCC rules when operating in the 60-meter band?
A. If you are using other than a dipole antenna, you must keep a record of the gain of your antenna
B. You must keep a record of the date, time, frequency, power level and stations worked
C. No records are required
D. You must keep a record of the manufacturer of your equipment and the antenna used

G2E Third-party communications; ITU Regions; VOX operation

G2E01
What type of messages may be transmitted to an amateur station in a foreign country?
A. Messages of any type
B. Messages that are not religious, political, or patriotic in nature
C. Messages of a technical nature or personal remarks of relative unimportance
D. Messages of any type, but only if the foreign country has a third-party communications agreement with the US

G2E02
Which of the following statements is true of VOX operation?
A. The received signal is more natural sounding
B. This mode allows "Hands Free' operation
C. Frequency spectrum is conserved
D. The duty cycle of the transmitter is reduced

G2E03
Which of the following user adjustable controls are usually associated with VOX circuitry?
A. Anti-VOX
B. VOX Delay
C. VOX Sensitivity
D. All of these choices are correct

| G2D11
| (C)
| Page 2-9

| G2D12
| (A)
| [97.303s]
| Page 1-7

| G2E01
| (C)
| [97.117]
| Page 1-11

| G2E02
| (B)
| Page 2-11

| G2E03
| (D)
| Page 2-12

G2E04
(C)
Page 2-12

G2E04
What is the purpose of the VOX sensitivity control?
A. To set the timing of transmitter activation
B. To set the audio frequency range at which the transmitter activates
C. To set the audio level at which the transmitter activates
D. None of these choices is correct

G2E05
(B)
Page 1-2

G2E05
In which International Telecommunication Union Region is the continental United States?
A. Region 1
B. Region 2
C. Region 3
D. Region 4

G2E06
(A)
Page 1-2

G2E06
In which International Telecommunication Union Region are Europe and Africa?
A. Region 1
B. Region 2
C. Region 3
D. Region 4

G2E07
(C)
Page 1-2

G2E07
In which International Telecommunication Union Region is Australia?
A. Region 1
B. Region 2
C. Region 3
D. Region 4

G2E08
(C)
Page 1-2

G2E08
Which of the following organizations is responsible for international regulation of the radio spectrum?
A. The International Regulatory Commission
B. The International Radio Union
C. The International Telecommunications Union
D. The International Frequency-Spectrum Commission

G2E09
(D)
Page 1-2

G2E09
What do the initials "ITU" stand for?
A. Interstate Telecommunications Union
B. International Telephony Union
C. International Transmission Union
D. International Telecommunications Union

G2E10
(B)
Page 2-11

G2E10
What is the circuit called that causes a transmitter to automatically transmit when an operator speaks into its microphone?
A. VXO
B. VOX
C. VCO
D. VFO

G2E11
What is the best reason to use a headset with an attached microphone and VOX control when using a mobile station?
A. For safer, handsfree operation
B. It allows you to make quicker transmissions
C. To eliminate ambient noise from your transmissions
D. To reduce outside distractions while operating

G2E11
(A)
Page 2-12

G2E12
What function does an anti-VOX circuit perform?
A. It prevents received audio from keying the transmitter
B. It prevents background noise from keying the transmitter
C. It prevents unauthorized persons from keying the transmitter
D. It prevents activation of the transmitter during CW operation

G2E12
(A)
Page 2-12

G2E13
Which of the following would indicate the completion of the transmitting of a formal message when using phone?
A. The phrase, "End of message"
B. The word "Break"
C. The Q-signal "QSL?"
D. The Q-signal "QRV"

G2E13
(A)
Page 2-2

G2F CW operating procedures, including procedural signals, Q signals and common abbreviations; full break-in; RTTY operating procedures, including procedural signals and common abbreviations and operating procedures for other digital modes, such as HF packet, AMTOR, PacTOR, G-TOR, Clover and PSK31

G2F01
Which of the following describes full break-in telegraphy (QSK)?
A. Breaking stations send the Morse code prosign BK
B. Automatic keyers are used to send Morse code instead of hand keys
C. An operator must activate a manual send/receive switch before and after every transmission
D. Incoming signals are received between transmitted key pulses

G2F01
(D)
Page 2-12

G2F02
In what segment of the 80-meter band do most data transmissions take place?
A. 3580 - 3620-kHz
B. 3500 - 3525-kHz
C. 3700 - 3750-kHz
D. 3775 - 3825-kHz

G2F02
(A)
Page 2-14

G2F03
(B)
Page 2-14

G2F03
In what segment of the 20-meter band do most RTTY transmissions take place?
A. 14.000 - 14.050 MHz
B. 14.070 - 14.095 MHz
C. 14.150 - 14.225 MHz
D. 14.275 - 14.350 MHz

G2F04
(D)
Page 2-13

G2F04
Which of the following is NOT correct?
A. ASCII is a 7-bit code, with start, stop and parity bits
B. The benefit of using AMTOR is its error detection and correction properties
C. Baudot is a 5-bit code, with additional start and stop bits
D. The two major AMTOR operating modes are SELCAL and LISTEN

G2F05
(B)
Page 2-13

G2F05
What is the most common frequency shift for RTTY emissions in the amateur HF bands?
A. 85 Hz
B. 170 Hz
C. 425 Hz
D. 850 Hz

G2F06
(A)
Page 2-14

G2F06
Why are the string of letters R and Y (sent as "RYRYRYRY...") occasionally used at the beginning of RTTY or other data transmissions?
A. This allows time to 'tune in' a station prior to the actual message being sent
B. To keep these commonly-used keys functional
C. These are the important mark and space keys
D. To make sure the transmitter is functional before sending a message

G2F07
(B)
Page 2-13

G2F07
What does the abbreviation "RTTY" stand for?
A. "Returning to you", meaning "your turn to transmit"
B. Radioteletype
C. A general call to all digital stations
D. Morse code practice over the air

G2F08
(C)
Page 2-13

G2F08
What prosign is sent using CW to indicate the end of a formal message?
A. SK - I acknowledge
B. BK - break
C. AR - end of message
D. KN - called station only, go ahead

G2F09
What character sequence is sent using RTTY or other data modes to indicate the end of a formal message?
A. CZCZ
B. KKKK
C. XXXXX
D. NNNN

G2F09
(D)
Page 2-14

G2F10
How many data bits are sent in a single PSK31 character?
A. The number varies
B. 5
C. 7
D. 8

G2F10
(A)
Page 2-16

G2F11
What part of a data packet contains the routing and handling information?
A. Directory
B. Preamble
C. Header
D. Footer

G2F11
(C)
Page 2-16

SUBELEMENT G3
RADIO-WAVE PROPAGATION
[3 EXAM QUESTIONS — 3 GROUPS]

Subelement
G3

G3A Ionospheric disturbances; sunspots and solar radiation

G3A01
What can be done at an amateur station to continue communications during a sudden ionospheric disturbance?
A. Try a higher frequency
B. Try the other sideband
C. Try a different antenna polarization
D. Try a different frequency shift

G3A01
(A)
Page 3-10

G3A02
What effect does a sudden ionospheric disturbance have on the daytime ionospheric propagation of HF radio waves?
A. It disrupts higher-latitude paths more than lower-latitude paths
B. It disrupts signals on lower frequencies more than those on higher frequencies
C. It disrupts communications via satellite more than direct communications
D. None, only areas on the night side of the earth are affected

G3A02
(B)
Page 3-10

G3A03
(C)
Page 3-10

G3A03
How long does it take the increased ultraviolet and X-ray radiation from solar flares to affect radio-wave propagation on the earth?
A. The effect is almost instantaneous
B. 1.5 minutes
C. 8 minutes
D. 20 to 40 hours

G3A04
(B)
Page 3-9

G3A04
What is solar flux?
A. The density of the sun's magnetic field
B. The radio energy emitted by the sun
C. The number of sunspots on the side of the sun facing the earth
D. A measure of the tilt of the earth's ionosphere on the side toward the sun

G3A05
(D)
Page 3-9

G3A05
What is the solar-flux index?
A. A measure of solar activity that is taken annually
B. A measure of solar activity that compares daily readings with results from the last six months
C. Another name for the American sunspot number
D. A measure of solar activity that is taken at a specific frequency

G3A06
(D)
Page 3-12

G3A06
What is a geomagnetic disturbance?
A. A sudden drop in the solar-flux index
B. A shifting of the earth's magnetic pole
C. Ripples in the ionosphere
D. A dramatic change in the earth's magnetic field over a short period of time

G3A07
(A)
Page 3-12

G3A07
At which latitudes are propagation paths more sensitive to geomagnetic disturbances?
A. Those greater than 45 degrees latitude
B. Those between 5 and 45 degrees latitude
C. Those near the equator
D. All paths are affected equally

G3A08
(B)
Page 3-12

G3A08
What can be the effect of a major geomagnetic storm on radio-wave propagation?
A. Improved high-latitude HF propagation
B. Degraded high-latitude HF propagation
C. Improved ground-wave propagation
D. Improved chances of UHF ducting

G3A09
When sunspot numbers are high, what is the effect on radio communications?
A. High-frequency radio signals are absorbed
B. Frequencies above 300 MHz become usable for long-distance communication
C. Long-distance communication in the upper HF and lower VHF range is enhanced
D. High-frequency radio signals become weak and distorted

G3A10
What is the sunspot number?
A. A daily index of sunspot activity
B. The number of sunspots observed during one solar rotation
C. The number of sunspots observed during a sunspot cycle
D. The number of sunspots observed averaged over a seven day period

G3A11
What is the sunspot cycle?
A. The 9- to 11-year periods when sunspots move from the sun's pole to its equatorial region
B. The 9- to 11-year periods when sunspots cause coronal holes to appear
C. The approximately 11year variation in the sunspot number
D. The approximately 11-year periods when sunspots combine to form flares

G3A12
What is the K-index?
A. A linear index of solar activity
B. A measure of geomagnetic stability
C. An index of solar flux measured at Boulder, Colorado
D. A daily value measured on a scale from 0 to 400 to express the range of disturbance of the geomagnetic field

G3A13
What is the A-index?
A. A monthly linear index of solar activity
B. An weekly index of solar flux measured at Boulder, Colorado
C. A daily value measured on a scale from 0 to 400 to express the range of disturbance of the geomagnetic field
D. An index used by NOAA to correlate the visual color wavelengths seen with Aurora Borealis (Northern Lights)

G3A09
(C)
Page 3-8

G3A10
(A)
Page 3-8

G3A11
(C)
Page 3-8

G3A12
(B)
Page 3-12

G3A13
(C)
Page 3-12

G3A14
How does solar coronal hole activity affect radio communications?
A. The activity emits charged particles that improve HF communications
B. The activity emits charged particles that improve VHF/UHF ducting
C. The activity emits charged particles that usually disrupt HF communications
D. The activity emits charged particles, but they never reach Earth's magnetosphere

G3A15
How long does it take charged particles from coronal mass ejections (CMEs) to affect radiowave propagation on the earth?
A. Almost instantaneously
B. About 5 minutes
C. About 8 minutes
D. 20 to 40 hours

G3A16
What might result during periods of high geomagnetic activity?
A. A visible aurora
B. Excellent highfrequency radio conditions
C. Poor 6-meter conditions
D. Flayer absorption

G3B Maximum usable frequency; propagation "hops"

G3B01
If the maximum usable frequency (MUF) on the path from Minnesota to France is 24 MHz, which band should offer the best chance for a successful contact?
A. 10 meters
B. 15 meters
C. 20 meters
D. 40 meters

G3B02
If the maximum usable frequency (MUF) on the path from Ohio to Germany is 17 MHz, which band should offer the best chance for a successful contact?
A. 80 meters
B. 40 meters
C. 20 meters
D. 2 meters

G3B03
If the HF radio-wave propagation (skip) is generally good on the 24-MHz and 28-MHz bands for several days, when might you expect a similar condition to occur?
A. 7 days later
B. 14 days later
C. 28 days later
D. 90 days later

G3B03
(C)
Page 3-9

G3B04
What is one way to determine if the maximum usable frequency (MUF) is high enough to support 28-MHz propagation between your station and western Europe?
A. Listen for signals on a 10-meter beacon frequency
B. Listen for signals on a 20-meter beacon frequency
C. Listen for signals on a 39-meter broadcast frequency
D. Listen for WWVH time signals on 20 MHz

G3B04
(A)
Page 3-10

G3B05
What usually happens to radio waves with frequencies below the maximum usable frequency (MUF) when they are sent into the ionosphere?
A. They are bent back to the earth
B. They pass through the ionosphere
C. They are completely absorbed by the ionosphere
D. They are bent and trapped in the ionosphere to circle the Earth

G3B05
(A)
Page 3-7

G3B06
Where would you tune to hear beacons that would help you determine propagation conditions on the 20-meter band?
A. 28.2 MHz
B. 21.1 MHz
C. 14.1 MHz
D. 18.1 MHz

G3B06
(C)
Page 3-7

G3B07
During periods of low solar activity, which frequencies are the least reliable for long-distance communication?
A. Frequencies below 3.5 MHz
B. Frequencies near 3.5 MHz
C. Frequencies on or above 10 MHz
D. Frequencies above 20 MHz

G3B07
(D)
Page 3-9

G3B08
At what point in the solar cycle does the 20-meter band usually support worldwide propagation during daylight hours?
A. At the summer solstice
B. Only at the maximum point of the solar cycle
C. Only at the minimum point of the solar cycle
D. At any point in the solar cycle

G3B08
(D)
Page 3-9

G3B09
(C)
Page 3-4

G3B09
What is the maximum distance along the Earth's surface that is normally covered in one hop using the F2 region?
A. 180 miles
B. 1200 miles
C. 2500 miles
D. None; the F2 region does not support radio-wave propagation

G3B10
(B)
Page 3-4

G3B10
What is the maximum distance along the Earth's surface that is normally covered in one hop using the E region?
A. 180 miles
B. 1200 miles
C. 2500 miles
D. None of these choices is correct

G3B11
(A)
Page 3-7

G3B11
What happens to HF propagation when the lowest usable frequency (LUF) exceeds the maximum usable frequency (MUF)?
A. No HF radio frequency will support communications along an ionospheric signal path
B. The lowest usable frequency can never exceed the maximum usable frequency
C. The ionospheric absorption of HF radio signals increases by 3 dB along every signal path
D. All ionospheric propagation paths are still usable, but the signal-to-noise ratio decreases

G3B12
(D)
Page 3-7

G3B12
What factors affect the maximum usable frequency (MUF)?
A. Path distance and locations
B. Time of day and season
C. Solar radiation and ionospheric disturbances
D. All of these choices are correct

G3B13
(D)
Page 3-6

G3B13
How might a skywave signal sound if it arrives at your receiver by both short path and long path propagation?
A. Periodic fading every 10 seconds
B. Signal strength increased by 3 dB
C. Signal strength decreased by 3 dB
D. A well-defined echo can be heard

G3B14
(A)
Page 3-7

G3B14
A short distance hop on 10 meters might indicate what conditions on 6 meters?
A. The MUF exceeds 50 MHz
B. Absolutely no chance of a skywave 6meter band opening
C. 6-meter ground waves will diminish
D. 10-meter propagation has no bearing on possible 6-meter propagation

G3C Height of ionospheric regions; critical angle and frequency; HF scatter

G3C01
What is the average height of maximum ionization of the E region?
A. 45 miles
B. 70 miles
C. 200 miles
D. 1200 miles

G3C02
When can the F2 region be expected to reach its maximum height at your location?
A. At noon during the summer
B. At midnight during the summer
C. At dusk in the spring and fall
D. At noon during the winter

G3C03
Why is the F2 region mainly responsible for the longest-distance radio-wave propagation?
A. Because it exists only at night
B. Because it is the lowest ionospheric region
C. Because it is the highest ionospheric region
D. Because it does not absorb radio waves as much as other ionospheric regions

G3C04
What is the "critical angle" as used in radio-wave propagation?
A. The lowest takeoff angle that will return a radio wave to the earth under specific ionospheric conditions
B. The compass direction of a distant station
C. The compass direction opposite that of a distant station
D. The highest takeoff angle that will return a radio wave to the earth under specific ionospheric conditions

G3C05
What is the main reason the 160-, 80- and 40-meter amateur bands tend to be useful only for short-distance communications during daylight hours?
A. Because of a lack of activity
B. Because of auroral propagation
C. Because of D-region absorption
D. Because of magnetic flux

G3C06
What is a characteristic of HF scatter signals?
A. High intelligibility
B. A wavering sound
C. Reversed modulation
D. Reversed sidebands

G3C01
(B)
Page 3-4

G3C02
(A)
Page 3-4

G3C03
(C)
Page 3-4

G3C04
(D)
Page 3-5

G3C05
(C)
Page 3-2

G3C06
(B)
Page 3-14

G3C07
(D)
Page 3-14

G3C07
What makes HF scatter signals often sound distorted?
A. Auroral activity and changes in the earth's magnetic field
B. Propagation through ground waves that absorb much of the signal
C. The state of the E-region at the point of refraction
D. Energy scattered into the skip zone through several radio-wave paths

G3C08
(A)
Page 3-14

G3C08
Why are HF scatter signals usually weak?
A. A part of the signal energy is propagated into the skip zone
B. Auroral activity absorbs most of the signal energy
C. Propagation through ground waves absorbs most of the signal energy
D. The F region of the ionosphere absorbs most of the signal energy

G3C09
(B)
Page 3-13

G3C09
What type of radio-wave propagation allows a signal to be detected at a distance too far for ground-wave propagation but too near for normal sky-wave propagation?
A. Ground wave
B. Scatter
C. Sporadic-E skip
D. Short-path skip

G3C10
(D)
Page 3-13

G3C10
When does scatter propagation on the HF bands most often occur?
A. When the sunspot cycle is at a minimum and D-region absorption is high
B. At night
C. When the F1 and F2 regions are combined
D. When communicating on frequencies above the maximum usable frequency (MUF)

G3C11
(A)
Page 3-7

G3C11
Which is true about ionospheric absorption near the maximum usable frequency (MUF)?
A. Absorption will be minimum
B. Absorption is independent of frequency
C. Absorption approaches maximum
D. There is no correlation between MUF and absorption

G3C12
(D)
Page 3-4

G3C12
Daylight fading on the 40-meter band is associated most with which ionospheric layer?
A. The F2 layer
B. The F1 layer
C. The E layer
D. The D layer

SUBELEMENT G4
AMATEUR RADIO PRACTICES
[5 EXAM QUESTIONS — 5 GROUPS]

G4A Twotone test; electronic TR switch; amplifier neutralization

G4A01
What kind of input signal is used to test the amplitude linearity of a single-sideband phone transmitter while viewing the output on an oscilloscope?
A. Normal speech
B. An audio-frequency sine wave
C. Two audio-frequency sine waves
D. An audio-frequency square wave

G4A02
When testing the amplitude linearity of a single-sideband transmitter, what kind of audio tones are fed into the microphone input and on what kind of instrument is the transmitter output observed?
A. Two harmonically related tones are fed in, and the output is observed on an oscilloscope
B. Two harmonically related tones are fed in, and the output is observed on a distortion analyzer
C. Two non-harmonically related tones are fed in, and the output is observed on an oscilloscope
D. Two non-harmonically related tones are fed in, and the output is observed on a distortion analyzer

G4A03
What audio frequencies are used in a two-tone test of the linearity of a single-sideband phone transmitter?
A. 20 Hz and 20 kHz tones must be used
B. 1200 Hz and 2400 Hz tones must be used
C. Any two audio tones may be used, but they must be within the transmitter audio passband, and must be harmonically related
D. Any two audio tones may be used, but they must be within the transmitter audio passband, and should not be harmonically related

G4A04
At what point in an HF transceiver block diagram would an electronic TR switch normally appear?
A. Between the transmitter and low-pass filter
B. Between the low-pass filter and antenna
C. At the antenna feed point
D. At the power supply feed point

G4A01
(C)
Page 4-6

G4A02
(C)
Page 4-6

G4A03
(D)
Page 4-6

G4A04
(A)
Page 4-13

G4A05
(C)
Page 4-14

G4A05
Why is an electronic TR switch preferable to a mechanical one?
A. It allows greater receiver sensitivity
B. Its circuitry is simpler
C. It has a higher operating speed
D. It allows cleaner output signals

G4A06
(A)
Page 4-9

G4A06
As a power amplifier is tuned, what reading on its grid-current meter indicates the best neutralization?
A. A minimum change in grid current as the output circuit is changed
B. A maximum change in grid current as the output circuit is changed
C. Minimum grid current
D. Maximum grid current

G4A07
(D)
Page 4-8

G4A07
Why is neutralization necessary for some vacuum-tube amplifiers?
A. To reduce the limits of loaded Q
B. To reduce grid-to-cathode leakage
C. To cancel AC hum from the filament transformer
D. To cancel oscillation caused by the effects of interelectrode capacitance

G4A08
(C)
Page 4-8

G4A08
In a properly neutralized RF amplifier, what type of feedback is used?
A. 5%
B. 10%
C. Negative
D. Positive

G4A09
(B)
Page 4-8

G4A09
What does a neutralizing circuit do in an RF amplifier?
A. It controls differential gain
B. It cancels the effects of positive feedback
C. It eliminates AC hum from the power supply
D. It reduces incidental grid modulation

G4A10
(B)
Page 4-8

G4A10
What is the reason for neutralizing the final amplifier stage of a transmitter?
A. To limit the modulation index
B. To eliminate self oscillations
C. To cut off the final amplifier during standby periods
D. To keep the carrier on frequency

G4A11
What type of transmitter performance does a two-tone test analyze?
A. Linearity
B. Carrier and undesired sideband suppression
C. Percentage of frequency modulation
D. Percentage of carrier phase shift

G4A11
(A)
Page 4-6

G4A12
What type of signals are used to conduct a two-tone test?
A. Two audio signals of the same frequency, but shifted 90degrees and are within the transmitter's modulation bandpass
B. Two non-harmonically related audio signals that are within the modulation bandpass of a transmitter
C. Any two audio frequency signals as long as they are within the transmitter's modulation bandpass
D. Two audio frequency range square wave signals of equal amplitude that are within the transmitter's modulation bandpass

G4A12
(B)
Page 4-6

G4A13
In what way is a diode like a switch?
A. It permits current flow at bias voltages less than its zener voltage and blocks current at bias voltages greater than its zener voltage
B. It permits current flow when reverse biased and blocks current when forward biased
C. The voltage drop across it increases as forward bias increases and decreases as reverse bias decreases
D. It permits current flow when forward biased and blocks current when reverse biased

G4A13
(D)
Page 4-14

G4B Test equipment: oscilloscope; signal tracer; antenna noise bridge; monitoring oscilloscope; field-strength meters

G4B

G4B01
What item of test equipment contains horizontal- and vertical-channel amplifiers?
A. An ohmmeter
B. A signal generator
C. An ammeter
D. An oscilloscope

G4B01
(D)
Page 4-2

G4B02
What is a digital oscilloscope?
A. An oscilloscope used only for signal tracing in digital circuits
B. An oscilloscope used only for troubleshooting computers
C. An oscilloscope used only for troubleshooting switching power supply circuits
D. An oscilloscope designed around digital technology rather than analog technology

G4B02
(D)
Page 4-3

G4B03
(D)
Page 4-4

G4B03
How would a signal tracer normally be used?
A. To identify the source of radio transmissions
B. To make exact drawings of signal waveforms
C. To show standing wave patterns on open-wire feed-lines
D. To identify an inoperative stage in a receiver

G4B04
(C)
Page 4-4

G4B04
How is a noise bridge normally used?
A. It is connected at an antenna's feed point and reads the antenna's noise figure
B. It is connected between a transmitter and an antenna and is tuned for minimum SWR
C. It is connected between a receiver and an antenna of unknown impedance and is tuned for minimum noise
D. It is connected between an antenna and ground and is tuned for minimum SWR

G4B05
(A)
Page 4-3

G4B05
What is the best instrument to use to check the signal quality of a CW or single-sideband phone transmitter?
A. A monitoring oscilloscope
B. A field-strength meter
C. A sidetone monitor
D. A signal tracer and an audio amplifier

G4B06
(D)
Page 4-3

G4B06
What signal source is connected to the vertical input of a monitoring oscilloscope when checking the quality of a transmitted signal?
A. The IF output of a monitoring receiver
B. The audio input of the transmitter
C. The RF signals of a nearby receiving antenna
D. The RF output of the transmitter

G4B07
(C)
Page 4-5

G4B07
What is the purpose of a field-strength meter?
A. To determine the standing-wave ratio on a transmission line
B. To check the output modulation of a transmitter
C. To monitor relative RF output
D. To increase average transmitter output

G4B08
(A)
Page 4-5

G4B08
What simple instrument may be used to monitor relative RF output during antenna and transmitter adjustments?
A. A field-strength meter
B. An antenna noise bridge
C. A multimeter
D. A metronome

G4B09
In order to raise the S-meter reading on a receiver from S8 to S9, how much must the power output of a transmitter be increased?
A. Approximately 2 times
B. Approximately 3 times
C. Approximately 4 times
D. Approximately 5 times

G4B09
(C)
Page 4-13

G4B10
What type of information does a field strength meter provide?
A. The gain in dBi of an antenna
B. The field pattern of an antenna
C. The presence and amount of phase distortion of a transmitter
D. The presence and amount of amplitude distortion of a transmitter

G4B10
(B)
Page 4-5

G4B11
For which of the following applications might you use a field strength meter?
A. Close-in RDF work
B. A modulation monitor for a frequency or phase modulation transmitter
C. An overmodulation indicator for a SSB transmitter
D. A keying indicator for a RTTY or packet transmitter

G4B11
(A)
Page 4-5

G4B12
What is one way a noise bridge might be used?
A. Determining an antenna's gain in dBi
B. Pre-tuning an antenna tuner
C. Determining the directivity of an antenna
D. Determining the line loss of the antenna system

G4B12
(B)
Page 4-4

G4B13
What information could a noise bridge directly provide about an unknown length and type of transmission line?
A. Its characteristic impedance
B. Its velocity factor
C. Its loss in dB per 100-feet
D. Its reflection coefficient

G4B13
(A)
Page 4-4

G4B14
How would you connect an oscilloscope to an AM or SSB transmitter to check transmitter modulation using double trapezoidal patterns?
A. Couple the detected RF output signal to the vertical plates; set the internal sweep to twice the modulating frequency
B. Couple the RF output signal to the vertical plates and external trigger; set the internal sweep to twice the modulating frequency
C. Couple the RF output signal to the vertical plates; apply the unmodulated RF drive signal to the horizontal plates
D. Couple the detected RF output signal to the vertical plates; apply a constant DC signal to the horizontal plates

G4B14
(B)
Page 4-3

G4C Audio rectification in consumer electronics; RF ground

G4C01
What devices would you install in home-entertainment systems to reduce or eliminate audio-frequency interference?
A. Bypass inductors
B. Bypass capacitors
C. Metal-oxide varistors
D. Bypass resistors

G4C02
What should be done if a properly operating amateur station is the cause of interference to a nearby telephone?
A. Make internal adjustments to the telephone equipment
B. Install RFI filters at the affected telephone
C. Stop transmitting whenever the telephone is in use
D. Ground and shield the local telephone distribution amplifier

G4C03
What sound is heard from a public-address system if audio rectification of a nearby single-sideband phone transmission occurs?
A. A steady hum whenever the transmitter's carrier is on the air
B. On-and-off humming or clicking
C. Distorted speech from the transmitter's signals
D. Clearly audible speech from the transmitter's signals

G4C04
What sound is heard from a public-address system if audio rectification of a nearby CW transmission occurs?
A. On-and-off humming or clicking
B. Audible, possibly distorted speech
C. Muffled, severely distorted speech
D. A steady whistling

G4C05
If your third-floor amateur station has a ground wire running 33 feet down to a ground rod, why might you get an RF burn if you touch the front panel of your HF transceiver?
A. Because the ground rod is not making good contact with moist earth
B. Because the transceiver's heat-sensing circuit is not working to start the cooling fan
C. Because of a bad antenna connection, allowing the RF energy to take an easier path out of the transceiver through you
D. Because the ground wire is a resonant length on several HF bands and acts more like an antenna than an RF ground connection

G4C06
Which of the following is NOT an important reason to have a good station ground?
A. To reduce the cost of operating a station
B. To reduce electrical noise
C. To reduce interference
D. To reduce the possibility of electric shock

G4C06
(A)
Page 4-18

G4C07
What is one good way to avoid stray RF energy in your amateur station?
A. Keep the station's ground wire as short as possible
B. Use a beryllium ground wire for best conductivity
C. Drive the ground rod at least 14 feet into the ground
D. Make a couple of loops in the ground wire where it connects to your station

G4C07
(A)
Page 4-18

G4C08
Which of the following statements about station grounding is NOT true?
A. Braid from RG-213 coaxial cable makes a good conductor to tie station equipment together into a station ground
B. Only transceivers and power amplifiers need to be tied into a station ground
C. According to the National Electrical Code, there should be only one grounding system in a building
D. The minimum length for a good ground rod is 8 feet

G4C08
(B)
Page 4-18

G4C09
Which of the following statements about station grounding is true?
A. The chassis of each piece of station equipment should be tied together with high-impedance conductors
B. If the chassis of all station equipment is connected with a good conductor, there is no need to tie them to an earth ground
C. RF hot spots can occur in a station located above the ground floor if the equipment is grounded by a long ground wire
D. A ground loop is an effective way to ground station equipment

G4C09
(C)
Page 4-18

G4C10
Which of the following is NOT covered in the National Electrical Code?
A. Minimum conductor sizes for different lengths of amateur antennas
B. The size and composition of grounding conductors
C. Electrical safety inside the ham shack
D. The RF exposure limits of the human body

G4C10
(D)
Page 4-21

G4C11
What can cause the unintended rectification of an RF signal?
A. Induced currents in conductors that are in poor electrical contact
B. Induced voltages in conductors that are in good electrical contact
C. Capacitive coupling of the RF signal to ground
D. Excessive standing wave ratio (SWR) of the transmission line system

G4C11
(A)
Page 4-16

G4C12
What is one cause of severe, broadband radio frequency noise at an amateur radio station?
A. Not using a balun or line isolator to feed balanced antennas
B. Lack of rectification of the transmitter's signal in power conductors
C. An intermittent RF ground
D. The use of horizontal, rather than vertical antennas

G4C13
How can a ground loop be avoided?
A. Series connect ("daisy chain") all ground conductors
B. Connect the AC neutral conductor to the ground wire
C. Avoid using lockwashers and star washers in making ground connections
D. Connect all ground conductors to a single point

G4D Speech processors; PEP calculations; wire sizes and fuses

G4D01
What is the reason for using a properly adjusted speech processor with a single-sideband phone transmitter?
A. It reduces average transmitter power requirements
B. It reduces unwanted noise pickup from the microphone
C. It improves voice-frequency fidelity
D. It improves signal intelligibility at the receiver

G4D02
If a single-sideband phone transmitter is 100% modulated, what will a speech processor do to the transmitter's power?
A. It will increase the output PEP
B. It will add nothing to the output PEP
C. It will decrease the peak power output
D. It will decrease the average power output

G4D03
What is the output PEP from a transmitter if an oscilloscope measures 200 volts peak-to-peak across a 50-ohm resistor connected to the transmitter output?
A. 100 watts
B. 200 watts
C. 400 watts
D. 1000 watts

G4D04
What is the output PEP from a transmitter if an oscilloscope measures 500 volts peak-to-peak across a 50-ohm resistor connected to the transmitter output?
A. 500 watts
B. 625 watts
C. 1250 watts
D. 2500 watts

G4D05
What is the output PEP of an unmodulated carrier transmitter if an average-reading wattmeter connected to the transmitter output indicates 1060 watts?
A. 530 watts
B. 1060 watts
C. 1500 watts
D. 2120 watts

G4D06
Which wires in a four-conductor line cord should be attached to fuses in a 240-VAC primary (single phase) power supply?
A. Only the "hot" (black and red) wires
B. Only the "neutral" (white) wire
C. Only the ground (bare) wire
D. All wires

G4D07
What size wire is normally used on a 20-ampere, 120-VAC household appliance circuit?
A. AWG number 20
B. AWG number 16
C. AWG number 14
D. AWG number 12

G4D08
What maximum size fuse or circuit breaker should be used in a household appliance circuit using AWG number 12 wiring?
A. 100 amperes
B. 60 amperes
C. 30 amperes
D. 20 amperes

G4D09
What operating benefit does properly adjusted speech clipping provide?
A. It removes any distortion in the audio waveform
B. Deep clipping restores the natural sound of the audio
C. It prevents overdriving the transmitter's modulator stage
D. It removes any AC hum and noise that might be in the audio

G4D10
What would be the voltage across a 50-ohm dummy load dissipating 1200 watts?
A. 173 volts
B. 245 volts
C. 346 volts
D. 692 volts

G4D05
(B)
Page 4-11

G4D06
(A)
Page 4-20

G4D07
(D)
Page 4-23

G4D08
(D)
Page 4-23

G4D09
(C)
Page 4-15

G4D10
(B)
Page 4-12

G4E Common connectors used in amateur stations: types; when to use; fastening methods; precautions when using; HF mobile radio installations; emergency power systems; generators; battery storage devices and charging sources including solar; wind generation

G4E01
(B)
Page 4-25

G4E01
Which of the following connectors is NOT designed for RF transmission lines?
A. PL-259
B. DB-25
C. Type N
D. BNC

G4E02
(D)
Page 4-22

G4E02
When installing a power plug on a line cord, which of the following should you do?
A. Twist the wire strands neatly and fasten them so they don't cause a short circuit
B. Observe the correct wire color conventions for plug terminals
C. Use proper grounding techniques
D. All of these choices

G4E03
(A)
Page 4-25

G4E03
Which of the following power connections would be the best for a 100-watt HF mobile installation?
A. A direct, fused connection to the battery using heavy gauge wire
B. A connection to the fuse-protected accessory terminal strip or distribution panel
C. A connection to the cigarette lighter
D. A direct connection to the alternator or generator

G4E04
(B)
Page 4-25

G4E04
Why is it best NOT to draw the DC power for a 100-watt HF transceiver from an automobile's cigarette lighter socket?
A. The socket is not wired with an RF-shielded power cable
B. The socket's wiring may not be adequate for the current being drawn by the transceiver
C. The DC polarity of the socket is reversed from the polarity of modern HF transceivers
D. The power from the socket is never adequately filtered for HF transceiver operation

G4E05
(C)
Page 4-25

G4E05
Which of the following most limits the effectiveness of an HF mobile transceiver operating in the 75-meter band?
A. The vehicle's electrical system wiring
B. The wire gauge of the DC power line to the transceiver
C. The HF mobile antenna system
D. The rating of the vehicle's alternator or generator

G4E06
Which of the following is true of both a permanent or temporary emergency generator installation?
A. The generator should be located in a well ventilated area
B. The installation should be grounded
C. Extra fuel supplies, especially gasoline, should not be stored in an inhabited area
D. All of these choices

G4E06
(D)
Page 4-23

G4E07
Which of the following is true of a lead-acid storage battery as it is being charged?
A. It tends to cool off
B. It gives off explosive oxygen gas
C. It gives off explosive hydrogen gas
D. It takes in oxygen from the surrounding air

G4E07
(C)
Page 4-24

G4E08
What is the name of the process by which sunlight is directly changed into electricity?
A. Photovoltaic conversion
B. Photosensitive conduction
C. Photosynthesis
D. Photocoupling

G4E08
(A)
Page 4-24

G4E09
What is the approximate open-circuit voltage from a modern, well illuminated photovoltaic cell?
A. 0.02 VDC
B. 0.5 VDC
C. 0.2 VDC
D. 1.38 VDC

G4E09
(B)
Page 4-24

G4E10
What determines the proper size solar panel to use in a solar-powered battery-charging circuit?
A. The panel's voltage rating and maximum output current
B. The amount of voltage available per square inch of panel
C. The panel's open-circuit current
D. The panel's short-circuit voltage

G4E10
(A)
Page 4-24

G4E11
What is the biggest disadvantage to using wind power as the primary source of power for an emergency station?
A. The conversion efficiency from mechanical energy to electrical energy is less that 2 percent
B. The voltage and current ratings of such systems are not compatible with amateur equipment
C. A large electrical storage system is needed to supply power when the wind is not blowing
D. All of these choices are correct

G4E11
(C)
Page 4-25

G4E12
(A)
Page 4-26

G4E12
What type of coaxial connector would be a good choice to use for 10 GHz feed-line connections?
A. Type N
B. Type BNC
C. Type UHF
D. Type F

G4E13
(A)
Page 4-23

G4E13
Where should you avoid placing a gasoline-fueled generator to power your station?
A. Inside a building or outside an open window
B. Close to cold water pipes or other grounded metal objects
C. Close to a driven ground
D. Downwind from your station

G4E14
(D)
Page 4-23

G4E14
What safety precaution should you observe when using a gasoline-fueled generator to power your home station?
A. Always ground the frame of the generator
B. Use only generators that produce a clean sine wave output
C. Make sure that the engine is well lubricated
D. All of these choices are correct

G4E15
(D)
Page 4-24

G4E15
During a commercial power outage, why would it be unwise to back feed the output of a gasoline generator into your house wiring by connecting the generator through an AC wall outlet?
A. It presents a hazard for electric company workers
B. You may draw too much current, overloading your generator
C. Power may be restored to your house, damaging your generator
D. All of these choices are correct

Subelement
G5

SUBELEMENT G5

ELECTRICAL PRINCIPLES

[2 EXAM QUESTIONS — 2 GROUPS]

G5A Impedance, including matching; resistance, including ohm; reactance; inductance; capacitance; and metric divisions of these values

G5A01
(C)
Page 5-19

G5A01
What is impedance?
A. The electric charge stored by a capacitor
B. The opposition to the flow of AC in a circuit containing only capacitance
C. The opposition to the flow of AC in a circuit
D. The force of repulsion between one electric field and another with the same charge

G5A02
What is reactance?
A. Opposition to DC caused by resistors
B. Opposition to AC caused by inductors and capacitors
C. A property of ideal resistors in AC circuits
D. A large spark produced at switch contacts when an inductor is de-energized

G5A03
In an inductor, what causes opposition to the flow of AC?
A. Resistance
B. Reluctance
C. Admittance
D. Reactance

G5A04
In a capacitor, what causes opposition to the flow of AC?
A. Resistance
B. Reluctance
C. Reactance
D. Admittance

G5A05
How does a coil react to AC?
A. As the frequency of the applied AC increases, the reactance decreases
B. As the amplitude of the applied AC increases, the reactance increases
C. As the amplitude of the applied AC increases, the reactance decreases
D. As the frequency of the applied AC increases, the reactance increases

G5A06
How does a capacitor react to AC?
A. As the frequency of the applied AC increases, the reactance decreases
B. As the frequency of the applied AC increases, the reactance increases
C. As the amplitude of the applied AC increases, the reactance increases
D. As the amplitude of the applied AC increases, the reactance decreases

G5A07
What happens when the impedance of an electrical load is equal to the internal impedance of the power source?
A. The source delivers minimum power to the load
B. The electrical load is shorted
C. No current can flow through the circuit
D. The source delivers maximum power to the load

G5A02
(B)
Page 5-17

G5A03
(D)
Page 5-18

G5A04
(C)
Page 5-17

G5A05
(D)
Page 5-18

G5A06
(A)
Page 5-17

G5A07
(D)
Page 5-20

G5A08
(A)
Page 5-20

G5A08
Why is impedance matching important?
A. So the source can deliver maximum power to the load
B. So the load will draw minimum power from the source
C. To ensure that there is less resistance than reactance in the circuit
D. To ensure that the resistance and reactance in the circuit are equal

G5A09
(B)
Page 5-17

G5A09
What unit is used to measure reactance?
A. Mho
B. Ohm
C. Ampere
D. Siemens

G5A10
(B)
Page 5-19

G5A10
What unit is used to measure impedance?
A. Volt
B. Ohm
C. Ampere
D. Watt

G5A11
(A)
Page 5-24

G5A11
Why should core saturation of a conventional impedance matching transformer be avoided?
A. Harmonics and distortion could result from saturation
B. Magnetic flux would increase with frequency
C. RF susceptance would increase
D. Temporary changes of the core permeability could result from saturation

G5B Decibel; Ohm's Law; current and voltage dividers; electrical power calculations and series and parallel components; transformers (either voltage or impedance); sine wave root-mean-square (RMS) value

G5B01
(B)
Page 5-10

G5B01
A two-times increase in power results in a change of how many dB?
A. 1 dB higher
B. 3 dB higher
C. 6 dB higher
D. 12 dB higher

G5B02
(B)
Page 5-4

G5B02
In a parallel circuit with a voltage source and several branch resistors, how is the total current related to the current in the branch resistors?
A. It equals the average of the branch current through each resistor
B. It equals the sum of the branch current through each resistor
C. It decreases as more parallel resistors are added to the circuit
D. It is the sum of each resistor's voltage drop multiplied by the total number of resistors

G5B03
How many watts of electrical power are used if 400 VDC is supplied
to an 800-ohm load?
A. 0.5 watts
B. 200 watts
C. 400 watts
D. 320,000 watts

G5B03
(B)
Page 5-8

G5B04
How many watts of electrical power are used by a 12-VDC light bulb
that draws 0.2 amperes?
A. 60 watts
B. 24 watts
C. 6 watts
D. 2.4 watts

G5B04
(D)
Page 5-8

G5B05
How many watts are being dissipated when 7.0 milliamperes flow
through 1.25 kilohms?
A. Approximately 61 milliwatts
B. Approximately 39 milliwatts
C. Approximately 11 milliwatts
D. Approximately 9 milliwatts

G5B05
(A)
Page 5-8

G5B06
What is the voltage across a 500-turn secondary winding in a
transformer if the 2250-turn primary is connected to 120 VAC?
A. 2370 volts
B. 540 volts
C. 26.7 volts
D. 5.9 volts

G5B06
(C)
Page 5-22

G5B07
What is the turns ratio of a transformer to match an audio amplifier
having a 600-ohm output impedance to a speaker having a 4-ohm
impedance?
A. 12.2 to 1
B. 24.4 to 1
C. 150 to 1
D. 300 to 1

G5B07
(A)
Page 5-22

G5B08
A DC voltage equal to what value of an applied sine-wave AC voltage
would produce the same amount of heat over time in a resistive
element?
A. The peak-to-peak value
B. The RMS value
C. The average value
D. The peak value

G5B08
(B)
Page 5-15

G5B09
(D)
Page 5-15

G5B09
What is the peak-to-peak voltage of a sine wave that has an RMS voltage of 120 volts?
A. 84.8 volts
B. 169.7 volts
C. 204.8 volts
D. 339.4 volts

G5B10
(B)
Page 5-16

G5B10
A sine wave of 17 volts peak is equivalent to how many volts RMS?
A. 8.5 volts
B. 12 volts
C. 24 volts
D. 34 volts

G5B11

G5B11
This question has been withdrawn.

G5B12
(C)
Page 5-20

G5B12
What causes a voltage to appear across the secondary winding of a transformer when a voltage source is connected across its primary winding?
A. Capacitive coupling
B. Displacement current coupling
C. Mutual inductance
D. Mutual capacitance

G5B13
(A)
Page 5-18

G5B13
What would be the capacitance and voltage rating of a series circuit consisting of two equal value capacitors with equal voltage ratings?
A. Total capacitance would be half that of each capacitor and maximum voltage would be twice that of each capacitor
B. Total capacitance would be half that of each capacitor and maximum voltage would be the same as each capacitor
C. Total capacitance and maximum voltage would be the same as each capacitor
D. Total capacitance and maximum voltage would be half that of each capacitor

G5B14
(D)
Page 5-12

G5B14
What percentage loss would result from a transmission line loss of 1 dB?
A. 16.6%
B. 12.5%
C. 14.7%
D. 20.6%

G5B15
If three equal resistors in parallel produce 50 ohms of resistance and the same resistors in series produce 450-ohms, what is the value of each resistor?
A. 1500-ohms
B. 90-ohms
C. 150-ohms
D. 175-ohms

G5B15
(C)
Page 5-7

SUBELEMENT G6 — CIRCUIT COMPONENTS
[1 EXAM QUESTION — 1 GROUP]

Subelement
G6

G6A Resistors; capacitors; inductors; rectifiers and transistors; etc.

G6A01
If a carbon resistor's temperature is increased, what will happen to the resistance?
A. It will increase by 20% for every 10 degrees centigrade
B. It will stay the same
C. It will change depending on the resistor's temperature coefficient rating
D. It will become time dependent

G6A01
(C)
Page 6-3

G6A02
What type of capacitor is often used in power-supply circuits to filter the rectified AC?
A. Disc ceramic
B. Vacuum variable
C. Mica
D. Electrolytic

G6A02
(D)
Page 6-7

G6A03
What function does a capacitor serve if it is used in a power-supply circuit to filter transient voltage spikes across the transformer's secondary winding?
A. Clipper capacitor
B. Trimmer capacitor
C. Feedback capacitor
D. Suppressor capacitor

G6A03
(D)
Page 6-6

G6A04
Where is the source of energy connected in a transformer?
A. To the secondary winding
B. To the primary winding
C. To the core
D. To the plates

G6A04
(B)
Page 6-9

G6A05
(A)
Page 6-10

G6A05
If no load is attached to the secondary winding of a transformer, what is current in the primary winding called?
A. Magnetizing current
B. Direct current
C. Excitation current
D. Stabilizing current

G6A06
(C)
Page 6-11

G6A06
What is the peak-inverse-voltage rating of a power-supply rectifier?
A. The maximum transient voltage the rectifier will handle in the conducting direction
B. 1.4 times the AC frequency
C. The maximum voltage the rectifier will handle in the non-conducting direction
D. 2.8 times the AC frequency

G6A07
(A)
Page 6-11

G6A07
What are the two major ratings that must not be exceeded for silicon-diode rectifiers used in power-supply circuits?
A. Peak inverse voltage; average forward current
B. Average power; average voltage
C. Capacitive reactance; avalanche voltage
D. Peak load impedance; peak voltage

G6A08
(A)
Page 7-6

G6A08
What is the output waveform of an unfiltered full-wave rectifier connected to a resistive load?
A. A series of pulses at twice the frequency of the AC input
B. A series of pulses at the same frequency as the AC input
C. A sine wave at half the frequency of the AC input
D. A steady DC voltage

G6A09
(B)
Page 7-2

G6A09
A half-wave rectifier conducts during how many degrees of each cycle?
A. 90 degrees
B. 180 degrees
C. 270 degrees
D. 360 degrees

G6A10
(D)
Page 7-4

G6A10
A full-wave rectifier conducts during how many degrees of each cycle?
A. 90 degrees
B. 180 degrees
C. 270 degrees
D. 360 degrees

G6A11
When two or more diodes are connected in parallel to increase the current-handling capacity of a power supply, what is the purpose of the resistor connected in series with each diode?
A. The resistors ensure the thermal stability of the power supply
B. The resistors regulate the power supply output voltage
C. The resistors ensure that one diode doesn't take most of the current
D. The resistors act as swamping resistors in the circuit

G6A11
(C)
Page 7-5

G6A12
Why would it not be a good idea to use a wire-wound resistor in a resonant circuit?
A. The resistor's tolerance value would not be adequate for such a circuit
B. The resistor's inductance would detune the circuit
C. The resistor would overheat
D. The resistor's internal capacitance would detune the circuit

G6A12
(B)
Page 6-3

G6A13
What is an advantage of ferrite toroidal inductors?
A. Large values of inductance may be obtained
B. The inductor may be used in applications where core saturation is desirable
C. Most of the magnetic field is contained in the core
D. All of these choices are correct

G6A13
(D)
Page 6-8

G6A14
Where would be the stable operating points for a bipolar transistor that is used as a switch in a logic circuit?
A. In its saturation and cutoff regions
B. In its active region (between cutoff and saturation regions)
C. Between its peak and valley current points
D. Between its enhancement and deletion modes

G6A14
(A)
Page 6-14

G6A15
How should two solenoid inductors be placed so as to minimize their mutual inductance?
A. In line with their winding axis
B. With their winding axis parallel to each other
C. At right angles to their winding axis
D. Within the same shielded enclosure

G6A15
(C)
Page 6-8

G6A16
Why might it be important to minimize the mutual inductance between two inductors?
A. To increase the energy transfer between both circuits
B. To reduce or eliminate stray coupling between RF stages
C. To reduce conducted emissions
D. To increase the self-resonant frequency of both inductors

G6A16
(B)
Page 6-8

SUBELEMENT G7 — PRACTICAL CIRCUITS
[1 EXAM QUESTION — 1 GROUP]

G7A Power supplies and filters; single-sideband transmitters and receivers

G7A01
(B)
Page 7-7

G7A01
What safety feature does a power-supply bleeder resistor provide?
A. It does not affect voltage regulation
B. It discharges the filter capacitors
C. It removes shock hazards from the induction coils
D. It eliminates ground-loop current

G7A02
(D)
Page 7-6

G7A02
What components are used in a power-supply filter network?
A. Diodes
B. Transformers and transistors
C. Quartz crystals
D. Capacitors and inductors

G7A03
(C)
Page 7-5

G7A03
What should be the minimum peak-inverse-voltage rating of the rectifier in a full-wave power supply?
A. One-quarter the normal output voltage of the power supply
B. Half the normal output voltage of the power supply
C. Double the normal peak output voltage of the power supply
D. Equal to the normal output voltage of the power supply

G7A04
(D)
Page 7-5

G7A04
What should be the minimum peak-inverse-voltage rating of the rectifier in a half-wave power supply?
A. One-quarter to one-half the normal peak output voltage of the power supply
B. Half the normal output voltage of the power supply
C. Equal to the normal output voltage of the power supply
D. One to two times the normal peak output voltage of the power supply

G7A05
(B)
Page 7-9

G7A05
What should be the impedance of a low-pass filter as compared to the impedance of the transmission line into which it is inserted?
A. Substantially higher
B. About the same
C. Substantially lower
D. Twice the transmission line impedance

G7A06

In a typical single-sideband phone transmitter, what circuit processes signals from the balanced modulator and sends signals to the mixer?

A. Carrier oscillator
B. Filter
C. IF amplifier
D. RF amplifier

G7A07

In a single-sideband phone transmitter, what circuit processes signals from the carrier oscillator and the speech amplifier and sends signals to the filter?

A. Mixer
B. Detector
C. IF amplifier
D. Balanced modulator

G7A08

In a single-sideband phone superheterodyne receiver, what circuit processes signals from the RF amplifier and the local oscillator and sends signals to the IF filter?

A. Balanced modulator
B. IF amplifier
C. Mixer
D. Detector

G7A09

In a single-sideband phone superheterodyne receiver, what circuit processes signals from the IF amplifier and the BFO and sends signals to the AF amplifier?

A. RF oscillator
B. IF filter
C. Balanced modulator
D. Detector

G7A10

What type of power supply circuit is often used to provide overvoltage protection at its output?

A. Crowbar
B. Circuit breaker
C. Ferrite transformer
D. Buck-out transistor

G7A11

What type of capacitors should be used to filter the rectified DC output of a switching power supply?

A. Capacitors with low equivalent series resistance
B. Ordinary, large value electrolytic capacitors
C. NP0-type ceramic disc or silver mica capacitors
D. Capacitors with high equivalent series inductance

G7A06
(B)
Page 7-11

G7A07
(D)
Page 7-11

G7A08
(C)
Page 7-12

G7A09
(D)
Page 7-13

G7A10
(A)
Page 7-8

G7A11
(A)
Page 7-8

G7A12
(C)
Page 7-8

G7A12
Which of the following is an advantage of a switched-mode power supply as compared to a linear power supply?
A. Higher output voltages are possible with the switched-mode supply
B. Fewer circuit components are required for the switched-mode supply
C. The relatively high frequency power oscillator allows the use of small, lightweight and lowcost transformers in the switched-mode supply
D. All of these choices are correct

G7A13
(A)
Page 7-7

G7A13
In a switched-mode power supply, what is the first step in converting the 120 volt AC input voltage to a 12 volt DC output voltage?
A. The 120 volt AC is first rectified and filtered
B. The 120 volt AC is first converted to 12 volt AC with a transformer
C. The 120 volt AC is switched off when the waveform exceeds 12 volts, and is switched on again when the waveform drops below 12 volts
D. An AC clamp is used to limit the input signal to no more than 20 volts DC

Subelement
G8

SUBELEMENT G8 —SIGNALS AND EMISSIONS
[2 EXAM QUESTIONS — 2 GROUPS]

G8A Signal information; AM; FM; single and double sideband and carrier; bandwidth; modulation envelope; deviation; overmodulation

G8A01
(D)
Page 8-2

G8A01
What type of modulation system changes the amplitude of an RF wave for the purpose of conveying information?
A. Frequency modulation
B. Phase modulation
C. Amplitude-rectification modulation
D. Amplitude modulation

G8A02
(B)
Page 8-9

G8A02
What type of modulation system changes the phase of an RF wave for the purpose of conveying information?
A. Pulse modulation
B. Phase modulation
C. Phase-rectification modulation
D. Amplitude modulation

G8A03
What type of modulation system changes the frequency of an RF wave for the purpose of conveying information?
A. Phase-rectification modulation
B. Frequency-rectification modulation
C. Amplitude modulation
D. Frequency modulation

G8A04
What emission is produced by a reactance modulator connected to an RF power amplifier?
A. Multiplex modulation
B. Phase modulation
C. Amplitude modulation
D. Pulse modulation

G8A05
In what emission type does the instantaneous amplitude (envelope) of the RF signal vary in accordance with the modulating audio?
A. Frequency shift keying
B. Pulse modulation
C. Frequency modulation
D. Amplitude modulation

G8A06
How much should the carrier be suppressed below peak output power in a properly designed single-sideband (SSB) transmitter?
A. No more than 20 dB
B. No more than 30 dB
C. At least 40 dB
D. At least 60 dB

G8A07
What is one advantage of carrier suppression in a double-sideband phone transmission?
A. Only half the bandwidth is required for the same information content
B. Greater modulation percentage is obtainable with lower distortion
C. More power can be put into the sidebands
D. Simpler equipment can be used to receive a double-sideband suppressed-carrier signal

G8A08
Which popular phone emission uses the narrowest frequency bandwidth?
A. Single-sideband
B. Double-sideband
C. Phase-modulated
D. Frequency-modulated

G8A03
(D)
Page 8-9

G8A04
(B)
Page 8-11

G8A05
(D)
Page 8-2

G8A06
(C)
Page 8-6

G8A07
(C)
Page 8-5

G8A08
(A)
Page 8-6

G8A09
(D)
Page 8-3

G8A09
What happens to the signal of an overmodulated single-sideband or double-sideband phone transmitter?
A. It becomes louder with no other effects
B. It occupies less bandwidth with poor high-frequency response
C. It has higher fidelity and improved signal-to-noise ratio
D. It becomes distorted and occupies more bandwidth

G8A10
(B)
Page 8-6

G8A10
How should the microphone gain control be adjusted on a single-sideband phone transmitter?
A. For full deflection of the ALC meter on modulation peaks
B. For slight movement of the ALC meter on modulation peaks
C. For 100% frequency deviation on modulation peaks
D. For a dip in plate current

G8A11
(C)
Page 8-6

G8A11
What is meant by flattopping in a single-sideband phone transmission?
A. Signal distortion caused by insufficient collector current
B. The transmitter's automatic level control is properly adjusted
C. Signal distortion caused by excessive drive
D. The transmitter's carrier is properly suppressed

G8A12
(A)
Page 8-10

G8A12
What happens to the RF carrier signal when a modulating audio signal is applied to an FM transmitter?
A. The carrier frequency changes proportionally to the instantaneous amplitude of the modulating signal
B. The carrier frequency changes proportionally to the amplitude and frequency of the modulating signal
C. The carrier amplitude changes proportionally to the instantaneous frequency of the modulating signal
D. The carrier phase changes proportionally to the instantaneous amplitude of the modulating signal

G8A13
(A)
Page 8-6

G8A13
What signal(s) would be found at the output of a properly adjusted balanced modulator?
A. Both upper and lower sidebands
B. Either upper or lower sideband, but not both
C. Both upper and lower sidebands and the carrier
D. The modulating signal and the unmodulated carrier

G8B Frequency mixing; multiplication; bandwidths; HF data communications

| G8B

G8B01
What receiver stage combines a 14.25-MHz input signal with a 13.795-MHz oscillator signal to produce a 455-kHz intermediate frequency (IF) signal?
A. Mixer
B. BFO
C. VFO
D. Multiplier

| G8B01
| (A)
| Page 8-7

G8B02
If a receiver mixes a 13.800-MHz VFO with a 14.255-MHz received signal to produce a 455-kHz intermediate frequency (IF) signal, what type of interference will a 13.345-MHz signal produce in the receiver?
A. Local oscillator
B. Image response
C. Mixer interference
D. Intermediate interference

| G8B02
| (B)
| Page 8-8

G8B03
What stage in a transmitter would change a 5.3-MHz input signal to 14.3 MHz?
A. A mixer
B. A beat frequency oscillator
C. A frequency multiplier
D. A linear translator

| G8B03
| (A)
| Page 8-8

G8B04
What is the name of the stage in a VHF FM transmitter that selects a harmonic of an HF signal to reach the desired operating frequency?
A. Mixer
B. Reactance modulator
C. Preemphasis network
D. Multiplier

| G8B04
| (D)
| Page 8-14

G8B05
Why isn't frequency modulated (FM) phone used below 29.5 MHz?
A. The transmitter efficiency for this mode is low
B. Harmonics could not be attenuated to practical levels
C. The bandwidth would exceed FCC limits
D. The frequency stability would not be adequate

| G8B05
| (C)
| Page 8-15

G8B06
What is the total bandwidth of an FM-phone transmission having a 5-kHz deviation and a 3-kHz modulating frequency?
A. 3 kHz
B. 5 kHz
C. 8 kHz
D. 16 kHz

| G8B06
| (D)
| Page 8-15

G8B07
(B)
Page 8-14

G8B07
What is the frequency deviation for a 12.21-MHz reactance-modu-
lated oscillator in a 5-kHz deviation, 146.52-MHz FM-phone transmit-
ter?
A. 41.67 Hz
B. 416.7 Hz
C. 5 kHz
D. 12 kHz

G8B08
(C)
Page 8-16

G8B08
How is frequency shift related to keying speed in an FSK signal?
A. The frequency shift in hertz must be at least four times the
   keying speed in WPM
B. The frequency shift must not exceed 15 Hz per WPM of keying speed
C. Greater keying speeds require greater frequency shifts
D. Greater keying speeds require smaller frequency shifts

G8B09
(B)
Page 8-15

G8B09
What do RTTY, Morse code, PSK31 and packet communications
have in common?
A. They are multipath communications
B. They are digital communications
C. They are analog communications
D. They are only for emergency communications

G8B10
(B)
Page 8-17

G8B10
When sending data modes, why is it important to know the duty cycle
of the mode you are using?
A. Your connectors, feed line or antenna may be rated for
   intermittent amateur service
B. To prevent damage to your transmitter's final output stage due to
   its inability to dissipate excess heat
C. To prevent blowing your power supply's fuse due to its inability to
   dissipate excess heat
D. All of these choices are correct

G8B11
(D)
Page 8-17

G8B11
In what segment of the 20-meter band are most PSK31 operations
found?
A. At the bottom of the slow-scan TV segment, near 14.230 MHz
B. At the top of the SSB phone segment, near 14.325 MHz
C. In the middle of the CW segment, near 14.100 MHz
D. Below the RTTY segment, near 14.070 MHz

G8B12
(A)
[97.303s]
Page 8-6

G8B12
What is the maximum bandwidth permitted by FCC rules for amateur
radio stations when operating on USB frequencies in the 60-meter
band?
A. 2.8 kHz
B. 5.6 kHz
C. ±2.8 kHz
D. 3 kHz

G8B13
What is another term for the mixing of two RF signals?
A. Heterodyning
B. Synthesizing
C. Cancellation
D. Multiplying

G8B13
(A)
Page 8-8

SUBELEMENT G9
ANTENNAS AND FEED LINES

[4 EXAM QUESTIONS —4 GROUPS]

Subelement
G9

G9A Yagi antennas — physical dimensions; impedance matching; radiation patterns; directivity and major lobes

G9A01
When designing a Yagi antenna, how can the SWR bandwidth be increased?
A. Use larger diameter elements
B. Use closer element spacing
C. Use traps on the elements
D. Use tapered-diameter elements

G9A01
(A)
Page 9-12

G9A02
Approximately how long is the driven element of a Yagi antenna for 14.0 MHz?
A. 17 feet
B. 33 feet
C. 35 feet
D. 66 feet

G9A02
(B)
Page 9-9

G9A03
Approximately how long is the director element of a Yagi antenna for 21.1 MHz?
A. 42 feet
B. 21 feet
C. 17 feet
D. 10.5 feet

G9A03
(B)
Page 9-9

G9A04
Approximately how long is the reflector element of a Yagi antenna for 28.1 MHz?
A. 8.75 feet
B. 16.6 feet
C. 17.5 feet
D. 35 feet

G9A04
(C)
Page 9-9

G9A05
(B)
Page 9-9

G9A05
Which statement about a three-element Yagi antenna is true?
A. The reflector is normally the shortest parasitic element
B. The director is normally the shortest parasitic element
C. The driven element is the longest parasitic element
D. Low feed-point impedance increases bandwidth

G9A06
(A)
Page 9-11

G9A06
What is one effect of increasing the boom length and adding directors to a Yagi antenna?
A. Gain increases
B. SWR increases
C. Weight decreases
D. Wind load decreases

G9A07
(C)
Page 9-10

G9A07
Why is a Yagi antenna often used for radio communications on the 20-meter band?
A. It provides excellent omnidirectional coverage in the horizontal Plane
B. It is smaller, less expensive and easier to erect than a dipole or vertical antenna
C. It helps reduce interference from other stations off to the side or behind
D. It provides the highest possible angle of radiation for the HF bands

G9A08
(C)
Page 9-11

G9A08
What does "antenna front-to-back ratio" mean in reference to a Yagi antenna?
A. The number of directors versus the number of reflectors
B. The relative position of the driven element with respect to the reflectors and directors
C. The power radiated in the major radiation lobe compared to the power radiated in exactly the opposite direction
D. The power radiated in the major radiation lobe compared to the power radiated 90 degrees away from that direction

G9A09
(D)
Page 9-10

G9A09
What is the "main lobe" of a Yagi antenna radiation pattern?
A. The direction of least radiation from the antenna
B. The point of maximum current in a radiating antenna element
C. The maximum voltage standing wave point on a radiating element
D. The direction of maximum radiated field strength from the antenna

G9A10
(A)
Page 9-11

G9A10
What is a good way to get maximum performance from a Yagi antenna?
A. Optimize the lengths and spacing of the elements
B. Use RG-58 feed-line
C. Use a reactance bridge to measure the antenna performance from each direction around the antenna
D. Avoid using towers higher than 30 feet above the ground

G9A11
Which of the following is NOT a Yagi antenna design variable that should be considered to optimize the forward gain, front-to-rear ratio and SWR bandwidth?
A. The physical length of the boom
B. The number of elements on the boom
C. The spacing of each element along the boom
D. The polarization of the antenna elements

G9B Loop antennas — physical dimensions; impedance matching; radiation patterns; directivity and major lobes

G9B01
Approximately how long is each side of a cubical-quad antenna driven element for 21.4 MHz?
A. 1.17 feet
B. 11.7 feet
C. 47 feet
D. 469 feet

G9B02
Approximately how long is each side of a cubical-quad antenna driven element for 14.3 MHz?
A. 17.6 feet
B. 23.4 feet
C. 70.3 feet
D. 175 feet

G9B03
Approximately how long is each side of a cubical-quad antenna reflector element for 29.6 MHz?
A. 8.23 feet
B. 8.7 feet
C. 9.7 feet
D. 34.8 feet

G9B04
Approximately how long is each leg of a symmetrical delta-loop antenna driven element for 28.7 MHz?
A. 8.75 feet
B. 11.7 feet
C. 23.4 feet
D. 35 feet

G9B05
Approximately how long is each leg of a symmetrical delta-loop antenna driven element for 24.9 MHz?
A. 10.99 feet
B. 12.95 feet
C. 13.45 feet
D. 40.36 feet

G9A11
(D)
Page 9-12

G9B

G9B01
(B)
Page 9-12

G9B02
(A)
Page 9-12

G9B03
(B)
Page 9-12

G9B04
(B)
Page 9-15

G9B05
(C)
Page 9-15

G9B06
(C)
Page 9-15

G9B06
Approximately how long is each leg of a symmetrical delta-loop antenna reflector element for 14.1 MHz?
A. 18.26 feet
B. 23.76 feet
C. 24.35 feet
D. 73.05 feet

G9B07
(A)
Page 9-16

G9B07
Which statement about two-element quad antennas is true?
A. They compare favorably with a three-element Yagi
B. They perform poorly above HF
C. They perform very well only at HF
D. They are effective only when constructed using insulated wire

G9B08
(D)
Page 9-16

G9B08
Compared to a dipole antenna, what are the directional radiation characteristics of a cubical-quad antenna?
A. The quad has more directivity in the horizontal plane but less directivity in the vertical plane
B. The quad has less directivity in the horizontal plane but more directivity in the vertical plane
C. The quad has less directivity in both horizontal and vertical planes
D. The quad has more directivity in both horizontal and vertical planes

G9B09
(D)
Page 9-14

G9B09
Moving the feed point of a multielement quad antenna from a side parallel to the ground to a side perpendicular to the ground will have what effect?
A. It will significantly increase the antenna feed-point impedance
B. It will significantly decrease the antenna feed-point impedance
C. It will change the antenna polarization from vertical to horizontal
D. It will change the antenna polarization from horizontal to vertical

G9B10
(D)
Page 9-16

G9B10
What does the term "antenna front-to-back ratio" mean in reference to a cubical-quad antenna?
A. The number of directors versus the number of reflectors
B. The relative position of the driven element with respect to the reflectors and directors
C. The power radiated in the major radiation lobe compared to the power radiated 90 degrees away from that direction
D. The power radiated in the major radiation lobe compared to the power radiated in exactly the opposite direction

G9B11
(C)
Page 9-16

G9B11
What is the "main lobe" of a cubical-quad antenna radiation pattern?
A. The direction of least radiation from an antenna
B. The point of maximum current in a radiating antenna element
C. The direction of maximum radiated field strength from the antenna
D. The maximum voltage standing wave point on a radiating element

G9C Random wire antennas — physical dimensions; impedance matching; radiation patterns; directivity and major lobes; feed point impedance of ¹/₂-wavelength dipole and ¹/₄-wavelength vertical antennas

G9C

G9C01
What type of multiband transmitting antenna does NOT require a feed-line?
A. An end-fed random-wire antenna
B. A triband Yagi antenna
C. A delta-loop antenna
D. A Beverage antenna

G9C01
(A)
Page 9-3

G9C02
What is one advantage of using a random-wire antenna?
A. It is more efficient than any other kind of antenna
B. It will keep RF energy out of your station
C. It doesn't need an impedance matching network
D. It is a multiband antenna

G9C02
(D)
Page 9-3

G9C03
What is one disadvantage of a random-wire antenna?
A. It must be longer than 1 wavelength
B. You may experience RF feedback in your station
C. It usually produces vertically polarized radiation
D. You must use an inverted-T matching network for multiband operation

G9C03
(B)
Page 9-3

G9C04
What is an advantage of downward sloping radials on a ground-plane antenna?
A. It lowers the radiation angle
B. It brings the feed-point impedance closer to 300 ohms
C. It increases the radiation angle
D. It brings the feed-point impedance closer to 50 ohms

G9C04
(D)
Page 9-5

G9C05
What happens to the feed-point impedance of a ground-plane antenna when its radials are changed from horizontal to downward-sloping?
A. It decreases
B. It increases
C. It stays the same
D. It approaches zero

G9C05
(B)
Page 9-5

G9C06
(A)
Page 9-5

G9C06
What is the low-angle radiation pattern of an ideal half-wavelength dipole HF antenna installed a half-wavelength high, parallel to the earth?
A. It is a figure-eight at right angles to the antenna
B. It is a figure-eight off both ends of the antenna
C. It is a circle (equal radiation in all directions)
D. It is two smaller lobes on one side of the antenna, and one larger lobe on the other side

G9C07
(C)
Page 9-5

G9C07
How does antenna height affect the horizontal (azimuthal) radiation pattern of a horizontal dipole HF antenna?
A. If the antenna is too high, the pattern becomes unpredictable
B. Antenna height has no effect on the pattern
C. If the antenna is less than one-half wavelength high, the azimuthal pattern is almost omnidirectional
D. If the antenna is less than one-half wavelength high, radiation off the ends of the wire is eliminated

G9C08
(D)
Page 9-11

G9C08
If the horizontal radiation pattern of an antenna shows a major lobe at 0 degrees and a minor lobe at 180 degrees, how would you describe the radiation pattern of this antenna?
A. Most of the signal would be radiated towards 180 degrees and a smaller amount would be radiated towards 0 degrees
B. Almost no signal would be radiated towards 0 degrees and a small amount would be radiated towards 180 degrees
C. Almost all the signal would be radiated equally towards 0 degrees and 180 degrees
D. Most of the signal would be radiated towards 0 degrees and a smaller amount would be radiated towards 180 degrees

G9C09
(D)
Page 9-7

G9C09
If a slightly shorter parasitic element is placed 0.1 wavelength away and parallel to an HF dipole antenna mounted above ground, what effect will this have on the antenna's radiation pattern?
A. The radiation pattern will not be affected
B. A major lobe will develop in the horizontal plane, parallel to the two elements
C. A major lobe will develop in the vertical plane, away from the ground
D. A major lobe will develop in the horizontal plane, toward the parasitic element

G9C10

If a slightly longer parasitic element is placed 0.1 wavelength away and parallel to an HF dipole antenna mounted above ground, what effect will this have on the antenna's radiation pattern?

A. The radiation pattern will not be affected
B. A major lobe will develop in the horizontal plane, away from the parasitic element, toward the dipole
C. A major lobe will develop in the vertical plane, away from the ground
D. A major lobe will develop in the horizontal plane, parallel to the two elements

G9C10
(B)
Page 9-7

G9C11

Where should the radial wires of a ground-mounted vertical antenna system be placed?

A. As high as possible above the ground
B. Parallel to the antenna element
C. On the surface or buried a few inches below the ground
D. At the top of the antenna

G9C11
(C)
Page 9-3

G9D Popular antenna feed lines — characteristic impedance and impedance matching; SWR calculations

G9D

G9D01

Which of the following factors help determine the characteristic impedance of a parallel-conductor antenna feed-line?

A. The distance between the centers of the conductors and the radius of the conductors
B. The distance between the centers of the conductors and the length of the line
C. The radius of the conductors and the frequency of the signal
D. The frequency of the signal and the length of the line

G9D01
(A)
Page 9-17

G9D02

What is the typical characteristic impedance of coaxial cables used for antenna feed-lines at amateur stations?

A. 25 and 30 ohms
B. 50 and 75 ohms
C. 80 and 100 ohms
D. 500 and 750 ohms

G9D02
(B)
Page 9-18

G9D03

What is the characteristic impedance of flat-ribbon TV-type twin-lead?

A. 50 ohms
B. 75 ohms
C. 100 ohms
D. 300 ohms

G9D03
(D)
Page 9-17

G9D04
(C)
Page 9-19

G9D04
What is the typical cause of power being reflected back down an antenna feed-line?
A. Operating an antenna at its resonant frequency
B. Using more transmitter power than the antenna can handle
C. A difference between feed line impedance and antenna feed-point impedance
D. Feeding the antenna with unbalanced feed-line

G9D05
(D)
Page 9-19

G9D05
What must be done to prevent standing waves of voltage and current on an antenna feed-line?
A. The antenna feed point must be at DC ground potential
B. The feed line must be cut to an odd number of electrical quarter-wavelengths long
C. The feed line must be cut to an even number of physical half wavelengths long
D. The antenna feed-point impedance must be matched to the characteristic impedance of the feed-line

G9D06
(C)
Page 9-21

G9D06
Under what conditions would you use an inductively coupled matching network with a dipole antenna fed with parallel-conductor feed line?
A. It would not normally be used with parallel-conductor feed lines
B. It would be used to increase the SWR to an acceptable level
C. It would be used to match the unbalanced transmitter output to the balanced parallel-conductor feed line
D. It would be used at the antenna feed point to tune out the radiation resistance

G9D07
(A)
Page 9-17

G9D07
If a 160-meter signal and a 2-meter signal pass through the same coaxial cable, how will the attenuation of the two signals compare?
A. It will be greater at 2 meters
B. It will be less at 2 meters
C. It will be the same at both frequencies
D. It will depend on the emission type in use

G9D08
(D)
Page 9-17

G9D08
In what values are RF feed line losses usually expressed?
A. Bels/1000 ft
B. dB/1000 ft
C. Bels/100 ft
D. dB/100 ft

G9D09
What standing-wave-ratio will result from the connection of a 50-ohm feed line to a resonant antenna having a 200-ohm feed-point impedance?
A. 4:1
B. 1:4
C. 2:1
D. 1:2

G9D10
What standing-wave-ratio will result from the connection of a 50-ohm feed line to a resonant antenna having a 10-ohm feed-point impedance?
A. 2:1
B. 50:1
C. 1:5
D. 5:1

G9D11
What standing-wave-ratio will result from the connection of a 50-ohm feed line to a resonant antenna having a 50-ohm feed-point impedance?
A. 2:1
B. 1:1
C. 50:50
D. 0:0

G9D12
What physical aspects of an air-insulated parallel-conductor transmission line determine its characteristic impedance?
A. The RF resistance of the conductors and the length of the conductors
B. The diameter of the conductors and the distance between their centers
C. The RF resistance of the conductors and the dielectric constant of the insulation
D. The resistance of each wire to RF ground and the antenna's impedance

G9D13
What would be the SWR if you feed a vertical antenna that has a 25-ohm feed-point impedance with 50-ohm coaxial cable?
A. 2:1
B. 2.5:1
C. 1.25:1
D. You cannot determine SWR from impedance values

G9D09
(A)
Page 9-19

G9D10
(D)
Page 9-19

G9D11
(B)
Page 9-19

G9D12
(B)
Page 9-17

G9D13
(A)
Page 9-19

G9D14
What would be the SWR if you feed a folded dipole antenna that has a 300-ohm feed-point impedance with 50-ohm coaxial cable?
A. 1.5:1
B. 3:1
C. 6:1
D. You cannot determine SWR from impedance values

SUBELEMENT G0 — RF SAFETY
[5 EXAM QUESTIONS — 5 GROUPS]

G0A RF Safety Principles

G0A01
Depending on the wavelength of the signal, the energy density of the RF field, and other factors, in what way can RF energy affect body tissue?
A. It heats body tissue
B. It causes radiation poisoning
C. It causes the blood count to reach a dangerously low level
D. It cools body tissue

G0A02
Which property is NOT important in estimating RF energy's effect on body tissue?
A. Its duty cycle
B. Its critical angle
C. Its power density
D. Its frequency

G0A03
Which of the following has the most direct effect on the permitted exposure level of RF radiation?
A. The maximum usable frequency of the ionosphere
B. The frequency (or wavelength) of the energy
C. The environment near the transmitter
D. The distance from the antenna

G0A04
What unit of measurement best describes the biological effects of RF fields at frequencies used by amateur operators?
A. Electric field strength (V/m)
B. Magnetic field strength (A/m)
C. Specific absorption rate (W/kg)
D. Power density (W/cm2)

G0A05
RF radiation in which of the following frequency ranges has the most effect on the human eyes?
A. The 3.5-MHz range
B. The 2-MHz range
C. The 50-MHz range
D. The 1270-MHz range

G0A05
(D)
Page 10-3

G0A06
What does the term "athermal effects" of RF radiation mean?
A. Biological effects from RF energy other than heating
B. Chemical effects from RF energy on minerals and liquids
C. A change in the phase of a signal resulting from the heating of an antenna
D. Biological effects from RF energy in excess of the maximum permissible exposure level

G0A06
(A)
Page 10-3

G0A07
At what frequencies does the human body absorb RF energy at a maximum rate?
A. The high-frequency (3-30-MHz) range
B. The very-high-frequency (30-300-MHz) range
C. The ultra-high-frequency (300-MHz to 3-GHz) range
D. The super-high-frequency (3-GHz to 30-GHz) range

G0A07
(B)
Page 10-3

G0A08
What does "time averaging" mean when it applies to RF radiation exposure?
A. The average time of day when the exposure occurs
B. The average time it takes RF radiation to have any long term effect on the body
C. The total time of the exposure, e.g. 6 minutes or 30 minutes
D. The total RF exposure averaged over a certain time

G0A08
(D)
Page 10-6

G0A09
What guideline is used to determine whether or not a routine RF evaluation must be performed for an amateur station?
A. If the transmitter's PEP is 50 watts or more, an evaluation must always be performed
B. If the RF radiation from the antenna system falls within a controlled environment, an evaluation must be performed
C. If the RF radiation from the antenna system falls within an uncontrolled environment, an evaluation must be performed
D. If the transmitter's PEP and frequency are within certain limits given in Part 97, an evaluation must be performed

G0A09
(D)
Page 10-10

G0A10
(A)
Page 10-10

G0A10
If you perform a routine RF evaluation on your station and determine that its RF fields exceed the FCC's exposure limits in human-accessible areas, what are you required to do?
A. Take action to prevent human exposure to the excessive RF fields
B. File an Environmental Impact Statement (EIS-97) with the FCC
C. Secure written permission from your neighbors to operate above the controlled MPE limits
D. Nothing; simply keep the evaluation in your station records

G0A11
(C)
Page 10-12

G0A11
At a site with multiple transmitters operating at the same time, how is each transmitter included in the RF exposure site evaluation?
A. Only the RF field of the most powerful transmitter need be considered
B. The RF fields of all transmitters are multiplied together
C. Transmitters that produce more than 5% of the maximum permissible power density exposure limit for that transmitter must be included
D. Only the RF fields from any transmitters operating with high duty-cycle modes (greater than 50%) need to be considered

G0A12
(D)
Page 10-2

G0A12
What factors can affect the thermal aspects of RF energy exposure to human body tissues?
A. The body part and duration of its exposure
B. Frequency and power density
C. Wave polarization
D. All of these choices are correct

G0B

G0B RF Safety Rules and Guidelines

G0B01
(C)
Page 10-7

G0B01
What are the FCC's RF-safety rules designed to control?
A. The maximum RF radiated electric field strength
B. The maximum RF radiated magnetic field strength
C. The maximum permissible human exposure to all RF radiated fields
D. The maximum RF radiated power density

G0B02
(A)
Page 10-12

G0B02
At a site with multiple transmitters, who must ensure that all FCC RF-safety regulations are met?
A. All licensees contributing more than 5% of the maximum permissible power density exposure for that transmitter are equally responsible
B. Only the licensee of the station producing the strongest RF field is responsible
C. All of the stations at the site are equally responsible, regardless of any station's contribution to the total RF field
D. Only the licensees of stations which are producing an RF field exceeding the maximum permissible exposure limit are responsible

G0B03
What effect does duty cycle have when evaluating RF exposure?
A. Low duty-cycle emissions permit greater short-term exposure levels
B. High duty-cycle emissions permit greater short-term exposure levels
C. The duty cycle is not considered when evaluating RF exposure
D. Any duty cycle may be used as long as it is less than 100 percent

G0B03
(A)
Page 10-3

G0B04
What is the threshold power used to determine if an RF environmental evaluation is required when the operation takes place in the 15-meter band?
A. 50 watts PEP
B. 100 watts PEP
C. 225 watts PEP
D. 500 watts PEP

G0B04
(B)
Page 10-10

G0B05
Why do the power levels used to determine if an RF environmental evaluation is required vary with frequency?
A. Because amateur operators may use a variety of power levels
B. Because Maximum Permissible Exposure (MPE) limits are frequency dependent
C. Because provision must be made for signal loss due to propagation
D. All of these choices are correct

G0B05
(B)
Page 10-7

G0B06
What is the threshold power used to determine if an RF environmental evaluation is required when the operation takes place in the 10-meter band?
A. 50 watts PEP
B. 100 watts PEP
C. 225 watts PEP
D. 500 watts PEP

G0B06
(A)
Page 10-10

G0B07
What is the threshold power used to determine if an RF environmental evaluation is required for transmissions in the amateur bands with frequencies less than 10 MHz?
A. 50 watts PEP
B. 100 watts PEP
C. 225 watts PEP
D. 500 watts PEP

G0B07
(D)
Page 10-10

G0B08
What amateur frequency bands have the lowest power limits above which an RF environmental evaluation is required?
A. All bands between 17 and 30 meters
B. All bands between 10 and 15 meters
C. All bands between 40 and 160 meters
D. All bands between 1.25 and 10 meters

G0B08
(D)
Page 10-11

G0B09
(C)
Page 10-10

G0B09
What is the threshold power used to determine if an RF safety
evaluation is required when the operation takes place in the 20-meter
band?
A. 50 watts PEP
B. 100 watts PEP
C. 225 watts PEP
D. 500 watts PEP

G0B10
(D)
Page 10-10

G0B10
Which of the following amateur radio stations are subject to routine
environmental evaluation?
A. Those stations that use gain-type antennas at HF frequencies
B. All except portable stations
C. All except those stations where no one is exposed to RF radiation
D. Those stations with transmitter output levels exceeding 500-watts
   PEP on the 40, 75/80 and 160 meter bands

G0C

**G0C   Routine Station Evaluation and Measurements
(FCC Part 97 refers to RF Radiation Evaluation)**

G0C01
(C)
Page 10-21

G0C01
If the free-space far-field strength of a 10-MHz dipole antenna
measures 1.0 millivolts per meter at a distance of 5 wavelengths,
what will the field strength measure at a distance of 10 wavelengths?
A. 0.10 millivolts per meter
B. 0.25 millivolts per meter
C. 0.50 millivolts per meter
D. 1.0 millivolts per meter

G0C02
(B)
Page 10-21

G0C02
If the free-space far-field strength of a 28-MHz Yagi antenna
measures 4.0 millivolts per meter at a distance of 5 wavelengths,
what will the field strength measure at a distance of 20 wavelengths?
A. 2.0 millivolts per meter
B. 1.0 millivolts per meter
C. 0.50 millivolts per meter
D. 0.25 millivolts per meter

G0C03
(A)
Page 10-21

G0C03
If the free-space far-field strength of a 1.8-MHz dipole antenna
measures 9 microvolts per meter at a distance of 4 wavelengths,
what will the field strength measure at a distance of 12 wavelengths?
A. 3 microvolts per meter
B. 3.6 microvolts per meter
C. 4.8 microvolts per meter
D. 10 microvolts per meter

G0C04
If the free-space far-field power density of a 18-MHz Yagi antenna measures 10 milliwatts per square meter at a distance of 3 wavelengths, what will it measure at a distance of 6 wavelengths?
A. 11 milliwatts per square meter
B. 5.0 milliwatts per square meter
C. 3.3 milliwatts per square meter
D. 2.5 milliwatts per square meter

G0C05
If the free-space far-field power density of an antenna measures 9 milliwatts per square meter at a distance of 5 wavelengths, what will the field strength measure at a distance of 15 wavelengths?
A. 3 milliwatts per square meter
B. 1 milliwatt per square meter
C. 0.9 milliwatt per square meter
D. 0.09 milliwatt per square meter

G0C06
What factors determine the location of the boundary between the near and far fields of an antenna?
A. Wavelength of the signal and physical size of the antenna
B. Antenna height and element material
C. Boom length and element material
D. Transmitter power and antenna gain

G0C07
Which of the following steps might an amateur operator take to ensure compliance with the RF safety regulations?
A. Post a copy of FCC Part 97 in the station
B. Post a copy of OET Bulletin 65 in the station
C. Nothing; amateur compliance is voluntary
D. Perform a routine RF exposure evaluation

G0C08
In the free-space far field, what is the relationship between the electric field (E field) and magnetic field (H field)?
A. The electric field strength is equal to the square of the magnetic field strength
B. The electric field strength is equal to the cube of the magnetic field strength
C. The electric and magnetic field strength has a fixed impedance relationship of 377 ohms
D. The electric field strength times the magnetic field strength equals 377 ohms

G0C09
What type of instrument can be used to accurately measure an RF field?
A. A receiver with an S meter
B. A calibrated field-strength meter with a calibrated antenna
C. A betascope with a dummy antenna calibrated at 50 ohms
D. An oscilloscope with a high-stability crystal marker generator

G0C04
(D)
Page 10-20

G0C05
(B)
Page 10-20

G0C06
(A)
Page 10-18

G0C07
(D)
Page 10-13

G0C08
(C)
Page 10-19

G0C09
(B)
Page 10-17

G0C10
(C)
Page 10-13

G0C10
If your station complies with the RF safety rules and you reduce its power output from 500 to 40 watts, how would the RF safety rules apply to your operations?
A. You would need to reevaluate your station for compliance with the RF safety rules because the power output changed
B. You would need to reevaluate your station for compliance with the RF safety rules because the transmitting parameters changed
C. You would not need to perform an RF safety evaluation, but your station would still need to be in compliance with the RF safety rules
D. The RF safety rules would no longer apply to your station because it would be operating with less than 50 watts of power

G0C11
(D)
Page 10-13

G0C11
If your station complies with the RF safety rules and you reduce its power output from 1000 to 500 watts, how would the RF safety rules apply to your operations?
A. You would need to reevaluate your station for compliance with the RF safety rules because the power output changed
B. You would need to reevaluate your station for compliance with the RF safety rules because the transmitting parameters changed
C. You would need to perform an RF safety evaluation to ensure your station would still be in compliance with the RF safety rules
D. Since your station was in compliance with RF safety rules at a higher power output, you need to do nothing more

G0D

G0D   Practical RFsafety applications

G0D01
(C)
Page 10-15

G0D01
Considering RF safety, what precaution should you take if you install an indoor transmitting antenna?
A. Locate the antenna close to your operating position to minimize feed line losses
B. Position the antenna along the edge of a wall where it meets the floor or ceiling to reduce parasitic radiation
C. Locate the antenna as far away as possible from living spaces that will be occupied while you are operating
D. Position the antenna parallel to electrical power wires to take advantage of parasitic effects

G0D02
(A)
Page 10-16

G0D02
Considering RF safety, what precaution should you take whenever you make adjustments to the feed line of a directional antenna system?
A. Be sure no one can activate the transmitter
B. Disconnect the antenna-positioning mechanism
C. Point the antenna away from the sun so it doesn't concentrate solar energy on you
D. Be sure you and the antenna structure are properly grounded

G0D03
What is the best reason to place a protective fence around the base of a ground-mounted transmitting antenna?
A. To reduce the possibility of persons being exposed to levels of RF in excess of the maximum permissible exposure (MPE) limits
B. To reduce the possibility of animals damaging the antenna
C. To reduce the possibility of persons vandalizing expensive equipment
D. To improve the antenna's grounding system and thereby reduce the possibility of lightning damage

G0D03
(A)
Page 10-15

G0D04
What RF-safety precautions should you take before beginning repairs on an antenna?
A. Be sure you and the antenna structure are grounded
B. Be sure to turn off the transmitter and disconnect the feed-line
C. Inform your neighbors so they are aware of your intentions
D. Turn off the main power switch in your house

G0D04
(B)
Page 10-16

G0D05
What precaution should be taken when installing a ground-mounted antenna?
A. It should not be installed higher than you can reach
B. It should not be installed in a wet area
C. It should be painted so people or animals do not accidentally run into it
D. It should be installed so no one can be exposed to RF radiation in excess of the maximum permissible exposure (MPE) limits

G0D05
(D)
Page 10-15

G0D06
What precaution should you take before beginning repairs on a microwave feed horn or waveguide?
A. Wear tight-fitting clothes and gloves to protect your body and hands from sharp edges
B. Be sure the transmitter is turned off and the power source is disconnected
C. Wait until the weather is dry and sunny
D. Be sure propagation conditions are not favorable for troposphere ducting

G0D06
(B)
Page 10-16

G0D07
Why should directional high-gain antennas be mounted higher than nearby structures?
A. To eliminate inversion of the major and minor lobes
B. So they will not damage nearby structures with RF energy
C. So they will receive more sky waves and fewer ground waves
D. So they will not direct excessive amounts of RF energy toward people in nearby structures

G0D07
(D)
Page 10-15

G0D08
(C)
Page 10-15

G0D08
For best RF safety, where should the ends and center of a dipole antenna be located?
A. Near or over moist ground so RF energy will be radiated away from the ground
B. As close to the transmitter as possible so RF energy will be concentrated near the transmitter
C. As far away as possible to minimize RF exposure to people near the antenna
D. Close to the ground so simple adjustments can be easily made without climbing a ladder

G0D09
(B)
Page 10-16

G0D09
What should you do to reduce RF radiation exposure when operating at 1270 MHz?
A. Make sure that an RF leakage filter is installed at the antenna feed point
B. Keep the antenna away from your eyes when RF is applied
C. Make sure the standing wave ratio is low before you conduct a test
D. Never use a shielded horizontally polarized antenna

G0D10
(A)
Page 10-16

G0D10
Considering RF safety, which of the following is the best reason to mount the antenna of a mobile VHF transceiver in the center of a metal roof?
A. The roof will greatly shield the driver and passengers from RF radiation
B. The antenna will be out of the driver's line of sight
C. The center of a metal roof is the sturdiest mounting place for an antenna
D. The wind resistance of the antenna will be centered between the wheels and not drag on one side or the other

G0D11
(A)
Page 10-15

G0D11
Why should you avoid using attic-mounted antennas?
A. They may expose people in the house to strong, near field RF energy
B. The attic may not have adequate thermal insulation for the antenna
C. People moving around in the house might detune the antenna
D. All of these choices are correct

G0D12
(D)
Page 10-15

G0D12
Why must you be careful when aiming EME (moonbounce) arrays toward the horizon?
A. Their high ERP may produce hazardous RF fields in uncontrolled areas
B. They could cause TVI/RFI for your neighbors
C. Reflections from nearby objects could detune the array
D. All of these choices are correct

G0E RFsafety solutions

G0E01
If you receive minor burns every time you touch your microphone
while you are transmitting, which of the following statements is true?
A. You need to use a low-impedance microphone
B. You and others in your station may be exposed to more than the
   maximum permissible level of RF radiation
C. You need to use a surge suppressor on your station transmitter
D. All of these choices are correct

G0E02
If measurements indicate that individuals in your station are exposed
to more than the maximum permissible level of radiation, which of
the following corrective measures would be effective?
A. Ensure proper grounding of the equipment
B. Ensure that all equipment covers are tightly fastened
C. Use the minimum amount of transmitting power necessary
D. All of these choices are correct

G0E03
If calculations show that you and your family may be receiving more
than the maximum permissible RF radiation exposure from your
20-meter indoor dipole, which of the following steps might be
appropriate?
A. Use RTTY instead of CW or SSB voice emissions
B. Move the antenna to a safe outdoor environment
C. Use an antenna-matching network to reduce your transmitted
   SWR
D. All of these choices are correct

G0E04
Considering RF exposure, which of the following steps should you
take when installing an antenna?
A. Install the antenna as high and far away from populated areas as
   possible
B. If the antenna is a gain antenna, point it away from populated
   areas
C. Minimize feed line radiation into populated areas
D. All of these choices are correct

G0E05
What might you do if an RF radiation evaluation shows that your
neighbors may be receiving more than the maximum RF radiation
exposure limit from your Yagi antenna when it is pointed at their
house?
A. Change from horizontal polarization to vertical polarization
B. Change from horizontal polarization to circular polarization
C. Use an antenna with a higher front to rear ratio
D. Take precautions to ensure you can't point your antenna at their
   house

G0E06
(A)
Page 10-14

G0E06
What might you do if an RF radiation evaluation shows that your neighbors may be receiving more than the maximum RF radiation exposure limit from your quad antenna when it is pointed at their house?
A. Reduce your transmitter power to a level that reduces their exposure to a value below the maximum permissible exposure (MPE) limit
B. Change from horizontal polarization to vertical polarization
C. Use an antenna with a higher front to side ratio
D. Use an antenna with a sharper radiation lobe

G0E07
(C)
Page 10-15

G0E07
Why does a dummy antenna provide an RF safe environment for transmitter adjusting?
A. The dummy antenna carries the RF energy far away from the station before releasing it
B. The RF energy is contained in a halo around the outside of the dummy antenna
C. The RF energy is not radiated from a dummy antenna, but is converted to heat
D. The dummy antenna provides a perfect match to the antenna feed impedance

G0E08
(A)
Page 10-16

G0E08
From an RF radiation exposure point of view, which of the following materials would be the best to use for your homemade transmatch enclosure?
A. Aluminum
B. Bakelite
C. Transparent acrylic plastic
D. Any nonconductive material

G0E09
(B)
Page 10-14

G0E09
From an RF radiation exposure point of view, what is the advantage to using a high-gain, narrow-beamwidth antenna for your VHF station?
A. High-gain antennas absorb stray radiation
B. The RF radiation can be focused in a direction away from populated areas
C. Narrow-beamwidth antennas eliminate exposure in areas directly under the antenna
D. All of these choices are correct

G0E10
From an RF radiation exposure point of view, what is the disadvantage in using a high-gain, narrow-beamwidth antenna for your VHF station?
A. High-gain antennas must be fed with coaxial cable feed-line, which radiates stray RF energy
B. The RF radiation can be better focused in a direction away from populated areas
C. Individuals in the main beam of the radiation pattern will receive a greater exposure than when a low-gain antenna is used
D. All of these choices are correct

G0E10
(C)
Page 10-14

G0E11
If your station is located in a residential area, which of the following would best help you reduce the RF exposure to your neighbors from your amateur station?
A. Use RTTY instead of CW or SSB voice emissions
B. Use top-quality coaxial cable to reduce RF losses in the feed-line
C. Install your antenna as high as possible to maximize the distance to nearby people
D. Use an antenna matching network to reduce your transmitted SWR

G0E11
(C)
Page 10-14

G0E12
What could be done to ensure greater RF safety near a ground mounted vertical antenna?
A. Construct fencing to exclude people from getting too close to the antenna
B. Avoid transmitter output power levels above 50 watts
C. Increase the gain of the antenna
D. Add a parasitic element to redirect RF energy away from uncontrolled area

G0E12
(A)
Page 10-14

Schematic Symbols Used in Circuit Diagrams

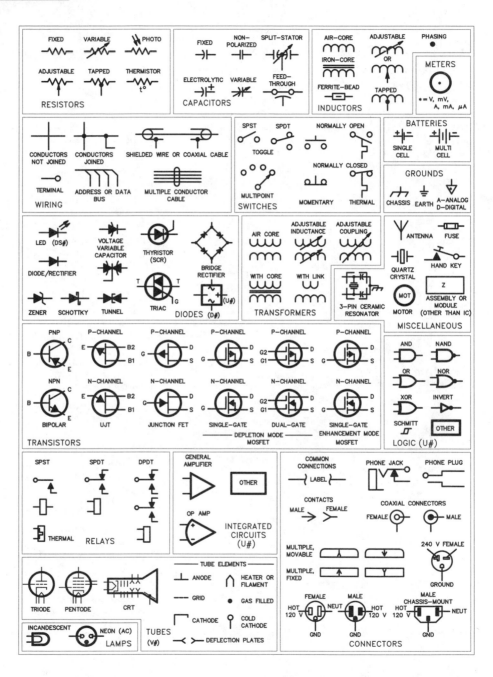

US Customary—Metric Conversion Factors

International System of Units (SI)—Metric Prefixes

Prefix	Symbol	Multiplication Factor	
exa	E	10^{18}	= 1 000 000 000 000 000 000
peta	P	10^{15}	= 1 000 000 000 000 000
tera	T	10^{12}	= 1 000 000 000 000
giga	G	10^{9}	= 1 000 000 000
mega	M	10^{6}	= 1 000 000
kilo	k	10^{3}	= 1 000
hecto	h	10^{2}	= 100
deca	da	10^{1}	= 10
(unit)		10^{0}	= 1
deci	d	10^{-1}	= 0.1
centi	c	10^{-2}	= 0.01
milli	m	10^{-3}	= 0.001
micro	μ	10^{-6}	= 0.000001
nano	n	10^{-9}	= 0.000000001
pico	p	10^{-12}	= 0.000000000001
femto	f	10^{-15}	= 0.000000000000001
atto	a	10^{-18}	= 0.000000000000000001

Linear

1 metre (m) = 100 centimetres (cm) = 1000 millimetres (mm)

Area

1 m^2 = 1 × 10^4 cm^2 = 1 × 10^6 mm^2

Volume

1 m^3 = 1 × 10^6 cm^3 = 1 × 10^9 mm^3
1 litre (l) = 1000 cm^3 = 1 × 10^6 mm^3

Mass

1 kilogram (kg) = 1 000 grams (g)
  (Approximately the mass of 1 litre of water)
1 metric ton (or tonne) = 1 000 kg

US Customary Units

Linear Units

12 inches (in) = 1 foot (ft)
36 inches = 3 feet = 1 yard (yd)
1 rod = $5\frac{1}{2}$ yards = $16\frac{1}{2}$ feet
1 statute mile = 1 760 yards = 5 280 feet
1 nautical mile = 6 076.11549 feet

Area

1 ft^2 = 144 in^2
1 yd^2 = 9 ft^2 = 1 296 in^2
1 rod^2 = $30\frac{1}{4}$ yd^2
1 acre = 4840 yd^2 = 43 560 ft^2
1 acre = 160 rod^2
1 $mile^2$ = 640 acres

Volume

1 ft^3 = 1 728 in^3
1 yd^3 = 27 ft^3

Liquid Volume Measure

1 fluid ounce (fl oz) = 8 fluidrams = 1.804 in^3
1 pint (pt) = 16 fl oz
1 quart (qt) = 2 pt = 32 fl oz = $57\frac{3}{4}$ in^3
1 gallon (gal) = 4 qt = 231 in^3
1 barrel = $31\frac{1}{2}$ gal

Dry Volume Measure

1 quart (qt) = 2 pints (pt) = 67.2 in^3
1 peck = 8 qt
1 bushel = 4 pecks = 2 150.42 in^3

Avoirdupois Weight

1 dram (dr) = 27.343 grains (gr) or (gr a)
1 ounce (oz) = 437.5 gr
1 pound (lb) = 16 oz = 7 000 gr
1 short ton = 2 000 lb, 1 long ton = 2 240 lb

Troy Weight

1 grain troy (gr t) = 1 grain avoirdupois
1 pennyweight (dwt) or (pwt) = 24 gr t
1 ounce troy (oz t) = 480 grains
1 lb t = 12 oz t = 5 760 grains

Apothecaries' Weight

1 grain apothecaries' (gr ap) = 1 gr t = 1 gr a
1 dram ap (dr ap) = 60 gr
1 oz ap = 1 oz t = 8 dr ap = 480 fr
1 lb ap = 1 lb t = 12 oz ap = 5 760 gr

Multiply —
Metric Unit = Conversion Factor × US Customary Unit

— Divide
Metric Unit ÷ Conversion Factor = US Customary Unit

	Conversion				Conversion	
Metric Unit =	Factor	× US Unit		Metric Unit =	Factor	× US Unit
(Length)				**(Volume)**		
mm	25.4	inch		mm³	16387.064	in³
cm	2.54	inch		cm³	16.387	in³
cm	30.48	foot		m³	0.028316	ft³
m	0.3048	foot		m³	0.764555	yd³
m	0.9144	yard		ml	16.387	in³
km	1.609	mile		ml	29.57	fl oz
km	1.852	nautical mile		ml	473	pint
				ml	946.333	quart
(Area)				l	28.32	ft³
mm²	645.16	inch²		l	0.9463	quart
cm²	6.4516	in²		l	3.785	gallon
cm²	929.03	ft²		l	1.101	dry quart
m²	0.0929	ft²		l	8.809	peck
cm²	8361.3	yd²		l	35.238	bushel
m²	0.83613	yd²				
m²	4047	acre		**(Mass)**	**(Troy Weight)**	
km²	2.59	mi²		g	31.103	oz t
				g	373.248	lb t
(Mass)	**(Avoirdupois Weight)**					
grams	0.0648	grains		**(Mass)**	**(Apothecaries' Weight)**	
g	28.349	oz		g	3.387	dr ap
g	453.59	lb		g	31.103	oz ap
kg	0.45359	lb		g	373.248	lb ap
tonne	0.907	short ton				
tonne	1.016	long ton				

Standard Resistance Values

Numbers in **bold** type are ± 10% values. Others are 5% values.

Ohms

1.0	3.6	**12**	43	**150**	510	**1800**	6200	**22000**
1.1	**3.9**	13	**47**	160	**560**	2000	**6800**	24000
1.2	4.3	**15**	51	**180**	620	**2200**	7500	**27000**
1.3	**4.7**	16	**56**	200	**680**	2400	**8200**	30000
1.5	5.1	**18**	62	**220**	750	**2700**	9100	**33000**
1.6	**5.6**	20	**68**	240	**820**	3000	**10000**	36000
1.8	6.2	**22**	75	**270**	910	**3300**	11000	**39000**
2.0	**6.8**	24	**82**	300	**1000**	3600	**12000**	43000
2.2	7.5	**27**	91	**330**	1100	**3900**	13000	**47000**
2.4	**8.2**	30	**100**	360	**1200**	4300	**15000**	51000
2.7	9.1	**33**	110	**390**	1300	**4700**	16000	**56000**
3.0	**10.0**	36	**120**	430	**1500**	5100	**18000**	62000
3.3	11.0	**39**	130	**470**	1600	**5600**	20000	**68000**

Megohms

75000	0.24	0.62	1.6	4.3	11.0
82000	**0.27**	**0.68**	**1.8**	**4.7**	**12.0**
91000	0.30	0.75	2.0	5.1	13.0
100000	**0.33**	**0.82**	**2.2**	**5.6**	**15.0**
110000	0.36	0.91	2.4	6.2	16.0
120000	**0.39**	**1.0**	**2.7**	**6.8**	**18.0**
130000	0.43	1.1	3.0	7.5	20.0
150000	**0.47**	**1.2**	**3.3**	**8.2**	**22.0**
160000	0.51	1.3	3.6	9.1	
180000	**0.56**	**1.5**	**3.9**	**10.0**	
200000					
220000					

Resistor Color Code

Color	Sig. Figure	Decimal Multiplier	Tolerance (%)	Color	Sig. Figure	Decimal Multiplier	Tolerance (%)
Black	0	1		Violet	7	10,000,000	
Brown	1	10		Gray	8	100,000,000	
Red	2	100		White	9	1,000,000,000	
Orange	3	1,000		Gold	—	0.1	5
Yellow	4	10,000		Silver	—	0.01	10
Green	5	100,000		No color	—		20
Blue	6	1,000,000					

Standard Values for 1000-V Disc-Ceramic Capacitors

pF	pF	pF	pF
3.3	39	250	1000
5	47	270	1200
6	50	300	1500
6.8	51	330	1800
8	56	360	2000
10	68	390	2500
12	75	400	2700
15	82	470	3000
18	100	500	3300
20	120	510	3900
22	130	560	4700
24	150	600	5000
25	180	680	5600
27	200	750	6800
30	220	820	8200
33	240	910	10000

Common Values for Small Electrolytic Capacitors

μF	V*	μF	V*
33	6.3	10	35
33	10	22	35
100	10	33	35
220	10	47	35
330	10	100	35
470	10	220	35
10	16	330	35
22	16	470	35
33	16	1000	35
47	16	1	50
100	16	2.2	50
220	16	3.3	50
470	16	4.7	50
1000	16	10	50
2200	16	33	50
4.7	25	47	50
22	25	100	50
33	25	220	50
47	25	330	50
100	25	470	50
220	25	10	63
330	25	22	63
470	25	47	63
1000	25	1	100
2200	25	10	100
4.7	35	33	100

*Working voltage

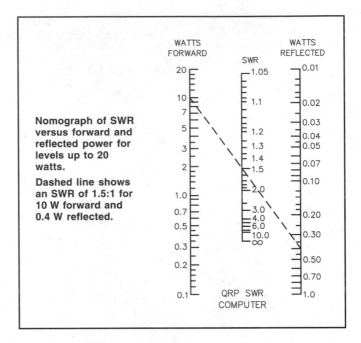

Nomograph of SWR versus forward and reflected power for levels up to 20 watts.

Dashed line shows an SWR of 1.5:1 for 10 W forward and 0.4 W reflected.

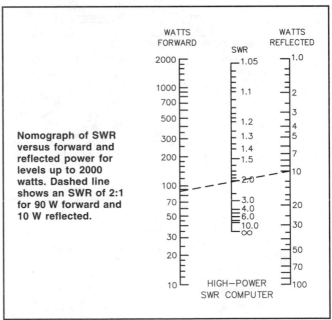

Nomograph of SWR versus forward and reflected power for levels up to 2000 watts. Dashed line shows an SWR of 2:1 for 90 W forward and 10 W reflected.

Fractions of an Inch with Metric Equivalents

Fractions Of An Inch		Decimals Of An Inch	Millimeters	Fractions Of An Inch		Decimals Of An Inch	Millimeters
	1/64	0.0156	0.397		33/64	0.5156	13.097
1/32		0.0313	0.794	17/32		0.5313	13.494
	3/64	0.0469	1.191		35/64	0.5469	13.891
1/16		0.0625	1.588	9/16		0.5625	14.288
	5/64	0.0781	1.984		37/64	0.5781	14.684
3/32		0.0938	2.381	19/32		0.5938	15.081
	7/64	0.1094	2.778		39/64	0.6094	15.478
1/8		0.1250	3.175	5/8		0.6250	15.875
	9/64	0.1406	3.572		41/64	0.6406	16.272
5/32		0.1563	3.969	21/32		0.6563	16.669
	11/64	0.1719	4.366		43/64	0.6719	17.066
3/16		0.1875	4.763	11/16		0.6875	17.463
	13/64	0.2031	5.159		45/64	0.7031	17.859
7/32		0.2188	5.556	23/32		0.7188	18.256
	15/64	0.2344	5.953		47/64	0.7344	18.653
1/4		0.2500	6.350	3/4		0.7500	19.050
	17/64	0.2656	6.747		49/64	0.7656	19.447
9/32		0.2813	7.144	25/32		0.7813	19.844
	19/64	0.2969	7.541		51/64	0.7969	20.241
5/16		0.3125	7.938	13/16		0.8125	20.638
	21/64	0.3281	8.334		53/64	0.8281	21.034
11/32		0.3438	8.731	27/32		0.8438	21.431
	23/64	0.3594	9.128		55/64	0.8594	21.828
3/8		0.3750	9.525	7/8		0.8750	22.225
	25/64	0.3906	9.922		57/64	0.8906	22.622
13/32		0.4063	10.319	29/32		0.9063	23.019
	27/64	0.4219	10.716		59/64	0.9219	23.416
7/16		0.4375	11.113	15/16		0.9375	23.813
	29/64	0.4531	11.509		61/64	0.9531	24.209
15/32		0.4688	11.906	31/32		0.9688	24.606
	31/64	0.4844	12.303		63/64	0.9844	25.003
1/2		0.5000	12.700	1		1.0000	25.400

Equations used in This Book

$P = I \times E$ (Equation 4-1)

$I = \dfrac{E}{R}$ (Equation 4-2)

$P = \dfrac{E}{R} \times E = \dfrac{E^2}{R}$ (Equation 4-3)

$PEP = \dfrac{[(0.707PEV)(0.707PEV)]}{R_L}$ (Equation 4-4)

$\text{Peak Envelope Voltage} = \dfrac{\text{Peak} - \text{to} - \text{Peak Voltage}}{2}$ (Equation 4-5)

$E = I \times R$ (Equation 5-1)

$I = \dfrac{E}{R}$ (Equation 5-2)

$R = \dfrac{E}{I}$ (Equation 5-3)

$I_1 + I_2 + I_3 + \ldots = I_T$ (Equation 5-4)

$R_{\text{Total in Series}} = R_1 + R_2 + R_3 + \cdots + R_N$ (Equation 5-5)

$R_{\text{Total in Parallel}} = \dfrac{1}{\dfrac{1}{R_1} + \dfrac{1}{R_2} + \dfrac{1}{R_3} + \cdots + \dfrac{1}{R_N}}$ (Equation 5-6)

$P = I \times E$ (Equation 5-7)

$$I = \frac{P}{E}$$ (Equation 5-8

$$E = \frac{P}{I}$$ (Equation 5-9)

$$dB = 10 \times \log_{10}\left(\frac{P_2}{P_1}\right)$$ (Equation 5-10)

$$V_{PEAK} = V_{RMS} \times \sqrt{2} = V_{RMS} \times 1.414$$ (Equation 5-11)

$$V_{RMS} = \frac{V_{PEAK}}{\sqrt{2}} = V_{PEAK} \times 0.707$$ (Equation 5-12)

$$X_C = \frac{1}{2\pi f C}$$ (Equation 5-13)

$$C_{Total\ in\ Series} = \frac{C_1 \times C_2}{C_1 + C_2}$$ (Equation 5-14)

$$X_L = 2\pi f L$$ (Equation 5-15)

$$E_S = \frac{N_S}{N_P} \times E_P$$ (Equation 5-16)

$$Z_P = Z_S\left(\frac{N_P}{N_S}\right)^2$$ (Equation 5-17)

$$\frac{N_P}{N_S} = \sqrt{\frac{Z_P}{Z_S}}$$ (Equation 5-18)

$$Z_P = Z_S \times \left(\frac{N_P}{N_S}\right)^2$$ (Equation 5-19)

$$P = I \times E$$ (Equation 6-1)

$$\text{Multiplication factor} = \frac{\text{Transmitter output frequency}}{\text{Oscillator frequency}}$$ (Equation 8-1)

$$\text{Oscillator deviation} \times \text{Multiplication factor} = \text{Transmitter deviation}$$
(Equation 8-2)

$$Bw = 2 \times (D + M) \qquad \text{(Equation 8-3)}$$

$$(\text{Yagi})\, L_{\text{Driven}}\,(\text{in feet}) = \frac{468}{f(\text{in MHz})} \qquad \text{(Equation 9-1)}$$

$$(\text{Yagi})\, L_{\text{Director}} = L_{\text{Driven}} \times 0.95 \qquad \text{(Equation 9-2)}$$

$$(\text{Yagi})\, L_{\text{Reflector}} = L_{\text{Driven}} \times 1.05 \qquad \text{(Equation 9-3)}$$

$$(\text{Quad})\, L_{\text{Driven}}\,(\text{in feet}) = \frac{1005}{f(\text{in MHz})} \qquad \text{(Equation 9-4)}$$

$$(\text{Quad})\, L_{\text{Director}}\,(\text{in feet}) = \frac{975}{f(\text{in MHz})} \qquad \text{(Equation 9-5)}$$

$$(\text{Quad})\, L_{\text{Reflector}}\,(\text{in feet}) = \frac{1030}{f(\text{in MHz})} \qquad \text{(Equation 9-6)}$$

$$SWR = \frac{E_{MAX}}{E_{MIN}} \qquad \text{(Equation 9-7)}$$

$$SWR = \frac{Z_0}{R} \ \text{ or } SWR = \frac{R}{Z_0} \qquad \text{(Equation 9-8)}$$

$$D \approx \frac{2L^2}{\lambda} \qquad \text{(Equation 10-1)}$$

$$\frac{E}{H} = \text{Impedance of Free Space} = 377\Omega \qquad \text{(Equation 10-2)}$$

$$\text{Power Density}_1 \times (\text{Distance}_1)^2 = \text{Power Density}^2 \times (\text{Distance}_2)^2 \qquad \text{(Equation 10-3)}$$

$$\text{Power Density}_2 = \text{Power Density}_1 \times \frac{(\text{Distance}_1)^2}{(\text{Distance}_2)^2} \qquad \text{(Equation 10-4)}$$

$$\text{Field Strength}_1 \times \text{Distance}_1 = \text{Field Strength}_2 \times \text{Distance}_2 \qquad \text{(Equation 10-5)}$$

$$\text{Field Strength}_2 = \text{Field Strength}_1 \times \frac{\text{Distance}_1}{\text{Distance}_2} \qquad \text{(Equation 10-6)}$$

Glossary of Keywords

A-Index — A daily measurement for the state of activity of the Earth's magnetic field. It is based on the eight **K-index** readings from the previous day, so the A-index tells you mainly how yesterday was. The A-index is given on a scale of values from 0 to 400, to indicate the range of geomagnetic field disturbance.

Alternating current (ac) — Electric current that flows first in one direction in a wire and then in the other direction. The applied voltage is changing polarity as the current direction changes. This direction reversal continues at a rate that depends on the frequency of the ac.

Amateur Auxiliary — A voluntary organization, administered by ARRL. The primary objectives are to foster amateur self-regulation and compliance with the rules.

Amateur Teleprinting Over Radio (AMTOR) — AMTOR provides error-correcting capabilities. See Automatic Repeat Request and Forward Error Correction.

American National Standard Code for Information Interchange (ASCII) — A seven-bit digital code used in computer and radioteleprinter applications.

Amplitude modulation (AM) — A method of combining an information signal and an RF carrier wave in which the amplitude of the RF envelope (carrier and sidebands) is varied in relation to the information signal strength.

Athermal effects of RF radiation — Health effects of RF radiation related to low-level energy fields that are insufficient to cause ionization or heating effects in body tissue. While research is ongoing, no conclusive evidence has been found to demonstrate that such fields cause serious health effects.

Audio rectification — Interference to electronic devices caused by a strong RF field that is rectified and amplified in the device.

Audio-frequency shift keying (AFSK) — A method of transmitting radioteletype information by switching between two audio tones fed into an FM transmitter microphone input. This is the RTTY mode most often used on VHF and UHF.

Automatic Repeat Request (ARQ) — One of two AMTOR communications modes. In ARQ, also called Mode A, the two stations are constantly confirming each other's transmissions. If information is lost, it is repeated until the receiving station confirms reception.

Azimuthal-equidistant projection map — A map made with its center at one geographic location and the rest of the continents projected from that point. Also called a great-circle map, this map is the most useful type for determining where to point a directional antenna to communicate with a specific location.

Balanced line — Feed line with two conductors having equal but opposite voltages, with neither conductor at ground potential.

Balanced modulator — A mixer circuit that combines an audio input signal with a carrier-oscillator signal. The output signal contains the two sidebands produced by this mixing, but does not include the original carrier-oscillator signal or the pure audio signal. A modulated RF signal contains some information to be transmitted. A circuit used in a single-sideband suppressed-carrier transmitter to combine a voice signal and the RF signal. The balanced modulator isolates the input signals from each other and the output, so that only the difference of the two input signals reaches the output.

Balun — A transformer used between a *bal*anced and an *un*balanced system, such as for feeding a balanced antenna with an unbalanced feed line.

Band plan – An agreement for operating within a certain portion of the radio spectrum. Band plans set aside certain frequencies for each different mode of amateur operation, such as CW, SSB, RTTY, SSTV, FM, repeaters and simplex.

Band-pass filter — A circuit that allows signals to go through it only if they are within a certain range of frequencies, and attenuates signals above and below this range.

Bandwidth — The frequency range (measured in hertz) over which a signal is stronger than some specified amount below the peak signal level. For example, if a certain signal is at least half as strong as the peak power level over a range of +3 kHz, the signal has a 3-dB bandwidth of 6 kHz.

Baudot — A five-bit digital code used in teleprinter application.

Beacon — An amateur station transmitting communications for the purposes of observing propagation and reception or other related experimental activities.

Beacon station — An amateur station transmitting communications for the purposes of observation of propagation and reception or other related experimental activities. [§97.3(a)(9)]

Bipolar transistor — A transistor made of two PN semiconductor junctions, using two layers of similar-type material (N or P) with a third layer of the opposite type between them.

Bleeder resistor — A large-value resistor connected to the filter capacitor in a power supply to discharge the filter capacitors when the supply is switched off.

Broadcasting — Transmissions intended to be received by the general public, either direct or relayed. [§97.3(a)(10)] Broadcasting is prohibited on Amateur Radio.

Capacitor — An electrical component composed of two or more conductive plates separated by an insulating material. A capacitor stores energy in an electric field.

Cathode-ray tube (CRT) — A vacuum tube with a phosphor coating on the inside of the face. CRTs are used in oscilloscopes and as the "picture tube" in television receivers.

Coaxial cable — Feed line with a central conductor surrounded by plastic, foam or gaseous insulation, which in turn is covered by a shielding conductor and the entire cable is covered with vinyl insulation.

Coil — A conductor wound into a series of loops. (Also see **inductor**.)

Critical angle — If radio waves leave an antenna at an angle greater than the critical angle for that frequency they will pass through the ionosphere instead of returning to Earth.

Critical frequency — The highest frequency at which a vertically incident radio wave will return from the ionosphere. Above the critical frequency radio signals pass through the ionosphere instead of returning to the Earth.

Cubical quad antenna — A full-wavelength loop antenna built with its elements in the shape of squares.

Current — A flow of electrons in an electric circuit.

D region — The lowest region of the ionosphere. The D region contributes very little to short-wave radio propagation, acting mainly to absorb energy from radio waves below about 7.5 MHz during daylight.

Decibel (dB) — The smallest change in sound level that can be detected by the human ear. Power gains and losses are also expressed in decibels.

Delta loop antenna — A full-wavelength loop antenna built with its elements in the shape of equilateral triangles.

Detector — A mixer circuit used to recover the information signal from a modulated RF signal.

Deviation ratio — The ratio between the maximum change in RF-carrier frequency and the highest modulating frequency used in an FM transmitter.

Digital oscilloscope — An oscilloscope that uses digital circuits to store, change and display waveforms, and compare a waveform with one stored earlier.

Direct current (dc) — Electric current that flows in one direction only.

Directional antenna — An antenna that concentrates more of the transmitted energy in a particular direction. A directional antenna must be turned to provide the best signal coverage in various directions. Such antennas also receive signals better from the direction they are pointed. Also called a beam antenna.

Director — A parasitic element in "front" of the driven element in a multielement antenna.

Driven element — The element connected directly to the feed line in a multielement antenna.

Duty cycle — The ratio between the actual RMS value of an RF signal and the RMS value of a continuous signal having the same PEP value, expressed as a percentage. A duty cycle of 100% corresponds to a continuous carrier, such as a test signal.

E region — The second lowest ionospheric region, the E region exists only during the day, and under certain conditions may refract radio waves enough to return them to Earth.

Effective radiated power (ERP) — The relative amount of power radiated in the direction of maximum signal by an antenna, as compared to a dipole. ERP takes system gains and losses into account.

Effective voltage — The value of a dc voltage that will heat a resistive component to the same temperature as the ac voltage that is being measured.

Electric field strength — This is the field resulting from the electric charge distribution present on a radiating element. Electric field strength is expressed in volts per meter (V / m).

Electromotive force (EMF) — The force or pressure that pushes a current through a circuit.

Emergency communications — Communications involving the immediate safety of human life and immediate protection of property when normal communication systems are not available.

Emission types — Any signals from a transmitter. Phone, data and CW are emission types.

F region – The highest ionospheric region, the F region refracts radio waves and returns them to Earth. The height of the F region varies greatly depending on the time of day, season of the year and amount of sunspot activity. During the day this region often splits into two regions, called the F1 and F2 regions.

Far field of an antenna — That region of the electromagnetic field surrounding an antenna where the field strength as a function of angle (the antenna pattern) is essentially independent of the distance from the antenna. In this region (also called the free-space region), the field has a predominantly plane-wave character. That is, locally uniform distributions of electric field strength and magnetic field strength are in a plane perpendicular to the direction of propagation.

Feed line — The wire or cable used to connect an antenna to the transmitter and receiver. (Also called *transmission line*.)

Field-strength meter — A simple test instrument used to show the presence of RF energy and the relative strength of the RF field.

Filter — A circuit that will allow some signals to pass through it but will greatly reduce the strength of others. In a power-supply circuit, a filter smoothes the ac ripple.

Flattopping — A distorted audio signal produced by an SSB transmitter with the microphone gain set too high. The peaks of the voice waveform are cut off in the transmitter because of overmodulation. Also called *clipping*.

Forward Error Correction (FEC) — One of two AMTOR communications modes. In the FEC mode, each character is sent twice. The receiving station checks for errors in the mark/space ratio. If an error is detected, a space is printed to show that an incorrect character was received. Also called Mode B.

Fox hunt — A friendly Amateur Radio competition to locate a hidden transmitter. Amateurs practice their direction-finding skills, which can be useful in tracking down interference sources.

Frequency deviation — The amount the carrier frequency in an FM transmitter changes as it is modulated.

Frequency modulation (FM) — The process of varying the frequency of an RF carrier in response to the instantaneous changes in an audible signal.

Frequency privileges — The specific band segments assigned to holders of each particular license. General operators have frequency privileges on the amateur high-frequency bands.

Frequency-shift keying (FSK) — A method of transmitting radioteletype information by switching an RF carrier between two separate frequencies. This is the RTTY mode most often used on HF.

Front-to-back ratio — The energy radiated from the front of a directive antenna divided by the energy radiated from the back of the antenna.

Full break-in (QSK) — With QSK, an amateur can hear signals between code characters. This allows another amateur to break into the communication without waiting for the transmitting station to finish.

Full-wave bridge rectifier — A full-wave rectifier circuit that uses four diodes and does not require a center-tapped transformer.

Full-wave rectifier — A circuit basically composed of two half-wave rectifiers. The full wave rectifier allows the full ac waveform to pass through; one half of the cycle is reversed in polarity. This circuit requires a center-tapped transformer.

Gain — An increase in the effective power radiated by an antenna in a certain desired direction, or an increase in received signal strength from a certain direction. This is at the expense of power radiated in, or signal strength received from, other directions.

Geomagnetic disturbance — A dramatic change in the Earth's magnetic field that occurs over a short time.

Great-circle path — Either one of two direct paths between two points on the surface of the Earth. One of the great-circle paths is the shortest distance between those two points. Great-circle paths can be visualized if you think of a globe with a rubber band stretched around it, connecting the two points.

Ground-plane antenna — A vertical antenna built with a central radiating element one-quarter-wavelength long and several radials extending horizontally from the base. The radials are slightly longer than one-quarter wave, and may droop toward the ground.

Half-wave rectifier — A circuit that allows only half of the applied ac waveform to pass through it.

Half-wavelength dipole antenna — A fundamental antenna one-half wavelength long at the desired operating frequency, and connected to the feed line at the center. This is a popular amateur antenna.

Harmful interference — Interference that seriously degrades, obstructs or repeatedly interrupts a radiocommunication service operating in accordance with the Radio Regulations. [§97.3(a)(23)]

High-pass filter — A filter that allows signals above the cutoff frequency to pass through, and attenuates signals below the cutoff frequency.

Horizontally polarized wave — An electromagnetic wave with its electric lines of force parallel to the ground.

Image response — A form of interference to received signals that is produced when a mixer responds to a signal frequency equal to the LO – the IF when the desired signal frequency is equal to the LO + the IF. Also when the mixer responds to a signal frequency equal to the LO + IF when the desired signal frequency is the LO – IF.

Impedance — A term used to describe a combination of reactance and resistance in a circuit.

Inductor — An electrical component usually composed of a coil of wire wound on a central core. An inductor stores energy in a magnetic field.

International Telecommunication Union (ITU) — The international organization with responsibility for dividing the range of communications frequencies between the various services for the entire world.

Ionosphere — A region in the atmosphere about 30 to 260 miles above the Earth. The ionosphere is made up of charged particles, or ions.

Junction diode — An electronic component formed by placing a layer of N-type semiconductor material next to a layer of P-type material. Diodes allow current to flow in one direction only.

K-Index — Readings of the Earth's geomagnetic field, updated every 3 hours at Boulder, Colorado. K-index values, given on a scale of 0 to 9, indicate the stability of the Earth's geomagnetic field.

Local oscillator (LO) — A receiver circuit that generates a stable, pure signal used to mix with the received RF to produce a signal at the receiver intermediate frequency (IF).

Long-path communication — Communication made by pointing beam antennas in the directions indicated by the longer **great-circle path** between the stations. To work each other by long-path, an amateur in Hawaii would point his antenna west and an amateur in Florida would aim east.

Lowest Usable Frequency (LUF) — The lower limit to the range of frequencies that will provide useful communications between two locations, using **sky-wave** propagation.

Low-pass filter — A filter that allows signals below the cutoff frequency to pass through and attenuates signals above the cutoff frequency.

Magnetic field strength — This is the field resulting from the currents on a radiating element. Magnetic field strength is expressed in amperes per meter (A / m).

Magnetizing current — A small current that flows in a transformer primary winding, even with no load connected to the secondary.

Main lobe — The direction of maximum radiated field strength from an antenna. (Also called *major lobe*.)

Maximum average forward current — The highest average forward current that can flow through a diode for a given junction temperature.

Maximum Permissible Exposure (MPE) limits — The electric field strength, magnetic field strength and plane-wave equivalent power density associated with a radiated electromagnetic wave to which a person may be exposed without harmful effect, and with an acceptable safety factor.

Maximum usable frequency (MUF) — The highest frequency that allows a radio wave to reach a desired destination.

Mixer — A circuit that takes two or more input signals, and produces an output that is the sum or difference of those signal frequencies.

Modulation — The process of varying some characteristic (amplitude, frequency or phase) of an RF carrier for the purpose of conveying information.

Modulation Index — The ratio between the maximum carrier frequency deviation and the audio modulating frequency at a given instant in an FM transmitter.

Monitor oscilloscope — A test instrument connected to an amateur transmitter and used to observe the shape of the transmitted-signal waveform.

Near field of an antenna — That region of the electromagnetic field immediately surrounding an antenna where the reactive field dominates and where the field strength as a function of angle (the antenna pattern) depends upon the distance from the antenna. It is a region in which the electric and magnetic fields do not have a substantial plane-wave character, but vary considerably from point to point.

Negative feedback — The process in which a portion of the amplifier output is returned to the input, 180° out of phase with the input signal.

Net — A group of amateurs who meet at regular times on a specific frequency to share common interests.

Neutralization — Feeding part of the output signal from an amplifier back to the input so it arrives out of phase with the input. This negative feedback neutralizes the effect of positive feedback caused by coupling between the input and output circuits in the amplifier.

Noise bridge — A test instrument used to determine the impedance of an antenna system.

Ohm — The basic unit of resistance, reactance and impedance.

Ohm's Law — A basic law of electronics, it gives a relationship between voltage, resistance and current (E = IR).

One-way communications — Radio signals not directed to a specific amateur station, or for which no reply is expected. The FCC Rules provide for limited types of one-way communications on the amateur bands. [§97.111(b)]

Oscilloscope — An electronic test instrument used to observe waveforms and voltages on a cathode-ray tube.

Parallel-conductor feed line — Feed line constructed of two wires held a constant distance apart; either encased in plastic or constructed with insulating spacers placed at intervals along the line.

Parasitic element — Part of a directive antenna that derives energy from mutual coupling with the driven element. Parasitic elements are not connected directly to the feed line.

Peak envelope power (PEP) — The average power of the RF cycle having the greatest amplitude. (This occurs during a modulation peak.)

Peak-inverse-voltage (PIV) — The maximum voltage a diode can withstand when it is reverse biased (not conducting).

Phase — The time interval between one event and another in a regularly recurring cycle.

Phase modulation — Varying the phase of an RF carrier in response to the instantaneous changes in an audio signal.

Photovoltaic cell – A wafer of semiconductor material that produces electricity when light shines on it. Sometimes called a *solar cell*, each cell produces about ½ volt when fully illuminated. Cells are connected in series to increase the voltage. The size or surface area of the cell determines the maximum current that the cell can supply. An array of cells forms a *solar panel* that can be used to charge a lead-acid storage battery.

Photovoltaic conversion – The process by which a semiconductor PN junction changes sunlight directly into electricity.

Polarization — The orientation of the electric lines of force in a radio wave, with respect to the surface of the Earth.

Power — The rate at which energy is consumed. In an electric circuit, power is found by multiplying the voltage applied to the circuit by the current through the circuit.

Power density — A measure of the power flow through a unit area normal (perpendicular) to the direction of propagation. It is usually expressed in watts per square meter (W / m^2)

Power supply — A device used to convert the available voltage and current source (often the 120-V ac household supply) to a form that is required for a specific circuit requirement. This will often be a higher or lower dc voltage.

Primary winding — The coil in a transformer that is connected to the energy source.

Propagation — The means by which radio waves travel from one place to another.

Pulsating dc — The output from a rectifier before it is filtered. The polarity of a pulsating dc source does not change, but the amplitude of the voltage changes with time.

Quarter-wavelength vertical antenna — An antenna constructed of a quarter-wavelength-long radiating element placed perpendicular to the Earth. (See **Ground-plane antenna**.)

Radio Amateur Civil Emergency Service (RACES) — Part of the amateur service that provides radio communications only for civil defense purposes.

Radio-Frequency Interference (RFI) — Interference to an electronic device (radio, TV, stereo) caused by RF energy from an amateur transmitter or other source.

Radioteletype (RTTY) — Radio signals sent from one teleprinter machine to another machine. Anything that one operator types on a teleprinter will be printed on the other machine when the two operators are communicating with each other. Also known as narrow-band direct-printing telegraphy.

Random-length wire antenna — A multiband antenna consisting of any convenient length of wire, connected directly to a transmitter or impedance-matching network without the use of feed line.

Reactance — The opposition to current that a capacitor or inductor creates in an ac circuit.

Reactance modulator — An electronic circuit whose capacitance or inductance changes in response to an audio input signal.

Rectifier — An electronic component that allows current to pass through it in only one direction.

Reflector — A parasitic element placed "behind" the driven element in a directive antenna.

Refract — To bend. Electromagnetic energy is refracted when it passes through a boundary between different types of material. Light is refracted as it travels from air into water or from water into air.

Repeater — An amateur station that automatically retransmits the signals of other amateur stations. [§97.3(a)(39)]

Resistance — The ability to oppose an electric current.

Resistor — Any material that opposes a current in an electrical circuit. An electronic component specifically designed to oppose current.

RF envelope — The shape of an RF signal as viewed on an oscilloscope.

Ripple — The amount of change between the maximum voltage and the minimum voltage in a pulsating dc waveform.

Root-mean-square (RMS) voltage — Another name for effective voltage. The term refers to the method of calculating the value.

Routine RF evaluation — A procedure required of some Amateur Radio stations, to determine that the station meets the maximum permissible exposure (MPE) limits established by FCC Rules.

S meter — A meter in a receiver that shows the relative strength of a received signal.

Scatter — Several factors may cause some energy from a radio signal to follow a path other than the idealized "straight line" shown on diagrams like Figure 3-1. Scattering can take place from Earth's ionospheric and atmospheric layers as well as objects in the wave path.

Secondary winding — The coil in a transformer that is connected to the load.

Short-path communication — Communication made by pointing beam antennas in the direction indicated by the shorter **great-circle path**.

Sideband — The sum or difference frequencies generated when an RF carrier is mixed with an audio signal.

Signal tracer — A test instrument that shows the presence of RF or AF energy in a circuit. The signal tracer is used to trace the flow of a signal through a multi-stage circuit.

Single sideband (SSB) — The type of voice (phone) operation used most often on the amateur high-frequency (HF) bands. One sideband and the carrier are removed from a double-sideband amplitude modulated (AM) signal, leaving only a single sideband to convey the voice information.

Single-sideband, suppressed-carrier signal — A radio signal in which only one of the two sidebands generated by amplitude modulation is transmitted. The other sideband and the RF carrier signal are removed before the signal is transmitted.

Single-sideband, suppressed-carrier, amplitude modulation (SSB) — A technique used to transmit voice information in which the amplitude of the RF carrier is modulated by the audio input, and the carrier and one sideband are suppressed.

Sky waves — Radio waves that travel from an antenna upward to the ionosphere, where they either pass through the ionosphere into space or are refracted back to Earth.

Solar flare — A large eruption of energy and solar material from the surface of the sun.

Solar flux — Radio energy coming from the sun.

Solar flux index — A measure of solar activity. The solar flux is a measure of the radio noise on 2800 MHz.

Specific absorption rate (SAR) — A measure of the rate at which RF energy is absorbed in body tissue. SAR is expressed in units of watts per kilogram (W / kg).

Speech processor — A device used to increase the average power contained in a speech waveform. Proper use of a speech processor can greatly improve the readability of a voice signal.

Splatter — Interference to adjacent signals caused by overmodulation of a transmitter.

Standing-wave ratio (SWR) — The ratio of maximum voltage to minimum voltage along a feed line. Also the ratio of antenna impedance to feed line impedance when the antenna is a purely resistive load.

Sudden Ionospheric Disturbance (SID) — A blackout of HF sky-wave communications that occurs after a solar flare.

Sunspots — Dark blotches that appear on the surface of the sun.

Superheterodyne receiver — A receiver that converts RF signals to an intermediate frequency before detecting them (converting the signals to audio).

Suppressor capacitor — A capacitor (often ceramic) connected across the transformer primary or secondary winding in a power supply. These capacitors are intended to suppress any transient voltage spikes, preventing them from getting through the power supply.

Temperature coefficient — A number used to show whether a component will increase or decrease in value as it gets warm.

Thermal effects of RF radiation — Body tissues that are subjected to *very high* levels of RF energy may suffer serious heat damage. These effects depend upon the frequency of the energy, the power density of the RF field that strikes the body and factors such as the polarization of the wave. For example, in an extreme case, RF heating of an eye can result in cataract formation or can even cause blindness.

Third-party communication — A message from the control operator (first party) of an amateur station to another amateur station (second party) on behalf of another person (third party). [§97.3(a)(46)]

Time averaging — Transmitter power is averaged over times of 6 minutes for controlled RF environments or 30 minutes for uncontrolled RF environments for power density calculations to determine exposure levels.

Transformer — Mutually coupled coils used to change the voltage level of an ac power source to one more suitable for a particular circuit.

Transmit-receive (TR) switch — A mechanical switch relay or electronic circuit used to switch an antenna between a receiver and transmitter in an amateur station.

Two-tone test — Problems in a sideband transmitter can be detected by feeding two audio tones into the microphone input of the transmitter and observing the output on an oscilloscope.

Unbalanced line — Feed line with one conductor at ground potential, such as coaxial cable.

Varactor diode — A component whose capacitance varies as the reverse bias voltage is changed.

Vertically polarized wave — A radio wave that has its electric lines of force perpendicular to the surface of the Earth.

Voice-Operated Transmit (VOX) — Circuitry that activates the transmitter when the operator speaks into the microphone.

Voltage — The EMF or pressure that causes electrons to move through an electric circuit.

Yagi antenna — A directive antenna made with a half-wavelength driven element, and two or more parasitic elements arranged in the same horizontal plane.

ABOUT THE ARRL

The seed for Amateur Radio was planted in the 1890s, when Guglielmo Marconi began his experiments in wireless telegraphy. Soon he was joined by dozens, then hundreds, of others who were enthusiastic about sending and receiving messages through the air—some with a commercial interest, but others solely out of a love for this new communications medium. The United States government began licensing Amateur Radio operators in 1912.

By 1914, there were thousands of Amateur Radio operators—hams—in the United States. Hiram Percy Maxim, a leading Hartford, Connecticut, inventor and industrialist saw the need for an organization to band together this fledgling group of radio experimenters. In May 1914 he founded the American Radio Relay League (ARRL) to meet that need.

Today ARRL, with approximately 170,000 members, is the largest organization of radio amateurs in the United States. The ARRL is a not-for-profit organization that:

- promotes interest in Amateur Radio communications and experimentation
- represents US radio amateurs in legislative matters, and
- maintains fraternalism and a high standard of conduct among Amateur Radio operators.

At ARRL headquarters in the Hartford suburb of Newington, the staff helps serve the needs of members. ARRL is also International Secretariat for the International Amateur Radio Union, which is made up of similar societies in 150 countries around the world.

ARRL publishes the monthly journal *QST*, as well as newsletters and many publications covering all aspects of Amateur Radio. Its headquarters station, W1AW, transmits bulletins of interest to radio amateurs and Morse code practice sessions. The ARRL also coordinates an extensive field organization, which includes volunteers who provide technical information for radio amateurs and public-service activities. In addition, ARRL represents US amateurs with the Federal Communications Commission and other government agencies in the US and abroad.

Membership in ARRL means much more than receiving *QST* each month. In addition to the services already described, ARRL offers membership services on a personal level, such as the ARRL Volunteer Examiner Coordinator Program and a QSL bureau.

Full ARRL membership (available only to licensed radio amateurs) gives you a voice in how the affairs of the organization are governed. ARRL policy is set by a Board of Directors (one from each of 15 Divisions). Each year, one-third of the ARRL Board of Directors stands for election by the full members they represent. The day-to-day operation of ARRL HQ is managed by an Executive Vice President and his staff.

No matter what aspect of Amateur Radio attracts you, ARRL membership is relevant and important. There would be no Amateur Radio as we know it today were it not for the ARRL. We would be happy to welcome you as a member! (An Amateur Radio license is not required for Associate Membership.) For more information about ARRL and answers to any questions you may have about Amateur Radio, write or call:

ARRL—The national association for Amateur Radio
225 Main Street
Newington CT 06111-1494
Voice: 860-594-0200
Fax: 860-594-0259
E-mail: **hq@arrl.org**
Internet: **www.arrl.org/**

Prospective new amateurs call (toll-free):
800-32-NEW HAM (800-326-3942)
You can also contact us via e-mail at **newham@arrl.org**
or check out *ARRLWeb* at **http://www.arrl.org/**

Index

FEEDBACK

Please use this form to give us your comments on this book and what you'd like to see in future editions, or e-mail us at **pubsfdbk@arrl.org** (publications feedback). If you use e-mail, please include your name, call, e-mail address and the book title, edition and printing in the body of your message. Also indicate whether or not you are an ARRL member.

Where did you purchase this book?
  ☐ From ARRL directly   ☐ From an ARRL dealer

Is there a dealer who carries ARRL publications within:
  ☐ 5 miles   ☐ 15 miles   ☐ 30 miles of your location?   ☐ Not sure.

License class: _____

Name _____

_____

Daytime Phone (   ) _____

Address _____

City, State/Province, ZIP/Postal Code _____

If licensed, how long? _____

Other hobbies_____

_____

Occupation _____

ARRL member? ☐ Yes ☐ No

Call Sign _____

Age _____

E-mail _____

For ARRL use only	GCLM
Edition	5 6 7 8 9 10 11 12
Printing	1 2 3 4 5 6 7 8 9 10 11 12

From _____

_____

_____

EDITOR, GENERAL CLASS LICENSE MANUAL
ARRL—THE NATIONAL ASSOCIATION FOR AMATEUR RADIO
225 MAIN STREET
NEWINGTON CT 06111-1494

— — — — — — — — — — — — — — — — please fold and tape — — — — — — — — — — — — — — — — —